Theory and Practice of Urban Sustainability Transitions

This series aims to provide timely coverage of the theory and practice of urban sustainability transitions. In modern societies, cities are centers of social and economic activity and are the places where the majority of the population lives. Cities are also at the center of the sustainability debate as the sites where many sustainability problems become apparent. At the same time cities are also regarded as providing a scale on which sustainability challenges can be addressed most efficiently. In addition to questions of conceptualization of transitions in an urban context, this series includes the quest of cities to accelerate and stimulate transitions to sustainability. The series thus is informed by transition thinking as it was developed in the last decade in Europe and as it is increasingly being applied worldwide. The aim of this series is to: 1. further the conceptualization and theorizing of urban sustainability transitions; 2. provide insights into how cities are addressing the sustainability challenge conceptually and practically; 3. learn from a comparison of and timely reflection on governance strategies in different countries, in different kinds of cities, as well as across policy domains; 4. provide case studies and contextualized tools for the governance of urban transitions. This book series will lead to compelling new insights for an international audience into how cities address the sustainability challenges they face by not repeating old patterns but by searching for new and innovative ways of thinking, valuing, and doing that are based on shared principles of a transitions approach. With a foundation of state-of-the-art research and ongoing practices, the series provides rich insights, new conceptualizations, and concrete, inspiring cases as well as practical methods and tools.

Niki Frantzeskaki • Magnus Moglia •
Peter Newton • Deo Prasad •
Melissa Pineda Pinto
Editors

Future Cities Making

Mission-oriented Research for Urban
Sustainability Transitions in Australia

 Springer

Editors
Niki Frantzeskaki
Geosciences Faculty
Utrecht University
Utrecht, Utrecht, The Netherlands

Magnus Moglia
Centre for Urban Transitions
Swinburne University of Technology
Hawthorn, VIC, Australia

Peter Newton
Centre for Urban Transitions
Swinburne University of Technology
Hawthorn, VIC, Australia

Deo Prasad
Built Environment
UNSW Sydney
Kensington, NSW, Australia

Melissa Pineda Pinto
The University of Melbourne
Melbourne, VIC, Australia

ISSN 2199-5508   ISSN 2199-5516   (electronic)
Theory and Practice of Urban Sustainability Transitions
ISBN 978-981-97-7670-2   ISBN 978-981-97-7671-9   (eBook)
https://doi.org/10.1007/978-981-97-7671-9

This Springer imprint is published by the registered company Springer Nature Singapore Pte Ltd.
The registered company address is: 152 Beach Road, #21-01/04 Gateway East, Singapore 189721, Singapore

If disposing of this product, please recycle the paper.

Foreword

I began writing this foreword right after co-authoring a comprehensive review on the future challenges and opportunities that humanity faces. That review noted three major vulnerabilities to be addressed if human civilisations as we know them are to survive in the twenty-first century (Cork et al. 2023). Firstly, there is overwhelming evidence that fundamental changes (i.e., transformations) are needed in many aspects of how human social, economic, political, legal, and other systems work if we have any hope of taking advantage of emerging opportunities and managing existential threats. Making those changes requires recognition that they are needed and require support across societies for those who will need to design and manage the changes. This leads to the second vulnerability. A large proportion of people, at societal to global scales, do not understand the situation humanity is in and the future implications of that situation. This is partly because most people lack the technical knowledge to interpret the complex information about human–environment interrelationships that is being provided by the sciences and humanities. It is also partly because our brains filter out information that we find uncomfortable or that does not match our existing views about the world. This is sometimes referred to as cognitive filtering, selective listening, or selective recall and is part of a suite of cognitive distortions that help humans simplify complexity but can get in the way of understanding communication between people and understanding what is really happening. And our lack of appreciation of our situation is also partly because those entrusted by society with the responsibility to make complex information understandable—scientists, those in government, and members of other professions—are struggling to communicate and retain that trust while false and misleading information distributed by vested interests flourishes.

The third vulnerability also relates to human cognitive processes. Faced with the need to recognise the need for fundamental change, and to make it, people in general struggle to imagine what sorts of hopeful futures we might be able to transition or transform into, which limits our options and enthusiasm for change. This dearth of imagination has complex and fascinating causes. Catastrophe stories are much more common than hopeful ones, partly because they are thought to make good media fodder and partly because they tend to be simpler and, therefore, more easily understood and believed than hopeful stories. But our still-evolving human brains also play another role—they tend to draw images of the future from the same region that houses memories of the past, which means we readily imagine futures that are repeats of what we have seen or heard of in the past. We can be blind to future possibilities that we have not experienced previously. This includes both possible future shocks and opportunities for creating better futures. The good news is that other parts of the brain that are less constrained by past experiences can be brought into play by creative stimuli, like inspirational ideas, images, or sounds.

The good news about these vulnerabilities is that they are being recognised across diverse professional disciplines and, as the review I refer to above noted, efforts to address all three challenges are emerging from diverse areas of scholarship and practical engagement with people in many different communities. Those efforts involve helping people across society understand present and possible futures; stimulating imagination and visions of new and positive possibilities; and exploring pathways for transitioning from the present to desirable futures. When I was invited to write this foreword, I hoped to discover that this book made a significant contributions to these efforts. I was not disappointed.

Since many of the biggest challenges and opportunities facing us arise from, or are influenced by, lifestyles and other activities of humans, and most people globally live in towns and cities, many of the most pressing needs for action surround settlements, especially large ones. As pointed out in several chapters of this book, the design and management of cities have the potential to mitigate many risks as well as improve human and other planetary life. But past investment of time, money, and effort in long-life infrastructure in cities, especially in wealthy countries, can make it difficult for them to change course quickly. This book takes a "mission-oriented" approach. That is, it seeks to identify the *big* issues that need to be addressed—issues that are often at scales beyond what politicians, planners, and administrators have felt comfortable with in the past—and proposes pathways for addressing these issues. These pathways involve large-scale multidisciplinary efforts like space missions (a metaphor used in this book) where advice is distilled to practical and achievable steps that can be implemented if appropriate governance arrangements are put in place.

Individually and collectively, the chapters tackle our dearth of imagination by reviewing current ideas about a wide range of "imaginaries" (ideas and visions about what cities might be) and build on these to help readers appreciate what might be possible and what might be needed to make those possibilities achievable. I have noted the following imaginaries while reading the drafts of this book (and have probably missed some): smart/digital, green, garden, biophilic, sustainable, ecologically just, low-carbon, happy, healthy, resilient, compact/walkable, cool, and sweet cities. Every chapter challenges past ways of thinking about cities. One example that stuck in my mind was the depiction of Brisbane's relationship with its eponymous river, not as "a city with a river problem" but as "a river with a city problem" (after historian Margaret Cook).

All authors agree that transformative change is needed in how cities and towns are designed and managed and they present data to demonstrate that need. One way in which this need is described is: "fundamental shifts in ways of thinking, organising, acting and knowing". There are consistent commonalities in transition pathways from the present to the future. All are based on considering the full breadth of systems that cities embody, including not just the technological aspects relating to built environments and physical infrastructure but also the social, economic, political, and legal aspects that are intrinsically coupled with the technologies and how they are used. There is a strong focus on convergence of science and technology disciplines to create new capabilities and new ideas and to drive innovation. The possibilities of new and increased data on which to base design and management of cities is another theme throughout. The technologies and social processes for collecting, storing, sharing, and interpreting those data are considered in depth and will encourage readers to stretch their imaginations beyond day-to-day comfortable limits but in an exciting and fascinating way.

Special importance is placed on more effective engagement of people across society to draw on diverse viewpoints and types of knowledge, identify what is wanted and needed in cities, and encourage the sorts of policies required to achieve those needs. There are calls in several chapters to not only acknowledge and respect indigenous belief and knowledge systems but to recognise their contributions to creating cities that harmonise relationships among their people and between people and their ecosystems. There is no shying away from the reality that ways will have to be found to manage competing values and objectives, but mechanisms are proposed to do this cooperatively and with recognition of, and respect for, those multiple values and their implications across all aspects of city systems. All of this might sound terribly complex but at the heart of all chapters is some core wisdom about adaptive governance that has been distilled from studies of societies that have managed comparable complexity successfully.

The title of this book focuses on Australian cities, and most case studies are from this continent. However, the authors are well connected with international urban research and make use of Australian case studies to both test ideas from elsewhere and generate new and challenging ideas that I am sure will be tested in other countries. It struck me that Australia is in a unique situation in being a developed country that is at an early stage of encroachment on natural environments compared with many others and yet that encroachment is becoming critically concerning. It is not so highly developed that it lacks agility to change direction, and yet it probably has less agility than many less developed countries (as highlighted, for example, by discussions in the book about slow uptake of the principles of circular economies). Australia's cities and towns face many of the climatic challenges that exist globally, which has provided opportunities to explore a wide range of mitigation and adaptation strategies across the continent.

Resilience of cities and their inhabitants (not just humans) is addressed directly or indirectly in most chapters. Resilience is a concept that interlaces inextricably with the need to transform, but that interrelationship is poorly understood among politicians, policymakers, planners, and administrators in my experience. As explained in the book, resilience is an attribute of systems and is neither good nor bad intrinsically. In popular media and policymaking, resilience is usually interpreted positively as the ability of coupled social and ecological systems to cope with shocks without losing important functions, values, and identity. But a system that is behaving in undesirable ways can also have the ability to preserve itself. Many of the pathways for transitioning from today's cities to future ones recognise that it will often be necessary to overcome this type of "undesirable resistance" to then create new functions and values and build resilience that will maintain them. Accordingly, there is a lot of focus on "tipping points" and "game-changing interventions"—the types of actions that can break through the processes holding cities and towns in the past and can allow them to progress towards new and hopeful futures.

A point that deserves reiteration here is that the key requirement for transformative change in past societies has been the ability to recognise the need for change. Without this recognition, transformation will not happen at the scales of time and space required to avoid significant societal and ecological damage. And yet it is not as simple as people recognising the need for change—and then it happens. In complex social systems, there are feedbacks and interplays such that recognition develops along with change, which is why the system-level thinking about transition pathways shown throughout this book is so important.

Late last century, leading futures-thinkers were considering what might be needed to achieve society-wide awareness and preparedness for alternative futures. Like those thinkers, this book recognises that societies have a long way to go to achieve these objectives. Most of the book's authors also recognise that no one knowledge system holds all of the answers. One of the most exciting aspects of the pathways towards achieving big missions is that they propose mechanisms for stimulating the innovation that will hopefully generate new knowledge, understanding, and solutions fast enough to meet upcoming challenges and opportunities. Frameworks like Three-Horizons Planning explore how pathways towards missions can be monitored and adjusted as they unfold through time. That means that designers, planners, and managers of cities should be able to find low-risk, high-possibility pathways to follow herein. I highly recommend this book to anyone who plans to design, build, live in or near, or just visit a city in the future.

Crawford School of Public Policy,   Steven Cork
Australian National University
Canberra, ACT, Australia
September 2023

Reference

Cork S, Alexandra C, Alvarez-Romero JG, Bennett EM, Berbés-Blázquez M, Bohensky E, Bok B, Costanza R, Hashimoto S, Hill R, Inayatullah S, Kok K, Kuiper JJ, Moglia M, Pereira L, Peterson G, Weeks R, Wyborn C (2023) Exploring alternative futures in the Anthropocene. Annu Rev Environ Resour 48(1):25–54. https://doi.org/10.1146/annurev-environ-112321-095011

Acknowledgements

Cities have been evolving for centuries as centres of culture, commerce, creativity, and change. They are now places where an increasing majority of the population in the twenty-first century live, work, and play. They are the world's economic engines, required to manage increasing flows of people, goods, information, and capital. While they are currently the source of most of the world's environmental problems, they also possess the capacity for transformation. This will require levels of innovation and integration at a scale yet to be realised, given that cities are increasingly large and complex places to plan and manage.

Future Cities Making is a project that has brought together a group of multi-disciplinary scholars to provide new visions and pathways for a sustainable transition in key sectors of Australian cities. The editors greatly appreciate their contributions to this curated collection of essays. They also thank Dr. Laura Goodin for her role as copyeditor of the book.

The editors also wish to acknowledge the financial support received from the Co-operative Research Centre for Low Carbon Living (CRCLCL) for this project. The research funded by CRCLCL Ltd. is provided by the Cooperative Research Centres Program, an Australian Government initiative.

Contents

About the Editors

Niki Frantzeskaki is a Chair Professor of Regional and Metropolitan Governance and Planning, Section Spatial Planning, Geosciences Faculty, Utrecht University, the Netherlands. Her expertise is on urban transitions and transformations, their governance and planning, with a focus on achieving climate change adaptation and mitigation, sustainability, and resilience. Her research also focuses on the governance and planning of nature-based solutions to enhance climate change resilience and promote more just urban futures. She has a rich international research experience with a portfolio of ongoing projects in Australia, Canada, and the USA. She has been a Highly Cited Researcher awardee from Clarivate Analytics in 2020 and 2021, placing her in the top 1% of researchers globally in the cross-field of social sciences and ecology. From 2019 to 2021, she has been a Research Professor and Director of the Centre for Urban Transitions at Swinburne University of Technology, Melbourne, Australia. From 2010 to 2019, she has been an Associate Professor at the Dutch Research Institute for Transitions, affiliated with Erasmus University Rotterdam. She has published over 100 peer-reviewed articles and 18 special issues and released four books on urban sustainability transitions in 2017, 2018, and 2020.

Magnus Moglia is an Associate Professor of Urban Sustainability at the Centre for Urban Transitions, Swinburne University of Technology, Australia. He has a PhD in environmental management and complex systems science, from the Australian National University's Crawford School for Public Policy, as well as degrees in engineering, physics, and mathematics from the Royal Institute of Technology in Sweden. From 2001 to 2020, he was a researcher at the CSIRO, Australia's National Science Agency, holding the positions of Principal Research Scientist and Research Team Leader in Urban Systems Science. With a background in both physical and social sciences and extensive experience in participatory and foresight methods, he is a transdisciplinary researcher, focusing on how solutions for sustainability can be equitably promoted and implemented in cities and regions. His research topics include water management, climate adaptation, freight and transport decarbonisation strategies, sustainable agriculture, circular economy, urban regeneration, working-from-home practices, and infrastructure resilience.

Much of his research has been in the international context, including in Vietnam, Laos, and Kiribati. He has published more than 70 peer-reviewed journal articles, as well as numerous book chapters and industry reports. He is currently the lead Chief Investigator for projects on circular economy, freight decarbonisation, intermodal rail, and sustainable aquaculture.

Peter Newton   is an Emeritus Professor in the Centre for Urban Transitions at Swinburne University of Technology in Melbourne, Australia. From 2007 to 2021, he held the position of Research Professor, following 15 years as Chief Research Scientist with the Commonwealth Scientific and Industrial Research Organisation (CSIRO). He has had a distinguished career in built environment research and was elected as a Fellow of the Academy of Social Sciences in Australia in 2014. He holds the distinction of being the only academic in Australia to have held senior leadership positions in all competitively funded national research centres: the Centre for Geographic Information Systems and Analysis, the Australian Housing and Urban Research Institute, the CRC for Construction Innovation, the CRC for Spatial Information, the CRC for Water Sensitive Cities, and the CRC for Low Carbon Living. He has also been a Member of the Board of The International Council for Research and Innovation in Building and Construction, as well as the Board of the Australian Urban Research Infrastructure Network, including a period as Interim Director in 2022. His principal fields of research have focused on the technology of planning, sustainability science, and urban transitions. He has published over 25 books, his most recent titles include *Greening the Greyfields: New Models for Regenerating the Middle Suburbs of Low-Density Cities*; *Migration and Urban Transitions in Australia*; *Decarbonising the Built Environment: Charting the Transition*, and *Resilient Sustainable Cities: A Future*.

Deo Prasad   is a Scientia Professor in the field of sustainable built environments at the University of New South Wales in Sydney, Australia. His expertise covers sustainable, low carbon, smart, resilient, and regenerative buildings and cities. He has published over 300 refereed publications including ten books. The last of his books on Climate Emergency: Towards a Net Zero Carbon Built Environment (Palgrave) was released early in 2023. He is currently the CEO of the NSW Decarbonisation Innovation Hub and previously served as the CEO of the Co-operative Research Centre for Low Carbon Living. Deo has received acknowledgement of his contributions from all levels of government in Australia, including an Order of Australia, a Fellowship of the Australian Academy of Technological Sciences and Engineering, Fellow of the Australian Institute of Architects, NSW Government's Green Globe Award, and the Global Impact Award as well as the National Leadership in Sustainability Prize from the Australian Institute of Architects. He is currently leading the commercialisation of proof-of-concept technologies and systems by creating an industry-government, research collaborative community at scale.

Melissa Pineda Pinto   is a postdoctoral researcher at Trinity College, Dublin. She works on the NovelEco project, which examines wild ecosystems in cities through forecasting methodologies and policy analysis. Melissa completed her PhD at Swinburne University of Technology, where she critically explored nature-based solutions through an ecological justice lens. She holds a Master of Environment from the University of Melbourne. This work draws on my previous experience in the architectural and planning industries and not-for-profit sectors. Her transdisciplinary collaborations and work span different geographies, including Australia, Europe, Latin America, and North America. Her academic experience and interests cut across social research methods, inter-transdisciplinary collaboration, and systems thinking in the context of urban ecosystems, justice, and ethics. Her research examines urban nature through diverse justice lenses for achieving sustainable futures. As an early-career researcher, Mel has published 10 peer-reviewed journal articles, book chapters, and popular science pieces. She will serve as the Chief Investigator in 2024 of a project exploring urban biodiversity and justice through community engagement in the cities of Melbourne, Australia and San José, Costa Rica.

Chapter 1
Future Cities and Their Transitions Ahead

Magnus Moglia, Niki Frantzeskaki, Peter Newton, Melissa Pineda Pinto, and Deo Prasad

Abstract Change is needed in how cities are designed, built, and managed to meet the grand challenges of the twenty-first century. In this book, we invited authors to report on their visions for cities, using a missions-oriented perspective on transformative innovations that support more liveable, sustainable, resilient, inclusive, and just futures. The resulting chapters have proposed a set of distinctive missions, providing what we think can provide the primary focus for future urban research and sustainability efforts. However, the chapters provide a mosaic rather than a single unified vision. To weave them together, this introductory chapter provides a conceptual framework for connecting and operationalising the mission-oriented approach for urban development research as a nexus of imaginaries, missions, pathways, and transformative urban innovations. This allows for orienting and bringing together contributions that represent a forward-looking collection for missions to guide and inform future city-making. In this chapter, we identify the pathways, game changers, and positive tipping points that can reshape future cities. This requires conceiving and activating multiple mission-scale programmes of intervention capable of step-change urban transitions.

M. Moglia (✉) · P. Newton
Centre for Urban Transitions, Swinburne University of Technology, Hawthorn, VIC, Australia
e-mail: mmoglia@swin.edu.au; pnewton@swin.edu.au

N. Frantzeskaki
Centre for Urban Transitions, Swinburne University of Technology, Hawthorn, VIC, Australia

Faculty of Geosciences, Human Geography and Spatial Planning Section, Utrecht University, Utrecht, The Netherlands
e-mail: n.frantzeskaki@uu.nl

M. Pineda Pinto
The University of Melbourne, Melbourne, VIC, Australia
e-mail: melissa.pinedapinto@unimelb.edu.au

D. Prasad
School of Built Environment, University of New South Wales, Kensington, NSW, Australia
e-mail: d.prasad@unsw.edu.au

© The Author(s) 2025
N. Frantzeskaki et al. (eds.), *Future Cities Making*, Theory and Practice of
Urban Sustainability Transitions, https://doi.org/10.1007/978-981-97-7671-9_1

Keywords Urban imaginaries · Mission-oriented planning · Sustainable urban development · Urban governance

1.1 Introduction to This Book

This book is based on the premise that transformative change is needed in how cities are designed, built, and managed if United Nations Sustainable Development Goals are to be realised by the mid-twentieth century (Newton and Bai 2008; Seto et al. 2012; Crane et al. 2021; de Sa et al. 2022; Rockström et al. 2023). Furthermore, to craft and navigate towards the types of cities that we need, it is necessary to move beyond traditional scientific forecasting efforts and engage with our imagination, and to tap into multiple knowledges and values so as to direct our minds to how the future of our cities should be: what our future cities could look like, and whose visions and imaginaries they will take up, enable, and propagate. So far, most urban research literature does not fully embrace futures-oriented perspectives, as it generally sits outside of most disciplines, and it is often used in more applied or transdisciplinary research projects.

In this book, we have asked the authors of chapters to draw on their breadth of experiences and research to outline how future cities could be imagined, designed, and built within a missions-oriented approach to address some of this century's grand challenges. By doing so, this book provides a comprehensive set of visions and urban imaginaries, offering a set of blueprints for problem-solving.

Given that most authors and editors of this book come from Australia, we need to caution that the word 'missions' refers to the objective-oriented and innovation-centric framework of mission-oriented innovation proposed by Mazzucato (2018) and should be neither conflated nor associated with the colonising meanings and history of the word. The choice of mission-oriented innovation and the idea of missions as a frame to guide and organise the ideas and proposals is our attempt to provide a new lens to orient our collective imagination for future cities.

In this chapter, we also identify the drivers as well as transformative propositions across existing and emerging urban imaginaries, such as human-healthy cities; nature-based cities, technology-driven cities; resilient cities; and just cities. We do this by reviewing prominent and emerging scientific literature to enrich our conceptual lens further and position the contributed chapters across the spectrum of future-focused urban imaginaries.

> **Box 1.1 Glossary**
> **Grand challenges**: Situations that create the need for transformative change. Examples include global climate change, biodiversity loss or pollution.
> **Imaginaries**: Visions for how urban areas need to develop and transform. This represents the end goal and typically has explicit normative judgements and justice implications.

Pathways: Sequences of actions implemented progressively, typically based on cross-sectoral responses to achieve a goal (such as a response to a grand challenge).

Solutions: The actions, such as implementation of technology, that will help address a problem or a grand challenge.

Mission areas: The combination of imaginaries, pathways, and solutions in response to identified grand challenges.

Cities are where sustainability tensions manifest, places where solutions are tested, evidence collected at a local scale, and at the same time, side effects and unintended consequences of interventions are revealed, opening the discourse to new debates and contestations. There is no solution applied to or emerging from cities that is uncontested, showing the dynamism as well as the immediacy of responses that urban environments can provide. This further supports the statement from Newton and Bai's (2008, p. 4) pioneering work that 'the challenge of achieving sustainable development in the 21st century will be won or lost in the world's urban areas'. As spaces and places of opportunity and tension, cities are often seen as locations for emerging or accelerating transformations. Recent examples include the electrification of cities (Griffith 2022a, b), the growing number of sharing and circular economy initiatives (Winslow and Coenen 2023), the declaration of climate emergency by cities (Howarth et al. 2021; Harvey-Scholes et al. 2023; Greenfield et al. 2022), and the mainstreaming of climate adaptation agendas, including greening the urban environment (Adams et al. 2023a, b); replacing grey infrastructure investments with nature-based solutions, including water sensitive urban design (Coutts et al. 2013) and low-impact urban solutions (Sharma et al. 2018); and hosting debates and agendas for just transition measures (Hughes and Hoffmann 2020). Progressing across all those challenges, city governments and urban planners, particularly, must consider multiple and often misaligned urban objectives, dealing with trade-offs and tensions that urban interventions need to navigate to facilitate the achievement of broader goals of sustainability, justice, liveability, and climate resilience (Frantzeskaki et al. 2021).

One way to source or co-create solutions for future cities is through future-oriented visions or urban imaginaries stimulated by questions such as: Why is it important to look into the future, and why are future-oriented narratives and images important to urban planners, practitioners, citizens, and all involved actors in cities? Imagining urban futures is a practice and process that goes hand in hand with the way we deal with urban complexity and progress our understanding of urban development trajectories. One important reason for this is that much of the uncertainty and complexity in cities is associated with human choices and actions; the uncertainty can be significantly reduced when shaped by collective visions. Imagination is a fundamental process of conceptualising, envisioning, anticipating, and executing visions and pathways that can help us project desired futures (Cork et al. 2023; Dunn 2018; Keith et al. 2023). Imagination leads to imaginaries that can be used for many different purposes, including.

- Seeds for a better future—providing a common goal for diverse stakeholders
- Boundary objects that allow us to identify where to intervene to transform a city
- Points of discussion and analysis to help us rethink how we make decisions about cities in a more systems-oriented, anticipatory, and adaptive manner

It is within this imagining of urban futures with a focus on innovation that these futures are defined and taken forward through mission-oriented approaches that include major intervention programmes. Mission-oriented approaches and policies represent the solutions driving innovation to address the 'grand challenges' that are complex, interconnected, and systemic, such as climate change and socio-economic inequalities (Mazzucato 2018).

When we do this, we also need to acknowledge both the diversity in society and our cultural and ideological biases. Only with more inclusive methods for thinking about and crafting more positive outcomes can we move towards such a future. There is a need to embed this type of futures-thinking and the missions-oriented research paradigm into societal futures thinking capability and governance systems. However, this is currently far from mainstream practice. This book attempts to break new ground in this space and offer innovative ways to address our grand challenges.

1.2 Imagination, Change, and Transformation

In recent times, humans have become the primary drivers of the planet's environmental systems to the extent that the current era in Earth's history has been termed the Anthropocene (Steffen et al. 2007). This has profound implications for how we as humans think about the future, and there is a need to move from being passive observers to acknowledging our role more actively in shaping the future. Whilst it is tempting to become fatalistic and pessimistic in viewing the future, considering our current trajectory, we need to acknowledge that, just as humans have shaped the past, we can also shape the future, and we have the agency to turn our trajectory towards more positive outcomes. Moglia et al. (2018) have highlighted six grand challenges that cities must deal with:

1. *Failure of planning for rapid urbanisation*, which is still occurring in many, if not most, cities around the world. This tends to lead to inadequate provision of services, congestion, inequality, crime, loss of agricultural lands, and ecosystem damage.
2. *Climate change* and its associated heat waves and natural disasters. This is connected with damage to infrastructure, loss of lives, and water and food shortages, as well as refugee flows and international instability.
3. *Economic boom-bust cycles*, which are associated with periods of rapid economic growth followed by rapid economic contraction. This leads to unemployment, household stress, accentuation of societal problems, and infrastructure deficits.

4. *Natural disasters*, such as storms, bushfires, cyclones, and flooding. These occur naturally but are being exacerbated by climate change. This leads to infrastructure damage, loss of life, and damage to the economy.
5. *Technology-based disruptions* associated with the unintended consequence of technological change such as automation and AI, potentially leading to issues like deskilling, disruptions to social and family structures, pressure on social services, unemployment, increased inequality, and stranded infrastructure.
6. *Failure of governance*, for example, through polarised political systems, and reduced trust in democratic government. This tends to reduce the capacity to protect social good and institutions, thus leading to increased inequality, lower social participation, environmental degradation, reduced productivity, and reduced attractiveness of the city.

In addition to these six specifically urban challenges, we can also add the urgent issue of biodiversity loss, as a critical challenge for staying within a safe operating space for the planet:

7. *Biodiversity loss*, as the planet increasingly loses its biosphere integrity that supports all known life. Whilst ecosystems in cities are smaller relative to other surface areas, recent research shows that urban areas are hotspots for biodiversity and need to be reimagined as linking spots or corridors to peri-urban and rural biotopes and ecosystems. Biodiversity loss is thus a challenge that also has an urban character, especially due to the dual role of cities (Simkin et al. 2022) as generators of consumption and pollution (Seto et al. 2012; Güneralp et al. 2013), as well as the potential of urban areas to contribute positively to biodiversity and liveability.

Since Moglia's identification of grand challenges, we have also been reminded of ever-present challenges such as the risk of *pandemics*, *war*, and *global conflict*. Importantly, we note that cities are responsible for much of the pressure on planetary support systems, through their material consumption, pollution of air, land, and water, and especially their greenhouse gas emissions. With the planet now in unsafe territory in relation to planetary boundaries (Richardson et al. 2023), safeguarding Earth's life support systems urgently requires a reduction in the damaging effect of cities and their populations on the environment.

1.2.1 Our Theory of Change

In this book, we do not address all these challenges. The primary focus is on climate change, but extends to biodiversity loss, technology-based disruption, failure of governance, and the capacity of planning to handle future threats and opportunities. We propose that such challenges can only be addressed by large-scale mission-oriented innovation, and urban imaginaries that focus on aspirational futures depicting what could be and should be. Urban imaginaries provide a common vision and goal that can become the focus of innovative, long-term, mission-oriented intervention programmes (Fig. 1.1).

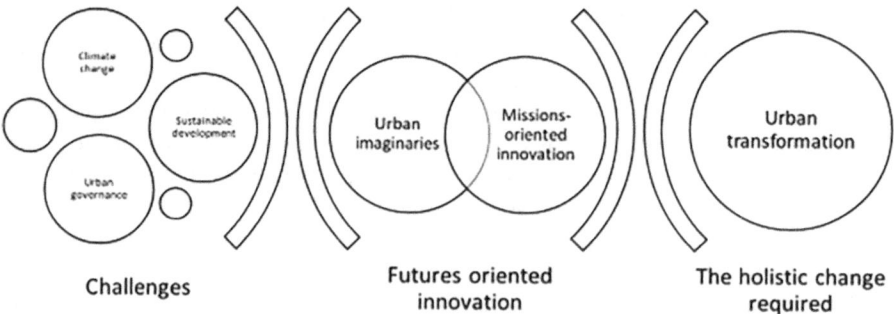

Fig. 1.1 The process of urban transformation to address urban challenges at the nexus of urban imaginaries and mission-oriented innovation

1.2.2   Urban Imaginaries of the Twenty-First Century

Achieving transformative change that addresses grand societal challenges is complex. To change intentionally, there must be both a recognition of the need for change and the imagination of a desired future. Imagination is therefore central to change processes.

> **Box 1.2a Examples of Urban Imaginaries Focused on Process**
> Smart City. Resilient City. Industrial City. Digital City. Global City. Neo-Liberal City. Regenerative City. Tactical Urbanism. Circular City. Cultural City. Inclusive City. Intelligent Urbanism. New Urbanism. Post Urbanism. Contested City.

Unsurprisingly, therefore, visioning and developing urban imaginaries is critical to the process of city-making. McPhearson et al. (2017, p. 6) remind us that 'positive visioning is a critical component to co-creating opportunities and generating realistic pathways for transformation toward sustainability. Research and practice are beginning to create positive visions, develop future scenarios, generate pathways, create plans, and initiate implementation projects for improving urban sustainability, resilience, and human livelihoods in cities'.

We note that while positive visioning is critical for place-making and enabling sustainability transitions, it also needs to be critically examined to ensure that an urban imaginary does not result in unintended consequences; for example, it is not reproducing inequalities or introducing narratives of unsustainable growth, or favouring the elites and certain cultural ideals that are underscored by wealth accumulation at the cost of marginalised groups and the environment (Bonakdar and Audirac 2021).

> **Box 1.2b Examples of Urban Imaginaries Focused on Urban Form and Transport**
> Garden City. Polycentric City. Compact City. Vertical City. Megacity. Megacity Region. Walking City. Happy City. Green City. Low-Carbon City. Transit City. Auto-City. Linear City. Edge City. Satellite City. Chrono-Urbanism. 30-Minute City. 15-Minute Neighbourhood.

In response to this need, 'Imaginaries' have emerged to stretch the boundaries of thinking across multiple disciplines and research areas. They are typically the result of a creative process involving often-speculative visions of some future state. Many are utopian, targeting new ideas and concepts that are conceived as transformative, involving some form of positive transition. Some are in response to grand challenges that need to be tackled to avoid or minimise negative (including catastrophic or dystopian) outcomes (Mazzucato 2018, 2020).

> **Box 1.2c Examples of Urban Imaginaries Focused on Outcomes**
> Green City. Healthy City. Low-Carbon City. Sustainable City. Competitive City. Productive City. Just City. Equitable City. Water-Sensitive City. Gentrified City. Liveable City. Safe City. Nature-Based City. Resilient City. Zero-Waste City.

Labelling of imaginaries varies, depending on whether the focus is on envisioning some 'end state' (e.g. smart city, garden city, etc.) or describing key drivers of change (e.g. globalisation, digitalisation, significant population upheaval, etc.). At a higher level, imaginaries can be clustered as social imaginaries (Taylor 2002), socio-technical imaginaries (Jasanoff et al. 2007), spatial imaginaries (Watkins 2015), and climate imaginaries (Nerlich and Morris 2015). Urban imaginaries can also be added to this list (Meissner and Lindner 2018, and earlier, Peter Hall's Cities of Tomorrow, whose 13 chapter titles are all replete with imaginaries). A compendium would be needed to capture all urban imaginaries that have emerged over the decades, so a representative collection is listed in Boxes 1.2a–1.2c. In this book, imaginaries will focus primarily on the topic of achieving/targeting 'sustainable urbanism' as a key goal of twenty-first-century city-making (here, sustainable urbanism is the effort to achieve the comprehensive set of sustainable development goals in cities).

1.2.3 Mission-Oriented Innovation

To move from imagination to action, Mazzucato (2018) outlined an approach that can collaboratively activate the key pillars of society, with governments in a leading role, but also applying the innovative capacity of industry and academia, to address big societal challenges. This is based on defining clear goals that can be collectively targeted, and that are based on a key set of principles that

- Are bold and inspirational with wide societal relevance
- Provide a clear direction by being targeted, measurable, and time bound
- Are linked with ambitious but realistic research and innovation activities
- Involve cross-disciplinary, cross-sectoral, and cross-actor innovation
- Operate across multiple scales
- Involve multiple, joined-up, bottom-up, and top-down solutions

This approach by Mazzucato has gathered traction, providing a guiding frame for EU and OECD innovation policy and frameworks (Mazzucato 2019; OECD 2022), as well as guiding the thinking of many universities globally (Broström et al. 2021).

1.3   Five Foci for Mission-Scale Urban Interventions

In this section, we identify several mission-critical areas for urban intervention. The goal: achieving critical sustainability outcomes, as represented by the UN SDGs and the UN's New Urban Agenda acting as a critical lever and implementation mechanism for the SDGs (United Nations 2017, 2020). They represent a nexus of missions, pathways, and transformative urban innovations. The goal of achieving sustainability outcomes, such as attending to issues of equality, health, well-being, and accessibility to services and infrastructure, and improving resilience, climate adaptation, and provision of ecosystem services and benefits, reflects the needs and aspirations that seek alternative futures.

1.3.1   Resilient Cities

The resilient cities imaginary describes a vision in which planning for, delivering, and maintaining infrastructure that is usually long-lived and expensive is done in a way that meets the rapidly changing priorities or needs of communities and economic activities in the future, in a world of shocks and stresses. An important assumption in resilience thinking, which originated in the study of social-ecological systems (Walker et al. 2004), is that systems like cities and the natural environment are dynamic, and constantly changing in interaction with each other; this realisation is moving resilience thinking away from a static or reductionist view of the world. Whilst there is no inherent value in the term resilience by itself, as it simply refers to the capacity to not change in response to stresses and shocks, urban resilience relates to a city's capacity to maintain key functions and thus meet the needs of its inhabitants as well as nature. In this context, this imaginary addresses the trifecta of the social, economic, and ecological domains.

Common critiques of the resilient city agenda argue that it is inadequate in dealing with power and politics, promotes the status quo, and aligns itself with the neoliberal politics field (Meerow and Newell 2019). In this book, we share this concern, and we argue that the resilient city imaginary needs to be radical and transformative, based on a paradigm shift of governance and decision-making (Chaps. 2 and 3), as well as radical and disruptive innovation (Chap. 4). Specifically, the resilient city imaginary needs to encompass notions of the just and nature-based city and needs to not simply reinforce poor outcomes for the natural environment, nor spatial or other types of inequalities for people, but rather build resilience for and with people and nature.

We note, then, that the resilient city imaginary (as a city that stays the same on some key aspect, regardless of shocks and stresses) relies on the social construction of what is considered a desirable city, and therefore it is important to consider who decides what a desirable urban system is, whose resilience is being prioritised, and where the boundaries of the city are (Meerow and Newell 2019). A good example to highlight this tension is the frequent debate about the role of private cars in cities. Although most cities have been built to support the widespread use of private cars, many argue for the extensive social and ecological benefits of reduced car use (Nieuwenhuijsen 2020; Nieuwenhuijsen and Khreis 2016). This creates a common dilemma, where decision-makers and the community tend to propagate a legacy urban form 'built for the car' whilst simultaneously recognising that reduced car use would benefit nearly everyone. Chapter 2 notes in particular the important role of allowing the community to be part of the process of socially constructing the goal of this imaginary inclusively and, in a way, open to systemic transformation. Chapter 3 notes the importance of a paradigm shift in governance based on intergenerational equity and resilience to climate change. Chapter 4 highlights the importance of radical and transformative innovation to address climate change threats, but specific innovations are currently meeting some resistance in governance regimes.

Resilience, as earlier noted, is a systems concept. While the resilience of systems, therefore, does not respect arbitrary administrative boundaries, interactions between scales need to be considered. On this note, Chap. 3 discusses the importance of considering such interactions when ensuring the resilience of water systems at precinct, city, and regional scales.

The focus of resilient cities is sometimes on general resilience, which is the capacity of social-ecological systems to adapt or transform in response to unfamiliar or unknown shocks (Carpenter et al. 2012), and sometimes on specific resilience, which is the capacity of social-ecological systems to adapt to specific shocks or threats. In this book, Chap. 2 focuses more on general resilience by addressing governance and decision-making capacity. Chapter 3 focuses on more specific threats to water systems, and Chap. 4 focuses on more specific climate change-related threats.

As outlined in Chap. 2, achieving resilient cities is more about the capacity and competencies for governing, adapting, and changing than it is about adopting any specific solution. The chapter argues that this involves building the capacity for a different decision-making process, which includes a better understanding of systems and anticipating challenges before they emerge, as well as incorporating more preventive strategies that help reduce risks in a more holistic sense. Expanding on this issue in relation to climate change, Chap. 4 stresses the importance of moving beyond simply coping with weather extremes and calls for more radical and disruptive innovation that speeds up the rate of change to rapidly reduce greenhouse gas emissions, and that better mitigates the negative impacts of climate change in a more proactive way.

In summary, the key to cities' resilience lies in the people and organisations having capacity, competencies, and governance for systemic interventions based on adaptive learning and collaborative decision-making. In terms of the solutions, they tend to relate to the challenges at hand, and in Chaps. 2–4, the following solutions are noted:

- Building with materials that reduce heat in the urban landscape
- Production and use of green energy, to reduce greenhouse gas emissions
- Nature-based solutions, like urban forests, for a range of benefits to people, biodiversity, and urban ecosystems
- Smart city technologies that promote more rapid and effective governance.
- Water-sensitive technologies such as water reuse, stormwater harvesting, and stormwater management, enhanced through green roofs or constructed wetlands
- Reduced private car use and the introduction of more public and active transport
- Provision of more affordable housing
- Sustainable urban re-development approaches like greening the greyfields (Newton et al. 2022; Chap. 7).

1.3.2 Low Carbon and Circular Cities

We currently inhabit a carbon- and resource-constrained world where population, consumption, and urbanisation are all on growth trajectories. In this context, the twenty-first-century goal of sustainable urban development urgently requires twin interlinked transitions to

- Renewable energy (from a long-established fossil fuel-based system where significant path dependencies exist) on a scale capable of halting and reversing rates of greenhouse gas emission and global warming, which is now reaching historic levels and heading higher (Copernicus Institute 2024)
- A circular economy, based on closed-loop systems linked to recycling, remanufacturing, reuse, repair, and sharing, which stand in stark contrast to currently dominant linear systems involving raw material extraction, manufacturing and distribution, consumption, and disposal (Kara et al. 2022; European Parliamentary Research Service 2024)

The chapters in this section of this book address these two grand challenges.

Decarbonisation of the built environment is the focus of Chap. 5. It draws heavily on an increasing body of research undertaken over the past half-century, and more recently by Australia's CRC for Low Carbon Living (Newton et al. 2019) and the United Nations Environment Programme (UNEP 2023a, b). The latter proposes a three-pillar pathway for creating clean, resource-efficient, green cities. Regarding resource efficiency improvement, there is prioritisation of measures to achieve a circular economy based on 3R (reduce, reuse, and recycle) principles, promoting lifecycle analysis of material and energy use and adoption of smart technologies. Suggestions about sustainable consumption and production, payment for pollution and waste, and accountability mechanisms represent solutions for making cities cleaner. Innovation in land-use planning, mobility management, and socio-economic equity improvement are challenges to be overcome, however, in realising the green growth opportunities that will underpin green cities and a green economy transition (see Newton and Newman 2015).

Chapter 6 focuses on the circular economy and its early stages of development in the state of Victoria, Australia. Circular economy concepts and their implementation are currently in their infancy in contemporary industrial and post-industrial societies (see McDonough 2008 for one of the early pioneering imaginaries), in contrast with agrarian and earlier industrial societies when circular economies were strong. This was before the Great Acceleration in manufacturing production associated with automation, global supply chains, and massive growth in a consumer society. Based on a survey targeting a wide range of businesses, the chapter found that there was no consistent or systemic understanding of the concept of the circular economy; rather, the narrative was narrower and revolved around waste management and recycling. Transition to a circular economy would require a systematic shift by industry and government involving a clear policy directive, financial outlays, advances in technical know-how, education, awareness, engagement, and collaboration across traditionally isolated sectors.

1.3.3   Nature-Based Regenerative Cities

At their most basic, cities are the habitats humans have created for themselves, but these habitats (like all habitats) are part of nature. Whilst in the modern post-industrial era there has been a tendency to consider cities as something different from nature, this view is now becoming less dominant. New imaginaries focus on regenerating ecosystems, waterways, and forests within cities, with the recognition that this provides benefits not just for nature but also for humans. The framework of ecosystem services may be simplistic in viewing nature as offering services to humans, but it highlights that ecosystems are of vital importance for life on our planet, and especially for humans, including for well-being, health, recreation, protection, and bolstering the earth's life support systems.

Along this line, the Green or Eco City is one of the oldest urban imaginaries to emerge, starting with ideals such as the garden city and affiliate imaginaries such as biophilic cities, in Beatley's coined term (Reeve et al. 2015; Kellert 2016; Lee and Kim 2021). Recent conceptualisations of the Green City envision nature, green spaces, and/or green infrastructure as essential components of the urban fabric. Biophilic cities, for example, focus on incorporating natural elements in urban design to improve inhabitants' sensory experiences and contact with nature to improve health and well-being, foster care and respect for nature, and foster resilience and adaptability, particularly in the face of climate change (Beatley and Newman 2013).

More recently, conceptualisations such as nature-based urbanism argue for designing with and for nature in cities to be integrated with processes and actions that seek justice for all species (Pineda-Pinto and Frantzeskaki 2023). Similarly, a regenerative urban imaginary seeks to go beyond sustainability and recognises the interconnections and interdependencies within and across urban boundaries. It seeks to minimise consumption, extraction, and impacts on ecosystems and the life that depends on them (Thomson and Newman 2018). Instead, it allows and supports the recovery of ecosystems to absorb, produce, and enhance the regenerative capacity of all interconnected systems. Three chapters in this book dive into the ideas behind the ideals of eco, green, nature-based, and regenerative cities to envision urban imaginaries with a mission to transform our current systems and transition to just and nature-positive futures.

For example, Chap. 8 positions integrating Indigenous knowledge as fundamental for planning nature-based cities. By proposing four pathways (thinking, organising, acting, and knowing), this chapter charts interlinked and interrelated priorities to effectively integrate nature in cities. Chapter 9 draws on the Three Horizons approach to put forward a paradigm shift to regenerative futures—one that seeks multispecies justice. In this chapter, planning and legal systems are reimagined through three horizons to achieve a future that recognises the rights of nature and plans for eco-commons through multispecies practices. Chapter 7 presents the efforts behind the model of greyfield-precinct regeneration. This model addresses the mission-scale challenge of regenerating the established, ageing, and occupied low-density greyfield suburbs of cities by re-developing at medium density with a careful integration of infrastructure retrofits with additional greenspace and services. At its core, this new model brings us an example currently being implemented in a Melbourne municipality that actively seeks to increase and enhance urban nature and provide greater access to local services by actively planning and implementing regenerative land-use and transport redevelopment at a precinct scale. A scale that is representative of master-planned greenfield and brownfield urban development rather than fragmented, piecemeal lot-by-lot redevelopment.

The three chapters in this section not only bring forward just transitions as a key element of building nature-based and regenerative futures, but also connect and transcend urban imaginaries of sustainable, resilient, circular, and healthy low-carbon cities.

1.3.4   Smart and Sustainable Cities

Cities are perhaps one of the most ingenious technologies that humans have invented, and this 'city technology' is still evolving. Viewing the city as a technology, or as home to technology, is a key theme in many urban imaginaries. The technology focus is important, because as humans are rapidly changing the planet that we live on and cities continue to adapt to the huge populations that few could have imagined at the beginning of the twentieth century, many believe that technology will help solve our current (and future) dilemmas.

According to this techno-optimist vision, new technologies—involving, for example, new materials, medical science, information technology, and artificial intelligence (AI)—are expected to solve many environmental, social, and economic problems. In the technologically driven Smart City vision (Chaps. 10 and 11), it is expected that technology will

- Monitor and optimise infrastructure performance, using smart sensors and AI
- Facilitate reduced and more efficient resource use
- Improve the breadth and speed of engagement among urban planners and managers and the public, creating more effective urban governance
- Optimise the transport of people and goods to reduce congestion and minimise costs and greenhouse emissions

This move towards a Smart City vision is part of the next wave of innovation. Several long waves of technological innovation have had a profound influence on industrial and urban development over time (Rodrigue 2020). Advances in transport technology have been associated with some of the most significant changes in urban form and fabric: from walking city to transit city (rail) to auto city and, most recently, the mega-city region, underpinned by high-speed rail, freeways, and broadband communications (Newton et al. 2024).

These significant reconfigurations of urban space and structure are explained by Marchetti's anthropological constant (Marchetti 1994), based on observations that time budgets for travel between home and place of work averaged an hour for a return journey and have remained so since the pre-industrial era. What changed was the mode and speed of travel, permitting the spread of built environments into the surrounding countryside and creating low-density suburban sprawl. This urban principle has been shaken by a surge in telecommuting by information workers, who now constitute the largest share of the modern industrial workforce, supplanting more 'hands-on' workers in agriculture, manufacturing, personal services, and retailing. Triggered by COVID but supported by advances in broadband communications technology and the Internet, telepresence has substituted for physical presence in most information economy-oriented workplaces, especially those in central business districts. This has been accompanied by significant population shifts to suburban and regional housing markets in search of greater living and working space. The planning and building sectors and related government agencies were unprepared for this shock. Indeed, the physical landscape of the urban agglomeration economy going forward is less clear, based as it is on the information economy—the principal outworking of the fifth wave of technological innovation, which is centred on digital networks, software, and new media.

Globally, cities are now experiencing the beginnings of the sixth long wave of innovation, where key drivers include renewable and distributed energy; a circular economy that includes water and sewage recycling and reuse and domestic, construction, and industrial waste recycling; electro-mobility; and increased automation, robotics, and AI. All are central to an urban sustainability transition, especially when combined with advances in digital information and communication technologies that can accelerate the needed change. This is critical as the window of opportunity for transition without significant economic and social dislocation is closing. Urban collaborations that can be both local and global in scope and operation and based on advanced digital infrastructure platforms (integrating IT, high-speed communications, data, and analytics) are illustrated in Chap. 10. The first of such innovation and creativity hubs, networked on digital platforms capable of linking geographically distributed groups of urban researchers and practitioners in real time, are now operating as prototypes and will become central to better facilitating the innovation that is core to the types of urban development projects envisioned in this book. New systems of governance are required to drive these new directions and rates of change, but the new infrastructure platforms are supportive of this evolutionary pathway.

1.3.5 Healthy Cities

What is it about cities that affect human health so profoundly? Since the beginning of the industrial era, human settlements have been faced with a series of health risks directly or indirectly associated with the type of built environment in which populations live and work. McMichael (2008) and Giles-Corti et al. (2016) have mapped these risks as infectious diseases (linked to poor sanitation and housing and lack of appropriate sewerage and water infrastructures); respiratory diseases (associated with air pollution from industry and transport); road trauma; obesity and associated non-communicable diseases (linked to lifestyle and the sedentary nature of urban living, working, and travelling); climate change-related health burdens (associated with increased temperatures and vector-borne diseases); and growing mental health and psychiatric disorders. While there has been significant progress in health outcomes over the past century, a range of urban-related health problems remain on the rise; for example, rates of diabetes and obesity are growing rapidly, while morbidity and mortality due to air pollution and increasing urban heat are challenges of growing concern. Importantly, many of the accelerating contemporary health problems are associated with the way that cities are built. Therefore, it is unsurprising that the World Health Organization is promoting an agenda of 'Healthy Cities'. The goal of healthy cities is also embedded in the United Nations Sustainable Development Goal 11.

The connections between urban planning, design, and public health are complex, being mediated by societal-scale political, economic, and social processes central to the development and management of built environments. Although it is less clear how these factors are causally related, the cost of not acting now on accumulated research evidence will be far greater in the future. Identifying built environment features associated with health outcomes can inform developments of future city

planning policies: 'designing-out' negative features and 'designing-in' good features that can promote health.

As built environments are socially produced, they represent modifiable determinants of urban health. The challenge is where best to intervene in the urban system
to deliver more sustainable outcomes involving human health. This requires visions
of key transformative interventions. The following two proposed city planning
imaginaries and pathways are not new, but have growing evidence-based support for
their efficacy from a health perspective:

- Make built environments more walkable and bikeable, with better public transport links connecting where people live with where they work, as well as better
  access to shops, services, and recreational places for undertaking more localised
  regular activities. This approach, the focus of Chap. 12, requires planning for
  more polycentric urban forms at the city level. Here there is a significant challenge for transport and land-use planning to be better integrated with
  twenty-first-century 'urban villages' concepts and designs (the 20-minute neighbourhood). Chapter 10 focuses on this municipal and neighbourhood scale,
  where there is a call for regenerative urban planning, requiring infrastructure
  retrofitting, redesign of neighbourhood road networks, urban greening, and
  increased mixed-use and medium-density residential redevelopment.
- Integrate nature-based services into urban design at all scales, ranging from lot
  to neighbourhood to city/catchment. These types of services are known to provide a range of health benefits to communities, including heat reduction, the
  opportunity to engage in physical activity, and well-being and mental health benefits. Therefore, the elimination of ecological inequalities needs to assume equal
  status to that of social inequality, especially in a future where urban heat is a
  leading economic and health threat. This introduces another hitherto intractable
  challenge: integrating strategic urban water planning with urban land-use and
  transport planning, a blue-green urban transition (a focus of Chap. 3).

1.4 Systemic Issues

In the mission-scale urban interventions that we have outlined in this book, the
focus has primarily been on achieving better human and environmental outcomes,
mostly in a focused way, but sometimes also in a more holistic way (e.g. in the
Resilient and Smart City imaginaries). With this diversity of imaginaries there is
both tension and contention between them, as well as within and between different
parts of affected communities. Therefore, here we also note that these tensions and
contentions need to be governed. Consequently, here we first introduce another
imaginary, the Just City, to note the importance of governing such tensions. Second,
we provide discussion on the synergies and trade-offs between the diverse imaginaries. Do we need to choose which imaginary to prioritise? Or do we need to develop

more comprehensive imaginaries that encompass a range of interventions? Or are we in fact enriched by the diversity of efforts and imaginaries?

1.4.1 Just Cities

Without consideration of justice, sustainability becomes impossible to achieve; thus, the notion of just transitions has become a key tenet of sustainability agendas (Bennett et al. 2019). Therefore, it is not surprising that notions of justice have been part of our urban imaginaries since cities and settlements began to take shape. Responding to calls for just processes, with attributes of democratic inclusivity, fairness, and equity, is key in attending to the vulnerabilities of marginalised populations, and the central prospect of 'the right to the city' has underlined this urban vision. Notions of justice in cities form a key part of urban resilience planning (Meerow et al. 2019), green city agendas (Cousins 2021), and healthy city planning (Corburn 2013). Justice forms part of each of these frameworks not only because it is normatively 'the right thing to do', but also because it makes decision-making more robust, provides a more integrative foundation of knowledge, reduces urban vulnerability and risks, and improves economic productivity. Importantly, growing social inequality represents both key pressure points and unintended consequences in cities, whilst at the same time being a key driver for societal transformation (Moglia et al. 2018).

Furthermore, with a backdrop of progressive agendas gathering strength in the last few decades, it is unsurprising that there has been a rapid growth of urban justice concerns. These are mainly driven by scholars and activists fighting for the recognition of diversity, inclusion, and equity in the way we envision change, create laws and regulations, and approach planning and design, and in the ways, we give shape to our urban environments. Nancy Fraser's (1998) approach, which is based on socio-economic redistribution, cultural recognition, and participatory equality as three tenets of justice, has provided a framework for activism, research, and counter-discourses, creating a diverse and pluralistic understanding of justice. Imaginaries like the *Just City* (Fainstein 2010) help us understand that principles of equity, democracy, and diversity can be aligned or conflict with each other, and that under the current neoliberal capitalist regime, policy reform for justice is possible, but requires political mobilisation. Harvey (2008), through the popularisation of his *Right to the City* ideal, instead argues that achieving urban justice will only be marginal under the capitalist system, and thus there is a need for social revolution. Very inspiringly, Harvey (2008, p. 23) asserts:

> The right to the city is far more than the individual liberty to access urban resources: it is a right to change ourselves by changing the city. It is, moreover, a common rather than an individual right since this transformation inevitably depends upon the exercise of a collective power to reshape the processes of urbanization. The freedom to make and remake our cities and ourselves is, I want to argue, one of the most precious yet most neglected of our human rights.

These stimulating frameworks and ideas have generated new ways of conceptualising, enacting, and questioning plural understandings of justice. Climate justice, environmental justice, ecological justice, and multispecies justice are some of the emerging themes. These new understandings highlight that inequalities, suffering, and invisibilities not only are represented through narrow socio-economic lenses, but also relate more broadly to our most pressing global challenges, such as climate change, biodiversity loss, worldwide social-ecological migration and displacement, and global pandemics.

These pluralities of justice apply to urban environments, and in many cases are born or are more pressing in these geographies. For instance, the environmental justice movement grew in numbers and strength in the USA as a movement against both toxic waste and environmental racism—that is, how issues of environmental pollution tend to fall on communities of colour, lower-income classes, and other minorities (Schlosberg 2007). Whilst these injustices are still present worldwide, environmental pollution is just another impact with which marginalised groups and communities are burdened. Climate change impacts will also increasingly affect these groups—cities with extreme heat, flooding, and other climatic events will experience the most damage and loss, particularly in those communities with fewer resources and capabilities to cope with them. More recent understandings of justice position urban environments as key places for imagining a just city for people, nature, and the planet. Ecological or multispecies justice argues for the recognition, fair treatment, and provisioning of habitat for all human and non-human species to flourish and exist in a state that allows for health and well-being outcomes for all (Celermajer et al. 2021; Pineda-Pinto et al. 2022).

All the chapters in this book bring to light issues of justice either as critical to achieving just outcomes for city-making processes or as a fundamental aspect for shifting existing narratives and paradigms. Planning for and shaping fairer, more accessible, and well-balanced urban environments have a strong connection to several sustainable development goals, including equity, biodiversity conservation, building resilience through climate change adaptation, and providing everyone with health and well-being through social, technological, and environmental improvements. Thus, most urban imaginaries intersect with one another. A resilient city is one that prepares for and can cope with current and future shocks. A just city seeks redistribution, multispecies recognition and participation, and the flourishing of all life through self-expression and self-determination on a shared common planet.

1.4.2   Trade-Offs and/or Synergies

This book has presented imaginaries of sustainable cities ranging from those associated with resilient cities to low-carbon circular cities, green and regenerative cities, those related to healthy cities, and everything in between. That said, somewhat unintentionally, nearly all chapters have addressed the different ways that cities need to adapt to climate change, and in some ways, this has become a common theme. There is also a strong overlap in the solutions and goals, with some solutions appearing across nearly all the imaginaries, such as nature-based solutions (incorporating urban forests, waterways, and urban parks). There have also been some themes that most, if not all, imaginaries have touched on. For example, the issue of equity and fairness is a common theme, as is the issue of human health. To some extent, this is unsurprising, given that most imaginaries have dealt with attempts to improve environmental outcomes, as they are recognised as an essential aspect of health and well-being.

Inherent in this, although it can sometimes be a false dichotomy, is a trade-off between human needs and the needs of other species and ecosystems. When this trade-off is present, human needs and environmental needs need to be balanced, which has infrequently been the case up to the present—and the results of this imbalance are now clear. Indeed, the Foresight, Research and Innovation team of ARUP, a large multinational consulting firm providing a range of services for the built environment, developed four plausible scenarios (ARUP 2023) that explore this trade-off and the delicate balance that needs to be found. Their scenarios highlight this powerfully, showing that the only viable path forward is to find this balance.

We also note that whilst balance is necessary, trade-offs are sometimes inevitable, and therefore occasionally the needs of some will have to be sacrificed. However, the needs that are sacrificed should be at first non-essential needs; in other words, we should strive to attend to the needs of as many as possible (human and nonhuman)—specifically, those needs that help us and other life forms lead a healthy, fulfilled, and good life, not the needs that enable destruction, consumption, and displacement. This is the basis of the new goal set out in Doughnut Economics (Raworth 2017): to stay within the safe and just space for humanity as defined through an ecological ceiling (nine dimensions) and a social foundation (12 dimensions), which is also closely aligned with the notion of missions-oriented innovation.

Along a similar vein, we note that the missions-oriented approach tends to require strong focus from governments and the community. This means that, at least in the short to medium term, there is limited bandwidth to achieve all the imaginaries that we have presented unless addressed collaboratively and holistically. There simply isn't sufficient funding and human capital to allow all the missions to be addressed simultaneously, and therefore there is a need to better engage and find a more unifying mission process that brings all the visions outlined in this book into a single coherent whole or sequenced according to some agreed level of importance. The emergent theme of climate adaptation and mitigation is a strong contender for a mission that brings together all the imaginaries in the book. That said, we also note that diversity and plurality of imaginations (here in terms of urban imaginaries) and the diversity of those involved in creating them is what will help us achieve the joint mission of resilient, healthy, just, and regenerative urban futures. Such tensions have been explored by Sharp et al. (2024), who note that the different framings at the same time enable pathways and collaborations, whilst at the same time obscuring other possibilities. Similarly, they note that a diversity in visions generates 'productive exchange between disconnected discourses and sheds light on possible blind spots' (Sharp et al. 2024, p. 13). This creative tension between the diversity of contributions and the need for coherence and prioritisation needs to be nurtured and carefully navigated.

Finally, another common theme is the recognition that to achieve the imaginaries presented, there needs to be the creation of a different type of governance and decision-making system—one that harnesses the increasingly powerful and ubiquitous digital platforms available for data collection, analysis, visualisation, communication, and engagement.

1.5 A Mission for Achieving Sustainable Development Goals in Cities

The big question addressed through the imaginaries in this book is: how can we adapt our cities, and by extension our societies, to meet the needs of the future? Eleven bold and ambitious urban imaginaries have been presented that stretch our imagination for what is possible whilst drawing on the latest available science. While they provide a diversity and mosaic of interdisciplinary solutions, ideas, and paradigm shifts, they also provide a window into what the future of cities may look like, albeit here framed in the context of Australian cities and perspectives. Dealing with the grand challenges that have been presented is not optional, but a necessity, so it is advisable that policymakers in government, industry, and the broader community take notice.

A major challenge that this book highlights is the need to rethink existing ways of organising and governing our cities, and many of the chapters propose or identify experimental actions. As Mazzucato (2018) also notes, progressing with missions requires the advancement of both basic research and innovations, as well as novel ways of combining existing innovations with each other (pp. 66, 74). From the sustainability-transitions literature, in the same vein, it is proposed that transformative change requires disruptive and conforming innovations that challenge the status quo in ways of thinking, organising, doing, and knowing (Frantzeskaki and Bush 2021; Loorbach et al. 2017). In Table 1.1, we list, by chapter, proposed experimental pathways that can advance a transformative agenda for future cities. Citing Mazzucato (2020, p. 105), 'Today's missions need to be nested on top of resilient systems and social and physical infrastructure'. As the contributions to this book note (again, Table 1.1), the needed transformations for creating future sustainable urban systems target ways of organising, knowing, and doing. The proposed experimentations and interconnected innovations are mostly in the areas of governance, planning, and knowledge. This is one of our concluding messages for planning, governing, and managing future cities: embrace new ways of thinking; develop new integrative, system-oriented, and transformative ways of organising and doing (linking tactical to strategic planning); and reform where we source our knowledge for future-oriented planning.

Table 1.1 Proposed transformative shifts and experimentations from this book's contributions to achieve a transformative mission for future cities in Australia

Chapter	Proposed shifts and experimentations	
2	Shift to an adaptive and anticipatory mode of urban governance that takes a systems-thinking approach that recognises uncertainties and opportunities Shift to solution-oriented thinking through an adaptive, integrative, and resilience-oriented urban-planning paradigm	Governance experimentation – Transforming ways of organising – Transforming ways of thinking.
3	Shift to a multi-scale urban water governance paradigm that embeds resilience and water-sensitive approaches	Governance experimentation – Transforming ways of organising and doing
4	Shift to understanding and designing climate innovations by also considering the temporal effects and scales in the context of climate change	Knowledge experimentation – Transforming ways of knowing and doing
5	Shift to interconnected systems thinking for transforming urban infrastructures towards low-carbon or net-zero	Infrastructure experimentation – Transforming ways of doing
6	Shift to a whole system thinking approach to chart innovations for circular economies in cities focusing on policy, practice, and infrastructures	Policy experimentation – Transforming ways of organising and doing
7	Shift to future-oriented planning by bringing evidence across sectors and objectives for greening greyfield precincts	Planning experimentation – Transforming ways of doing
8	Shift to strengthening people, nature, and place relations in cities by aligning planning, knowledge streams, and inclusive governance	Governance experimentation – Transforming ways of organising
9	Shift to new approaches for understanding, experiencing, and knowing nature in cities through a multispecies lens	Knowledge experimentation – Transforming ways of knowing and doing
10	Shift to collaborative governance for knowledge and innovation systems to be in synergy towards urban transformations	Governance experimentation – Transforming ways of organising and doing
11	Shift to policy innovation thinking and formulating for enabling and sustaining experimentation	Policy experimentation – Transforming ways of organising and doing
12	Shift to polycentric urban planning to enable active living as a cross-cutting way for urban planning (form, place-making, and people–place connections)	Planning experimentation – Transforming ways of doing

Finally, across all the urban imaginaries in this book, there is a call for rethinking the current paradigm and culture of urban planning and development; nested within the broader call for new systems of city governance. While we have often escaped

Fig. 1.2 Key arenas of applied research and innovation capable of mission-critical transitions for twenty-first-century cities

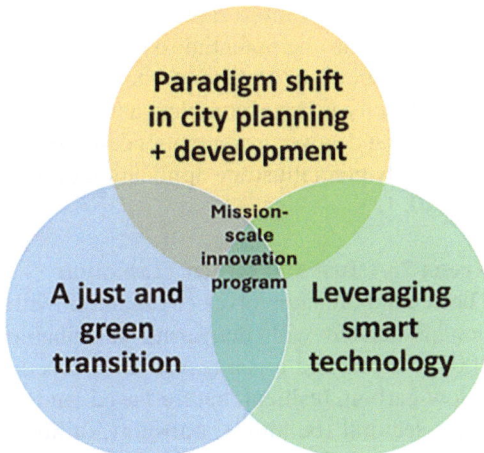

with poor practices in the past, from a social good perspective, the now-mounting problems and concerns demand a new paradigm—a new *kind* of city paradigm.

New paradigms, such as the New Urban Agenda, have been proposed, although it is alarming that there has been very little progress (United Nations 2020). Achieving this type of ambitious change, as already noted, requires a complete rethink about 'how we do cities'. What could our cities look like if we chose to comprehensively address current challenges? This book is not able to fully formulate what a future city should look like, but it outlines key pieces of the puzzle. Putting that puzzle together, and developing the final pieces of the puzzle, we argue, requires mission-oriented innovations.

1.5.1 Three Urban Innovation Arenas for Focusing Our Efforts

Taking a helicopter view on the challenge of future city-making suggests that there are three principal arenas where applied research and innovation need to be focused as illustrated by contributions to this book. They are sketched in Fig. 1.2. Each of the research arenas has been explored in this book.

Arena 1. Leveraging Smart Technology The first urban innovation arena involves taking advantage of the multiple technological advances occurring in what Batty (2018) has termed the fifth long wave of global innovation in information technology and digitalisation—the *digital transition*. Central to this transition is high-speed broadband communications delivering telepresence, the Internet and its IoT including synchronous collaboration platforms, an increasing array of sensing systems both ground-and-satellite-based contributing to big data and associated AI systems, and digital infrastructure platforms enabling interoperability of data and analytics and assembling a more rapid evidence base for urban decision-making. This new technology has the potential to overcome a key problem in urban governance, that is, the difficulty in observing, monitoring, and evaluating the impacts of interventions (the city is large, complex, and dynamic). This has the potential to make urban planning and governance evidence-based in a way it has rarely been in the past. Several chapters illustrate application of important new digital pathways (Chaps. 3, 5, 7, 10, 12).

Arena 2. A Just and Green Transition

The second arena involves engagement with the sixth long wave of global innovation associated with delivering sustainable development (Hargroves and Smith 2005)—a just and *green transition* focused on sustainable urban development that is low-carbon, resilient, nature-based, and healthy (Chaps. 2–11). This transition is multi-sectoral (buildings, transport, utilities, manufacturing, domestic) and multi-scale (household-precinct-city-region) and operates on green economy principles, for example, circular economy, regenerative development—an even greater challenge for applied research than sustainable development (Girardet 2015; Newton et al. 2022).

The challenge before us (as outlined in Chap. 10) is exploring the potential for increased integration and an accelerated convergence between fields of research associated with digitalisation and sustainable urban development. The challenges are formidable but need to be articulated as a focus for a mission-oriented response. OECD (2023) is the first global organisation to begin pursuing this goal. Convergence research has been identified as a fundamental underlying principle of scientific progress that assembles and integrates all relevant capabilities to answer contemporary grand challenges (Bainbridge and Roco 2016). It is a critical arena for applied research central to a much-needed accelerated transition to smart and sustainable urban development.

Arena 3. Paradigm Shift in City Planning and Development

The third innovation arena involves *transforming city governance and institutions* to enable smart sustainable cities. As outlined in Chap. 2, the engine that drives changes in the city is governance and decision-making. As also outlined throughout the chapters in this book, these decisions include how to accommodate growth in population through urban development, policy, and planning decisions in relation to transport systems, decisions about how to source energy for the city, how to create more affordable and accessible housing systems, or how to support nature in cities. It is recognised that to support cities that are more likely to achieve all sustainable development goals, there is a need to change the decision-making paradigm that supports city governance and urban development. A key focus in the paradigm needs to be how to support 'the commons' (i.e. shared community resources, culture, and institutions) and social good, in a planning and decision context that is commonly dominated by financial and development priorities.

To change the city, therefore, there is a need to reform city governance and associated institutions. To achieve the visions outlined in each of the chapters, as per

Table 1.1, a range of such reforms have been proposed, specifically ways of learning and knowing (as discussed in Chap. 2 on resilient cities, as well as described in Chap. 11 on experimentation), as well as changing the values and institutional rules that provide a basis and context for decision-making. Given the high levels of socio-technical complexity, and need to adapt to local contexts, experimentation is warranted as a guide for all types of reforms, including governance, knowledge, infrastructure, policy, and planning experimentation.

But how do governance and institutional reform happen? The answer to this is still an open question as the science and understanding of how reform happens is still evolving. A key emerging concept in relation to this question is the notion of 'mainstreaming'. As argued by Adams et al. (2024), mainstreaming is a process-oriented way of thinking and should be 'considered a strategy for doing sustainability transitions', involving the purposeful and cumulative actions of multiple actors. It needs to consider involving actors in a diverse set of roles, as well as the institutional spaces as they evolve, and their various mechanisms. Urban innovation research and development should have a capacity for mainstreaming. A key mechanism in mainstreaming is the process of experimentation which has been explored in Chap. 11. Chapter 10 further explores the role of digital innovations in the governance of cities, and to support mainstreaming.

1.5.2   An Innovation System to Support an Urban Transformation Mission

This book has provided critical inputs to help guide an urban transformation mission for the twenty-first century that both acknowledges the critical challenges that humanity is facing with the urgent environmental and ecological crisis, as well as the opportunities and risks presented by new technology. Most of the grand urban challenges and pressures faced by cities and their residents in advanced post-industrial Western societies are global in nature and are common to those identified in Australia. Notwithstanding, all chapters in this book are largely based on the Australian context, so the question arises, how well is the Australian innovation and governance system positioned to support a mission-scale *response* to the identified key challenges?

Until the release of a draft National Urban Policy in May 2024 (Australian Government 2024), there had been a gap of 13 years since its predecessor had been introduced (Department of Infrastructure and Transport 2011), A lack of continuity in national leadership in urban policy and settlement planning and development has characterised the last half century in Australia. Labor governments have launched several mission-scale urban programmes over this period, The decentralisation programme of the Whitlam Labor Government (1972–1975) targeted specific regional cities for Growth Centre investment (Bolleter et al. 2021), as well as provisioning sewerage for the outer suburbs of the larger cities and reinvesting in public housing construction. This administration was short-lived and the Liberal Government that succeeded it for the next decade, in common with most conservative governments in Australia, abrogated leadership in this space by pointing to the constitutional role of state governments in relation to city development and land use planning. It was not until the election of the next federal Labor government (Hawke-Keating, 1983–1996) that the next mission-scale urban programme emerged: *Building Better Cities*. Focused on the challenge of brownfield redevelopment, this programme's partners in state and local government in collaboration with the property development and construction sector developed a new model capable of successful implementation and replication in all major cities of Australia (Newton and Thomson 2017), continuing to the present with incremental enhancements to sustainability performance. The City Deals programme instituted in 2016 by the Liberal (Turnbull-Morrison) government became an imported UK initiative that has been widely criticised as 'ad hoc' and lacking in an overarching urban policy strategy (Burton and Nicholls 2019); and not supported by the incoming Albanese Labor government in 2022. Over the past decade there have been persistent calls for a national vision and plan for Australia's future settlement system such as that revealed in the Federal Inquiry into the Australian government's role in the development of cities *Building Up and Moving Out*. This report identified twin mission-scale challenges: creating both sustainable densification programmes for the major cities and effective decentralisation programmes targeting regional cities (Commonwealth of Australia 2018). Early reviews of the draft National Urban Policy have highlighted key weaknesses. Apart from it identifying five high-level goals (liveable, equitable, productive, sustainable, resilient) and six principles necessary for successful city planning, Freestone and Webb (2024, p. 1) considered 'there are no specifics in the national objectives, challenges and possible responses ... on how the desired urban transformation will be achieved in practice'.

It was also in 2023 that the federal government updated its 2015 National Science and Research Priorities (Australian Government 2023), which should desirably mesh with its national urban policies, also seeking to be innovative and evidence based. In Australia, urban research and innovation is reliant on comparatively narrow sources of funding and resourcing provided by key federal agencies such as Cooperative Research Centres (CRCs) and the Australian Research Council (ARC). Established in 1990, CRCs are a unique model of large, multi-year applied research initiatives supporting industry–government–research collaborations (Newton et al. 2009). Industry increasingly dictates the agenda of research projects in CRCs, although such projects tend to aim for a mix of economic and social outcomes and are often aligned with National Priorities, which currently lack any built environment directive. Two of the chapters in this book present urban innovation that has emerged from CRCs, that is, the Water Sensitive Cities CRC (Chap. 3) and the Low Carbon Living CRC (Chap. 5). However, from an urban transformation mission perspective, it could be said that CRCs tend to focus on sectoral problems that limit the opportunities for systems-oriented innovation needed for city-scale impact. This means that, for CRCs to become a suitable home for an urban transformation mission, they would need to expand their scope, level, and duration of funding (beyond the typical 7 years), and model of governance.

Research funded through the ARC, on the other hand, provides funding for a wide spread of smaller scale research on a case-by-case basis. Depending on the scheme, projects are either industry-led (Linkage) or academic-led (Discovery). This shorter-term and smaller-scale focus means that it again is not the appropriate mechanism for supporting an urban transformation mission. Furthermore, ARC projects tend to be more narrowly targeted, discipline-oriented, and based on academic credentials, thus often lacking a necessary transdisciplinary approach, and tending to be less applied than is required for an urban transformations mission. Recent media coverage also highlights several problematic issues with the ARC funding, including its politicisation, not covering the full costs of university research (let alone for the applications), the complexity of the application process, and inadequate evaluation metrics and processes (Message 2024; McCarthy 2023).

A scan of international programmes in these areas suggests that it is necessary to look to the European Union for leadership and a blueprint for a mission-scale research, innovation, and implementation programme capable of delivering more sustainable urban development at scale. *Horizon Europe* is such a programme (European Parliament 2023). A successor to previous Horizon 2020 and earlier Framework Programs extending back 30 years (European Commission 2015), it provides significant levels of financial support (currently 95 billion euros) and continuity for research innovation and collaboration unequalled in Australia. Horizon Europe (2021–2027) explicitly targets mission-scale initiatives that have been strongly influenced by the research and writings of Professor Mariana Mazzucato who likewise stimulated the ideas behind this book. Two of the five mission-scale programmes targeted in Horizon Europe in addition to cancer, oceans, and soil are climate and cities. These require the twin transitions of climate change mitigation, adaptation, and resilience and the digital innovations underpinning smart cities of the future with a capacity to achieve climate neutrality. These are two of the three mission-scale arenas of applied research and innovation identified by contributors to this book (again, see Fig. 1.2). The third involves a paradigm shift in city planning and development processes, largely focussed on new systems of governance capable of realising the six key planning principles outlined in the 2024 National Urban Policy. *Together* they can deliver the scale of step change transformative innovations necessary for delivering sustainable urban development this century.

Noteworthy, from the perspective of this book, one of the missions supported Horizon Europe is the 'Climate-neutral and Smart Cities' mission (Beretta and Bracchi 2023). As part of this mission, 100 European cities were selected to become climate-neutral by 2030, highlighting bold and ambitious, yet widely supported innovation missions that both industry and academia can collaborate on. The focus on tangible outcomes, embedded in legislative requirements through climate City Contracts that are co-created with local stakeholders and citizens, and a focus on place-based experimentation and innovation hubs serve as examples of the scale and systemic focus that is required.

The editors of this book argue that this type of ambitious and comprehensive mission, connected in a binding fashion into civil society and business, as we have seen in the EU, is needed in Australia, and indeed globally, to help achieve the cities of the future that are resilient, climate-neutral, climate-adapted, nature-based, regenerative, smart, sustainable, and healthy. If we can achieve this, the future for humanity looks every bit so much brighter.

References

Adams C, Frantzeskaki N, Moglia M (2023a) Mainstreaming nature-based solutions in cities: a systematic literature review and a proposal for facilitating urban transitions. Land Use Policy 130

Adams C, Frantzeskaki N, Moglia M (2023b) Space for mainstreaming? Learning from the implementation of urban forest strategies in metropolitan Melbourne. Aust Plan 59:154–169

Adams C, Frantzeskaki N, Moglia M (2024) Actors mainstreaming nature-based solutions in cities: a case study of Melbourne's change agents and pathways for urban sustainability transformations. Environ Sci Policy 155:103723

ARUP (2023) Four plausible futures: 2050 scenarios—ARUP. https://youtu.be/waeysF6h6po?si=_kIfPLQb8Sr7mpNL

Australian Government (2023) Australia's draft National Science and Research Priorities. Commonwealth of Australia, Canberra. https://consult.industry.gov.au/sciencepriorities2

Australian Government (2024) National Urban Policy. Consultation Draft, May. Canberra. https://www.infrastructure.gov.au/sites/default/files/documents/draft-national-urban-policy.pdf

Bonakdar A, Audirac I (2021) City planning, urban imaginary, and the branded space: untangling the role of city plans in shaping Dallas's urban imaginaries. Cities 117:103315. https://doi.org/10.1016/j.cities.2021.103315

Bainbridge WS, Roco MC (2016) Handbook of science and technology convergence, Springer Cham. https://doi.org/10.1007/978-3-319-07052-0

Batty M (2018) Inventing future cities. MIT Press, Cambridge, MA

Beatley T, Newman P (2013) Biophilic cities are sustainable, resilient cities. Sustainability. 5(8):3328–3345

Bennett NJ, Blythe J, Cisneros-Montemayor AM, Singh GG, Sumaila UR (2019) Just transformations to sustainability. Sustainability 11:3881

Beretta I, Bracchi C (2023) Climate-neutral and smart cities: a critical review through the lens of environmental justice. Front Sociol 8. https://doi.org/10.3389/fsoc.2023.1175592

Bolleter J, Freestone R, Cameron R, Wilkinson G, Hooper P (2021) Revisiting the Australian Government's Growth Centres programme 1972–1975. Plan Perspect 36(5):999–1023. https://doi.org/10.1080/02665433.2021.1885479

Broström A, Buenstorf G, McKelvey M (2021) The knowledge economy, innovation and the new challenges to universities: introduction to the special issue. Innovation 23:145–162

Burton P, Nicholls L (2019) A patchwork of City Deals or a national settlement strategy: what's best for our growing cities? The Conversation, Jul 25

Carpenter SR, Arrow KJ, Barrett S, Biggs R, Brock WA, Crépin A-S, Engström G, Folke C, Hughes TP, Kautsky N, Li C-Z, McCarney G, Meng K, Mäler K-G, Polasky S, Scheffer M, Shogren J, Sterner T, Vincent JR, Walker B, Xepapadeas A, Zeeuw AD (2012) General resilience to cope with extreme events. Sustainability 4:3248–3259

Celermajer D, Schlosberg D, Rickards L, Stewart-Harawira M, Thaler M, Tschakert P, Verlie B, Winter C (2021) Multispecies justice: theories, challenges, and a research agenda for environmental politics. Environ Polit 30:119–140

Commonwealth of Australia (2018) Building up & moving out: inquiry into the Australian Government's role in the development of cities. Canberra Australia. https://www. aph.gov.au/Parliamentary_Business/Committees/House/Former_Committees/ITC/ DevelopmentofCities/Report

Copernicus Institute (2024) 2023 is the hottest year on record, with global temperatures close to the 1.5 degrees limit. https://climate.copernicus.eu/copernicus-2023-hottest-year-record#:~:text=2023%20is%20confirmed%20as%20the,highest%20annual%20value%20 in%202016

Corburn J (2013) Healthy city planning: From neighbourhood to national health equity. Routledge, London

Cork S, Alexandra C, Alvarez-Romero JG, Bennett EM, Berbés-Blázquez M, Bohensky E, Bok B, Costanza R, Hashimoto S, Hill R, Inayatullah S, Kok K, Kuiper JJ, Moglia M, Pereira L, Peterson G, Weeks R, Wyborn C (2023) Exploring alternative futures in the Anthropocene. Annu Rev Environ Resour 48

Cousins JJ (2021) Justice in nature-based solutions: research and pathways. Ecol Econ 180:106874

Coutts AM, Tapper NJ, Beringer J, Loughnan M, Demuzere M (2013) Watering our cities: the capacity for water sensitive urban design to support urban cooling and improve human thermal comfort in the Australian context. Prog Phys Geogr Earth Environ 37:2–28

Crane M, Lloyd S, Haines A, Ding D, Hutchinson E, Belesova K, Davies M, Osrin D, Zimmermann N, Capon A, Wilkinson P, Turcu C (2021) Transforming cities for sustainability: a health perspective. Environ Int 147:106366

De Sa TH, Mwaura A, Vert C, Mudu P, Roebbel N, Tran N, Neira M (2022) Urban design is key to healthy environments for all. Lancet Glob Health 10:e786–e787

Department of Infrastructure and Transport (2011) Our cities, our future: a national urban policy for a productive, sustainable and liveable future. Australian Government

Dunn N (2018) Urban imaginaries and the palimpsest of the future. In: Lindner C, Meissner M (eds) The Routledge companion to urban imaginaries. Routledge, New York

European Commission (2015) Horizon Magazine Special Issue. EU Research Framework Programmes 1984–2014. https://sfe.lnl.infn.it/wp-content/uploads/2015/11/Horizon-Magazine-EU-Research-Framework-Programmes-1984-2014.pdf

European Parliament (2023) Horizon Europe research and innovation missions. State of play. European Parliamentary Research Service. PE 698.915—October. https://www.europarl. europa.eu/RegData/etudes/BRIE/2022/698915/EPRS_BRI(2022)698915_EN.pdf

European Parliamentary Research Service (2024) Circular economy. https://www.europarl.europa. eu/thinktank/infographics/circulareconomy/public/index.html

Fainstein S (2010) The just city. Cornell University Press, Ithaca, NY

Frantzeskaki N, Bush J (2021) Governance of nature-based solutions through intermediaries for urban transitions—a case study from Melbourne, Australia. Urban For Urban Green 64:127262

Frantzeskaki N, McPhearson T, Kabisch N (2021) Urban sustainability science: prospects for innovations through a system's perspective, relational and transformations approaches. Ambio 50:1650–1658

Fraser N (1998) Social justice in the age of identity politics: redistribution, recognition, participation. In: Peterson GB (ed) The Tanner lectures on human values. Stanford University, Stanford, CA

Freestone R, Webb B (2024) Labor's new National Urban Policy is welcome. But will it be transformative?. The Conversation, May 30

Giles-Corti B, Vernez-Moudon A, Reis R, Turrell G, Dannenberg AL, Badland H, Foster S, Lowe M, Sallis JF, Stevenson M, Owen N (2016) City planning and population health: a global challenge. Lancet 388:2912–2924

Girardet H (2015) Creating regenerative cities. Routledge, New York

Greenfield A, Moloney S, Granberg M (2022) Climate emergencies in Australian local governments: from symbolic act to disrupting the status quo? Climate 10

Griffith S (2022a) The Big Switch: Australia's electric future. Black Inc., Melbourne

Griffith S (2022b) Electrify everything: an optimist's playbook for our clean energy future. MIT Press, Cambridge, MA

Güneralp B, Seto KC, Ramachandran M (2013) Evidence of urban land teleconnections and impacts on hinterlands. Curr Opin Environ Sustain 5:445–451

Hargroves K, Smith M (eds) (2005) The natural advantage of nations: business opportunities, innovation and governance in the 21st century. Routledge, London

Harvey D (2008) The right to the city. https://newleftreview.org/issues/ii53/articles/david-harvey-the-right-to-the-city

Harvey-Scholes C, Mitchell C, Britton J, Lowes R (2023) Citizen policy entrepreneurship in UK local government climate emergency declarations. Rev Policy Res 40:950–971

Howarth C, Lane M, Fankhauser S (2021) What next for local government climate emergency declarations? The gap between rhetoric and action. Clim Change 167

Hughes S, Hoffmann M (2020) Just urban transitions: toward a research agenda. WIREs Climate Chang 11:e640

Jasanoff S, Kim S-H, Sperling S (2007) Sociotechnical imaginaries and science and technology policy: a cross-national comparison. Harvard STS Program, Cambridge, MA

Kara S, Hauschild M, Sutherland J, McAloone T (2022) Closed-loop systems to circular economy: a pathway to environmental sustainability? CIRP Ann 71:505–528

Keith M, Birch E, Buchoud NJA, Cardama M, Cobbett W, Cohen M, Elmqvist T, Espey J, Hajer M, Hartmann G, Matsumoto T, Parnell S, Revi A, Roberts DC, Saiz E, Schwanen T, Seto KC, Tuts R, Van Der Pütten M (2023) A new urban narrative for sustainable development. Nat Sustain 6:115–117

Kellert S (2016) Biophilic urbanism: the potential to transform. Smart and Sustainable. Built Environ 5

Lee S, Kim Y (2021) A framework of biophilic urbanism for improving climate change adaptability in urban environments. Urban For Urban Green 61:127104

Loorbach D, Frantzeskaki N, Avelino F (2017) Sustainability transitions research: transforming science and practice for societal change. Annu Rev Environ Resour 42:599–626

Marchetti C (1994) Anthropological invariants in travel behavior. Technol Forecast Soc Change 47:75–88

Mazzucato M (2018) Mission-oriented innovation policies: challenges and opportunities. Ind Corp Change 27:803–815

Mazzucato M (2019) Governing missions: governing missions in the European Union. European Commission, Brussels

Mazzucato M (2020) Mission economy: a moonshot guide to changing capitalism. Allen Lane

McCarthy GM (2023) A major review has recommended more independence for decisions about research funding in Australia. The Conversation. https://theconversation.com/a-major-review-has-recommended-more-independence-for-decisions-about-research-funding-in-australia-204184. Accessed 13 May 2024

McDonough W (2008) Cradle to cradle: remaking the way we make things. Henry Holt, New York

McMichael AJ (2008) The urban environment and health. In: Newton PW (ed) Transitions: pathways towards sustainable urban development in Australia. Springer, Dordrecht

McPhearson T, Iwaniec DM, Bai X (2017) Positive visions for guiding urban transformations toward sustainable futures. Curr Opin Environ Sustain 22:33–40

Meerow S, Newell JP (2019) Urban resilience for whom, what, when, where, and why? Urban Geogr 40:309–329

Meerow S, Pajouhesh P, Miller TR (2019) Social equity in urban resilience planning. Local Environ 24:793–808

Meissner M, Lindner C (2018) The Routledge companion to urban imaginaries. Routledge

Message K (2024) Research funding has been politicised and universities are losing public trust—is this the year that will reverse those trends? Australian Broadcasting Corporation. https://www.abc.net.au/religion/university-research-funding-and-declining-public-trust/103293026. Accessed 13 May 2024

Moglia M, Cork SJ, Boschetti F, Cook S, Bohensky E, Muster T, Page D (2018) Urban transformation stories for the 21st century: Insights from strategic conversations. Glob Environ Change 50:222–237

Nerlich B, Morris C (2015) Imagining imaginaries. University of Nottingham. https://blogs.nottingham.ac.uk/makingsciencepublic/2015/04/23/imagining-imaginaries/

Newton P, Bai X (2008) Transitioning to sustainable urban development. In: Newton PW (ed) Transitions: pathways towards sustainable urban development in Australia. Springer, Dordrecht

Newton P, Newman P (2015) Critical connections: the role of the built environment sector in delivering green cities and a green economy. Sustainability (Switzerland) 7:9417–9443

Newton P, Thomson G (2017) Urban regeneration in Australian cities. In: Roberts P, Sykes H, Granger R (eds) Urban regeneration: a handbook, revised edition. Sage, London

Newton P, Hampson K, Drogemuller R (2009) Transforming the built environment through construction innovation. In: Newton PW (ed) Technology, design and process innovation in the built environment, Spon research series. Taylor & Francis, London

Newton P, Prasad D, Sproul A, White S (2019) Decarbonising the built environment. Charting the transition. Palgrave Macmillan/Springer, Singapore

Newton PW, Newman PWG, Glackin S, Thomson G (2022) Greening the greyfields: new models for regenerating the middle suburbs of low-density cities. Springer, Singapore

Newton P, Whitten J, Glackin S, Reynolds M, Moglia M (2024) Prospects for a megacity-region transition in Australia: a preliminary examination of transport and communication drivers. Sustainability 16:3712. https://doi.org/10.3390/su16093712

Nieuwenhuijsen MJ (2020) Urban and transport planning pathways to carbon neutral, liveable and healthy cities; a review of the current evidence. Environ Int 140

Nieuwenhuijsen MJ, Khreis H (2016) Car free cities: pathway to healthy urban living. Environ Int 94:251–262

OECD (2022) Tackling policy challenges through public sector innovation.

OECD (2023) Navigating green and digital transitions: five imperatives for effective STI policy. OECD Science, Technology and Industry Policy Papers, No. 162, Paris

Pineda-Pinto M, Frantzeskaki N (2023) Chapter 15: Nature-based urbanism: designing for and with nature for sustainable cities and communities. In: McCormick K, Evans J, Voytenko Palgan Y, Frantzeskaki N (eds) A research agenda for sustainable cities and communities. Edward Elgar Publishing, Cheltenham

Pineda-Pinto M, Frantzeskaki N, Chandrabose M, Herreros-Cantis P, McPhearson T, Nygaard CA, Raymond CM (2022) Planning ecologically just cities: a framework to assess ecological injustice hotspots for targeted urban design and planning of nature-based solutions. Urban Policy Res 40:206–222

Raworth K (2017) Doughnut economics: Seven ways to think like a 21st century economist. Penguin, London

Reeve AC, Desha C, Hargreaves D, Hargroves K (2015) Biophilic urbanism: contributions to holistic urban greening for urban renewal. Smart Sustain Built Environ 4:215–233

Richardson K, Steffen W, Lucht W, Bendtsen J, Cornell SE, Donges JF, Drüke M, Fetzer I, Bala G, Von Bloh W, Feulner G, Fiedler S, Gerten D, Gleeson T, Hofmann M, Huiskamp W, Kummu M, Mohan C, Nogués-Bravo D, Petri S, Porkka M, Rahmstorf S, Schaphoff S, Thonicke K, Tobian A, Virkki V, Wang-Erlandsson L, Weber L, Rockström J (2023) Earth beyond six of nine planetary boundaries. Sci Adv 9:eadh2458

Rockström J, Gupta J, Qin D, Lade SJ, Abrams JF, Andersen LS, Armstrong McKay DI, Bai X, Bala G, Bunn SE, Ciobanu D, Declerck F, Ebi K, Gifford L, Gordon C, Hasan S, Kanie N, Lenton TM, Loriani S, Liverman DM, Mohamed A, Nakicenovic N, Obura D, Ospina D, Prodani K, Rammelt C, Sakschewski B, Scholtens J, Stewart-Koster B, Tharammal T, Van Vuuren D, Verburg PH, Winkelmann R, Zimm C, Bennett EM, Bringezu S, Broadgate W, Green PA, Huang L, Jacobson L, Ndehedehe C, Pedde S, Rocha J, Scheffer M, Schulte-Uebbing L, De Vries W, Xiao C, Xu C, Xu X, Zafra-Calvo N, Zhang X (2023) Safe and just Earth system boundaries. Nature 619:102–111

Rodrigue J-P (2020) Transportation, economy and society. In: Rodrigue J-P, Comtois C, Slack B (eds) The geography of transport systems. Routledge, London

Schlosberg D (2007) Defining environmental justice: theories, movements, and nature. Oxford University Press, Oxford

Seto KC, Güneralp B, Hutyra LR (2012) Global forecasts of urban expansion to 2030 and direct impacts on biodiversity and carbon pools. Proc Natl Acad Sci U S A 109:16083–16088

Sharma AK, Gardner T, Begbie D (2018) Approaches to water sensitive urban design: potential, design, ecological health, urban greening, economics, policies, and community perceptions. Elsevier, Amsterdam

Sharp D, Raven R, Farrelly M (2024) Pluralising place frames in urban transition management: net-zero transitions at precinct scale. Environ Innov Soc Transit 50:100803

Simkin RD, Seto C, McDonald RI, Jetz W (2022) Biodiversity impacts and conservation implications of urban land expansion projected to 2050. Proc Natl Acad Sci U S A 119:e2117297119

Steffen W, Crutzen PJ, McNeill JR (2007) The Anthropocene: are humans now overwhelming the great forces of nature? Ambio 36:614–621

Taylor C (2002) Modern social imaginaries. Duke University Press, Durham, NC

Thomson G, Newman P (2018) Urban fabrics and urban metabolism—from sustainable to regenerative cities. Resour Conserv Recycl 132:218–229

UNEP (2023a) Emissions gap report 2023. https://www.unep.org/resources/emissions-gap-report-2023

UNEP (2023b) Sustainable cities. Home regions Asia and the Pacific regional initiatives supporting resource efficiency. https://www.unep.org/regions/asia-and-pacific/regional-initiatives/supporting-resource-efficiency/sustainable-cities

United Nations (2017) New urban Agenda. Habitat III Secretariat, New York

United Nations (2020) The sustainable development goals report 2020. UNDESA, New York

Walker B, Holling CS, Carpenter SR, Kinzig A (2004) Resilience, adaptability and transformability in social-ecological systems. Ecol Soc 9

Watkins J (2015) Spatial imaginaries research in geography: synergies, tensions, and new directions. Geogr Compass 9:508–522

Winslow J, Coenen L (2023) Sustainability transitions to circular cities: experimentation between urban vitalism and mechanism. Cities 142:104531

Magnus Moglia is an Associate Professor of Urban Sustainability at the Centre for Urban Transitions, Swinburne University of Technology, Australia. He has a PhD in environmental management and complex systems science, from the Australian National University's Crawford School for Public Policy, as well as degrees in engineering, physics, and mathematics from the Royal Institute of Technology in Sweden. From 2001 to 2020, he was a researcher at the CSIRO, Australia's National Science Agency, holding the positions of Principal Research Scientist and Research Team Leader in Urban Systems Science. With a background in both physical and social sciences and extensive experience in participatory and foresight methods, he is a transdisciplinary researcher, focusing on how solutions for sustainability can be equitably promoted and implemented in cities and regions. His research topics include water management, climate adaptation, freight and transport decarbonisation strategies, sustainable agriculture, circular economy, urban regeneration, working-from-home practices, and infrastructure resilience. Much of his research has been in the international context, including in Vietnam, Laos, and Kiribati. He has published more than 70 peer-reviewed journal articles, as well as numerous book chapters and industry reports. He is currently the lead Chief Investigator for projects on circular economy, freight decarbonisation, intermodal rail, and sustainable aquaculture.

Niki Frantzeskaki is a Chair Professor of Regional and Metropolitan Governance and Planning, Section Spatial Planning, Geosciences Faculty, Utrecht University, the Netherlands. Her expertise is on urban transitions and transformations, their governance and planning, with a focus on achieving climate change adaptation and mitigation, sustainability, and resilience. Her research also focuses on the governance and planning of nature-based solutions to enhance climate change resilience and promote more just urban futures. She has a rich international research experience with a portfolio of ongoing projects in Australia, Canada, and the USA. She has been a Highly Cited Researcher awardee from Clarivate Analytics in 2020 and 2021, placing her in the top 1% of researchers globally in the cross-field of social sciences and ecology. From 2019 to 2021, she has been a Research Professor and Director of the Centre for Urban Transitions at Swinburne University of Technology, Melbourne, Australia. From 2010 to 2019, she has been an Associate Professor at the Dutch Research Institute for Transitions, affiliated with Erasmus University Rotterdam. She has published over 100 peer-reviewed articles and 18 special issues and released four books on urban sustainability transitions in 2017, 2018, and 2020.

Peter Newton is an Emeritus Professor in the Centre for Urban Transitions at Swinburne University of Technology in Melbourne, Australia. From 2007 to 2021, he held the position of Research Professor, following 15 years as Chief Research Scientist with the Commonwealth Scientific and Industrial Research Organisation (CSIRO). He has had a distinguished career in built environment research and was elected as a Fellow of the Academy of Social Sciences in Australia in 2014. He holds the distinction of being the only academic in Australia to have held senior leadership positions in all competitively funded national research centres: the Centre for Geographic Information Systems and Analysis, the Australian Housing and Urban Research Institute, the CRC for Construction Innovation, the CRC for Spatial Information, the CRC for Water Sensitive Cities, and the CRC for Low Carbon Living. He has also been a Member of the Board of The International Council for Research and Innovation in Building and Construction, as well as the Board of the Australian Urban Research Infrastructure Network, including a period as Interim Director in 2022. His principal fields of research have focused on the technology of planning, sustainability science, and urban transitions. He has published over 25 books, his most recent titles include *Greening the Greyfields: New Models for Regenerating the Middle Suburbs of Low-Density Cities*; *Migration and Urban Transitions in Australia*; *Decarbonising the Built Environment: Charting the Transition*, and *Resilient Sustainable Cities: A Future*.

Melissa Pineda Pinto is a postdoctoral researcher at Trinity College, Dublin. She works on the NovelEco project, which examines wild ecosystems in cities through forecasting methodologies and policy analysis. Melissa completed her PhD at Swinburne University of Technology, where she critically explored nature-based solutions through an ecological justice lens. She holds a Master of Environment from the University of Melbourne. This work draws on my previous experience in the architectural and planning industries and not-for-profit sectors. Her transdisciplinary collaborations and work span different geographies, including Australia, Europe, Latin America, and North America. Her academic experience and interests cut across social research methods, inter-transdisciplinary collaboration, and systems thinking in the context of urban ecosystems, justice, and ethics. Her research examines urban nature through diverse justice lenses for achieving sustainable futures. As an early-career researcher, Mel has published 10 peer-reviewed journal articles, book chapters, and popular science pieces. She will serve as the Chief Investigator in 2024 of a project exploring urban biodiversity and justice through community engagement in the cities of Melbourne, Australia and San José, Costa Rica.

Deo Prasad is a Scientia Professor in the field of sustainable built environments at the University of New South Wales in Sydney, Australia. His expertise covers sustainable, low carbon, smart, resilient, and regenerative buildings and cities. He has published over 300 refereed publications including ten books. The last of his books on Climate Emergency: Towards a Net Zero Carbon Built Environment (Palgrave) was released early in 2023. He is currently the CEO of the NSW Decarbonisation Innovation Hub and previously served as the CEO of the Co-operative Research

Centre for Low Carbon Living. Deo has received acknowledgement of his contributions from all levels of government in Australia, including an Order of Australia, a Fellowship of the Australian Academy of Technological Sciences and Engineering, Fellow of the Australian Institute of Architects, NSW Government's Green Globe Award, and the Global Impact Award as well as the National Leadership in Sustainability Prize from the Australian Institute of Architects. He is currently leading the commercialisation of proof-of-concept technologies and systems by creating an industry-government, research collaborative community at scale.

Part I
Resilient Cities

Chapter 2
A New Paradigm for Resilient Urban Infrastructure Planning: Game-Changing Interventions, Tipping Points, and Capacities

Magnus Moglia, Russell M. Wise, and Seona Meharg

Abstract What infrastructure do Australian cities need over the next century? Planning for, delivering, and maintaining infrastructure that is usually long-lived and expensive in a rapidly changing environment is difficult. Complexity and uncertainties are at play, with potentially serious consequences to be considered. Specifically, current infrastructure-investment risks are not fit for purpose and would fail to meet the rapidly evolving needs of communities and economic activities. This may create lock-in situations that are difficult to adapt to or reverse; they therefore close down opportunities for the transformation needed to reduce systemic risks. In this chapter, we argue that the key to cities' resilience lies in the people and organisations having capacity, competencies, and governance for systemic interventions based on adaptive learning and decision-making. We outline a method for changing the planning of infrastructure to meet urgent urban challenges, such as climate change, rapid technological change, and pandemics. This is based on a mission-oriented programme of innovation that guides, underpins, and supports inclusive and robust infrastructure decisions. Finally, we introduce principles that can trigger a set of tipping points that can promote necessary shifts in infrastructure planning, policies, and practices that encourage more resilient, sustainable and equitable outcomes.

Keywords Social-ecological systems · Urban transformation · Foresight · Adaptive Governance · Infrastructure resilience

M. Moglia (✉)
Centre for Urban Transitions, Swinburne University of Technology, Hawthorn, VIC, Australia
e-mail: mmoglia@swin.edu.au

R. M. Wise
Centre for Urban Transitions, Swinburne University of Technology, Hawthorn, VIC, Australia

CSIRO Environment, Canberra, ACT, Australia

S. Meharg
CSIRO Environment, Canberra, ACT, Australia

© The Author(s) 2025
N. Frantzeskaki et al. (eds.), *Future Cities Making*, Theory and Practice of
Urban Sustainability Transitions, https://doi.org/10.1007/978-981-97-7671-9_2

2.1 Introduction

We live in a time of uncertainty and large-scale shocks and stresses, sometimes referred to as the Anthropocene (Steffen et al. 2015). The increasing economic costs of climate change are very significant in Australia (Deloitte Access Economics 2021). From a human perspective, this new epoch is characterised by complex and overlapping dynamics and processes of both natural and human-made systems; for example, climate change, technological innovation, economic boom-bust cycles, natural disasters, pandemics, resource depletion, and an increasingly unstable global production system. Specifically, the characteristics of the Anthropocene (Table 2.1) mean that we need to recognise rapid, large-scale, and uncertain dynamics, shocks, and stresses, which are now complicating prevailing infrastructure planning, delivery, and maintenance decisions that are critical for creating desirable future cities (Infrastructure Australia 2021; The Resilience Shift 2022). To do this requires a new approach to decision-making.

This chapter outlines a missions-based innovation strategy to '*build infrastructure resilience by shifting the decision-making paradigm and governance*'. We first provide a description of key terms and a conceptualisation of infrastructure resilience. This is followed by a section discussing the current paradigm for planning infrastructure, and the case for a shift based on missions-oriented innovation. Finally, we describe three game-changing interventions that promote resilient infrastructure planning.

2.2 Conceptualising Resilient Infrastructure

To build the foundation of a new decision-making paradigm for planning resilient infrastructure, we first need to define the terms and clarify the language we use. Therefore, we here define what we mean by infrastructure, resilient infrastructure, and outline the key shocks and stresses that warrant this paradigm shift.

2.2.1 What Is Infrastructure?

An important ontological question, which is sometimes ignored, is 'What do we mean by infrastructure?' We consider infrastructure to comprise built, natural, and social forms of infrastructure. **Built infrastructure** provides the physical and typically engineered backbones of our cities, providing critical services including water, energy, roads, transport, housing, waste management, telecommunications, and food. This infrastructure is often extremely long-lasting, with decisions made now likely to have impacts for decades, or possibly centuries, to come. **Natural infrastructure**, often referred to as blue-green infrastructure, includes green spaces,

Table 2.1 Challenges to risk and economic assessments and decision-making with long-lived consequences

Decision-making dilemma in the Anthropocene	Challenges to assessment and decision-making	References
The magnitude and speed of the changes create systemic risks characterised by high complexity. Complexity obscures the workings of systemic risks and inhibits extrapolation from the past to the future	Established methods of science cannot accurately or confidently determine the probability of occurrence or the extent of the damage. This means that systemic risks and costs tend to be underestimated by all decision-makers due to the inherent uncertainties about the likelihoods of occurrence and magnitude of impact. Instead, approaches to assessing systemic risks need to use a range of knowledge types, methodologies, and models to underpin scenarios that map the stochastic and deeply uncertain nature of systemic risks	Stirling and Scoones (2009), Midgley and Rajagopalan (2021)
The drivers of change and their consequences typically originate in one subsystem of society (e.g. a jurisdiction or economic sector) and create ripple effects that affect many sectors, ecosystems, or jurisdictions regionally or globally	Such characteristics confound reductionist approaches to risk and economic analysis (i.e. neo-classical economics). Additionally, the causes of disaster risks are generally outside the influence or control of any single decision-maker and thus require new governance arrangements that are multi-level and have many centres of authority or control	Schweizer and Renn (2019)
Systemic risks are characterised by dynamic relationships between causes and effects, which are changing in uncertain ways (i.e. deterministic relationships between cause and effect are non-existent, and yet this is a widespread starting assumption underpinning most analyses)	Systemic issues are major challenges to risk diagnosis, communication, and management because identical causes can lead to diverging results, which creates uncertainty in knowledge about how systems work, the effectiveness of alternative actions or options, and what the objectives of intervention ought to be. This leads to contestation and conflict	Voß et al. (2007)
Almost all systems — Social, economic, and environmental— Have absorptive (coping, adaptive) capacities that allow them to continue functioning and providing services while experiencing shocks and stresses. This capacity often hides the inherent (generally unknown and unpredictable) limits to these capacities, at which point a system irreversibly crosses a threshold or tipping point into a new state	Governance arrangements are not designed to accommodate the possibility or eventuality of tipping points (unpredictable thresholds) due to their perceived low probability and high uncertainty. This is because considering these comes with additional management costs, whereas the benefits (such as avoided costs) are delayed, uncertain, and intangible, and therefore heavily discounted by prevailing economic-valuation approaches and the short-term economic incentives faced by decision-makers	Patterson et al. (2017), Schweizer (2019)

(continued)

Table 2.1 (continued)

Decision-making dilemma in the Anthropocene	Challenges to assessment and decision-making	References
The magnitude of the changes to natural, agricultural, and built environments, and the catastrophic impacts of disasters are driving unavoidable re-evaluations of preferences, priorities, and values of individuals, communities, governments, and businesses, and are challenging the feasibility of many existing organisations' and governments' policy, planning, and investment objectives	This creates new tensions, trade-offs, and potential conflicts as values and priorities need to be revisited, elicited, and renegotiated. These situations then need to be reflected in updated policies and regulations, which in turn require anticipatory competencies and adaptive governance that enable inclusive processes of continually eliciting and deliberating about changing values, and mechanisms for accommodating these new values in policy, regulations, legislation, and management	O'Connell et al. (2018), Muiderman et al. (2022)

waterways, ecosystems (Newton et al. 2020; Pereira and Baró 2022), and the broad family of nature-based solutions (Raymond et al. 2017) that provide a broad set of social and physical functions to support cities. *Social infrastructure* includes social services, networks of actors, relationships, places for social interaction, and human and social capital (Cuthill 2010). It is generally the case that these forms of infrastructure, when present, are highly interdependent in form and function. Therefore, investing in and managing urban infrastructure for resilience involves contending with interconnected sets of challenges, as discussed in Moglia et al. (2021).

2.2.2 What Is Resilient Infrastructure?

The Anthropocene is an era dominated by the environmental impacts caused by human activity. In this context, an important question is, 'What do we mean by resilient infrastructure?' To answer this question, we need to consider that decisions about where we invest and build infrastructure today will have an impact far into the future, and therefore it would be negligent not to have a long-term focus. Therefore, current infrastructure decisions should factor in future uncertain possibilities in ways that support cities to perform well into the future. Simultaneously, these decisions should protect and support human populations, given that they need to thrive, ensure they collectively limit pressures on Earth's life-supporting systems to within safe levels, and be able to reduce and accommodate (resist or adapt to) large and uncertain shocks and stresses.

The question about resilience also raises the issue about what type of systems we are dealing with. As we have moved into the Anthropocene, human impacts on planetary systems at both a local and global scale mean that all decisions need to consider human systems within a wider context of ecological systems. Therefore, we consider the term 'resilience' in a way that draws on socio-ecological systems theory (Walker 2020), which describes resilient system properties of providing the ability to bounce back from disturbance or shock, as well as the capacity to reorganise and change so that the system maintains its key features and properties. Thus, while resilience is neither a positive nor a negative term, we believe that when it is applied in association with cities, it takes on a normative property that is associated with cities being the primary human habitat. Drawing on previous studies of the meaning of a resilient city (Meerow et al. 2016; Boschetti et al. 2017; Moglia et al. 2018, 2021), we adopt the following definition:

> A resilient city is a city that can equitably meet the needs and desires of its inhabitants and visitors into uncertain and risky futures, and that allows itself and its inhabitants (permanent or temporary) to be good neighbours and custodians of the land and ecosystems upon which they are located and depend, along with those surrounding it, as well as Earth's life-supporting systems.

For the purposes of this book, then, urban resilience is normative and associated with the inclusive and shared vision of a desired better-off configuration that does not erode the inherent and surrounding environmental and social foundations.

2.2.3   What Are the Key Shocks and Stresses?

Resilience can be general (to any shock or stress) or specific (to a specific shock or stress). Building on exploratory scenarios of the future of several Australian cities, Moglia and colleagues identified several urban mega-challenges (Moglia et al. 2018) that included climate change, population growth, technological disruption, and natural disasters such as drought and bushfires. In related studies, pandemics were also raised as perhaps one of the greatest challenges for cities (Newton and Doherty 2014; Moglia et al. 2019). Resilience also needs to be found in the context of the cascading impacts of hazards, such as those that followed in the wake of Australia's 2019/2020 Black Summer bushfires (Kemter et al. 2021). However, it is suggested by many that specific resilience to high-risk challenges is the priority in the short term, and that general resilience to other types of shocks is important, but not as urgent or just too hard to realise. 'Low-hanging fruit' is the term often used for these incremental quick wins that build specific resilience. Often-stated examples of 'low-hanging fruit' that builds specific resilience are 'early warning systems', 'capability-building exercises', and 'maintenance of buildings or roads'. This perspective and strategic prioritisation, it is argued, ensures targeted interventions to achieve specific resilience that buys time to implement the less tangible and often more difficult interventions required for general resilience. Concerns raised about this line of argument are that these 'low-hanging fruits' are used as displacement activities by vested interests in the status quo or that these reinforce maladaptive practices which increase exposure and vulnerabilities to future climate risks (Schipper 2020).

2.3   The Need for a New Paradigm in Planning for Resilient Infrastructure

Here we argue that the current paradigm for planning infrastructure in Australia is problematic and leads to highly risky decisions and highly vulnerable societal outcomes. Therefore, there is a need for a paradigm shift, although we acknowledge that this shift requires considerable innovation for it to be realised.

2.3.1   Current Paradigm for Planning Infrastructure in Australia

Currently, the predominant ways that infrastructure decisions are made in Australia do not consider resilience, uncertainty, large-scale or transformative change, or complexity, and these omissions come at a high cost (Infrastructure Australia 2021). Infrastructure Australia (2018) has suggested that a new set of principles be used to help decision-makers account for these factors. However, infrastructure investment decisions globally continue to be made predominantly on limited and optimistic projections of the future, the evaluation of an insufficient variety of options, and both non-transparent and questionable assumptions (National Infrastructure Commission 2020; Crona et al. 2021). Current decision-making may lend itself well to optimising costs or profits in highly regulated and controlled environments where the dynamics of the systems are known and can be assumed to remain stable into the future. However, in situations of large and uncertain change and complexity, and where the stakes are high (i.e. where consequences of social, economic, or environmental impacts are potentially catastrophic), this is problematic for two main reasons.

First, it is risky if the future turns out very differently to what was 'expected' and planned for, such as in the case of highly uncertain, 'unprecedented' catastrophic events, which tend to be ignored in decision-making processes based on expected net present values. Second, at some point, the negative 'external' (non-market) effects of such decisions (e.g. environmental degradation, inequality, declining mental health) accumulate—such as in the cases of climate change and biodiversity loss—threatening the stability of entire social and economic systems.

This indicates that new thinking and practices in infrastructure decision-making are required, with particular focus on interpersonal, learning, and adaptive competencies, including systems thinking. Yet it is known from our experience and the literature (Smith and Stirling 2010; Stirling 2014; Jorgenson and Stephens 2022) that changes of practice across an industry, sector, or organisation are difficult, as they challenge prevailing ways of thinking, people's identity, decision-making norms, and existing power structures. In other words, the personal stakes of change are often high for incumbent players, and therefore there are strong economic, political, and psychological reasons for them to maintain the status quo. So how can new competencies and practices be built within organisations where funding flows and power dynamics are defined based on an old paradigm?

We argue there is a need for a paradigm shift to enable new competencies and approaches to survive and take root so they can transform organisations and the wider governance system to demonstrably benefit all. Such a paradigm would promote decisions and outcomes that are more suited to building and enabling people, communities, economies, and infrastructure to better prevent, cope, respond to, adapt to, or transform local and global changes, and demonstrate strong economic and financial reasons for doing so.

2.3.2 The Case for a Shift Based on Missions-Based Innovation

Recognising that this new way of planning and investing in infrastructure in cities is currently far from current practice, there is a need for disruptive innovation to allow the shift in infrastructure planning to occur. In this chapter we propose 'mission-directed innovation' as an effective framing and approach capable of catalysing and enabling coordinated actions to shift prevailing infrastructure decision-making competencies and governance towards their normative best-practice characteristics. We do this in a way that draws upon lessons from their adoption by the Organisation for Economic Co-operation and Development (OPSI 2022) and the European Union (Mazzucato 2019), and from recently published critiques or evaluations of mission-directed innovation programmes (Janssen et al. 2021).

Mission-directed innovation is framed around clear targets with societal importance, direction, and community support, and enabled through well-resourced collaboration and agreements across governments, communities, and the private sector. Under different guises, this approach has been successfully implemented in a range of areas, including telecommunications, space exploration, defence technology, energy systems, and agriculture (Klerkx and Rose 2020), as well as in the context of various environmental and social-good challenges (Mazzucato 2018; Janssen et al. 2021).

Here, we propose themes and elements based on which missions-based innovation could catalyse and drive these shifts in decision-making in the built environment. These are proposed as the domains and approaches to support the step changes away from the current predominantly technocratic and economic-centric approaches towards something that is more inclusive of plural values and diverse knowledge types (traditional, experiential, and scientific), based on transdisciplinary systems perspectives that acknowledge complexity and deep uncertainty, and which have at their core systemic changes that can deliver more equitable and sustainable outcomes.

2.4 Game-Changing Interventions That Enable a Paradigm Shift

In this section, we describe elements and themes of the missions-based innovation agenda we have argued is needed to achieve a new paradigm for resilient infrastructure decision-making. This is based on three systemic and game-changing interventions that, when implemented strategically and in a coordinated fashion, can shift the paradigm for planning infrastructure resilience in Australia.

2.4.1 Intervention 1: Adopt an Adaptive Governance Approach for Managing Cities as Systems

This intervention aims for inclusive, systemic, and anticipatory decisions and governance that enable robust, low-regrets outcomes, consider externalities, and build resilience. Such a paradigm shift in decision-making and governance requires that decision-makers purposefully interrogate and diagnose the contexts in which they make decisions—defined by the interacting systems of rules, values, and knowledge (Gorddard et al. 2016)—and to strategically invest efforts into shifting these to create environments that more effectively enable the accomplishment of what is required (i.e. making the appropriate decisions) to deliver more resilient and sustainable infrastructure and cities.

The speed, pervasiveness, complexity, impacts, and uncertainties of current and future changes to cities and societies, which the scale of human activity has accelerated, have also been further enhanced by innovation occurring at a faster rate than imagined. From a decision-making perspective, this poses significant challenges, as outlined in Table 2.1, that materialise in terms of implications for a number of areas (Wise et al. 2022):

- Methodology: The current practices, tools, processes, competencies, and governance are not adequately dealing with uncertainties and ambiguities generated by the speed and complexity of changes and the systemic risks being created as systems become increasingly interconnected; nor with cascading causes and effects and unpredictable tipping points and thresholds.
- Temporal dimensions: Unavoidable tensions and trade-offs exist between the longer lifetimes of infrastructure-related decisions; for example, extended lead, lag, or consequence times (as per Stafford-Smith et al. 2011). These are particularly pertinent relating to built and blue-green infrastructure, relative to the shorter-term decisions associated with economic and political cycles and the highly dynamic natural and social environments within which infrastructure needs to operate.

- Uncertainty is hidden: The likelihoods and outcomes of major hazards are highly uncertain, but the resulting uncertainty in risk estimates is rarely considered in assessments or decisions.
- New dynamics and tensions in values: The collective 'successes', which are now commonly measured in singular and short-term economic terms in Australia, crowd out the opportunity for addressing systemic and long-term issues. Values are often ambiguous and contested, creating decision inertia and inaction. Additionally, the large-scale changes and impacts of climate change on existing social, cultural, and environmental systems mean that many things currently valued will be inevitably threatened or lost, which will necessitate a re-evaluation of priorities and policy and management objectives.
- Spatially constrained mandates: The causes and effects of the drivers of change and impacts on infrastructure are often outside the control or mandate of any single decision-maker, organisation, or jurisdiction, necessitating new governance arrangements that enable cross-organisational and cross-jurisdictional coordination and collaboration.

To address these issues and avert or minimise conflicts and undesirable futures, we need new ways of thinking, assessing risks and trade-offs, and making decisions. In short, what has worked in the past will not adequately address the challenges of the future.

To better address these issues, it is suggested that a significant component of decision-making is based on reframing existing decision contexts that currently prevent decision-makers from effectively considering and acting on the systemic risks to urban communities and infrastructure. This reframing requires individuals and organisations to invest efforts into diagnosing and reforming the prevailing societal and organisational values (such as goals, objectives, and key performance indicators), rules (such as practices, policies, and regulations), and knowledge. Without such reframing, they cannot credibly, legally, and legitimately adopt the adaptive decision-making cycle (Fig. 2.1a) needed to tackle systemic risks and build resilience (Gorddard et al. 2016). Examples of what these changes in values, rules, and knowledge (Fig. 2.1b) might involve are provided in Box 2.1. Such changes are not trivial, especially if initiated or driven bottom-up by employees who are already overwhelmed trying to meet business-as-usual priorities, who have limited time, competencies, and resources, and who are highly constrained by organisational policies and processes. This challenge highlights the important roles that leaders of organisations (who may also need to enhance their competencies) need to create the authorising and enabling environments for employees to do things differently to tackle systemic risks.

Figure 2.1a shows an adaptive learning and decision-making cycle—based on stakeholder engagement, ongoing monitoring and evaluation, and systems understanding—that is,

Futures-oriented and anticipatory; that is, the best practice is to anticipate emerging issues before they become a significant problem (Moglia et al. 2018). This

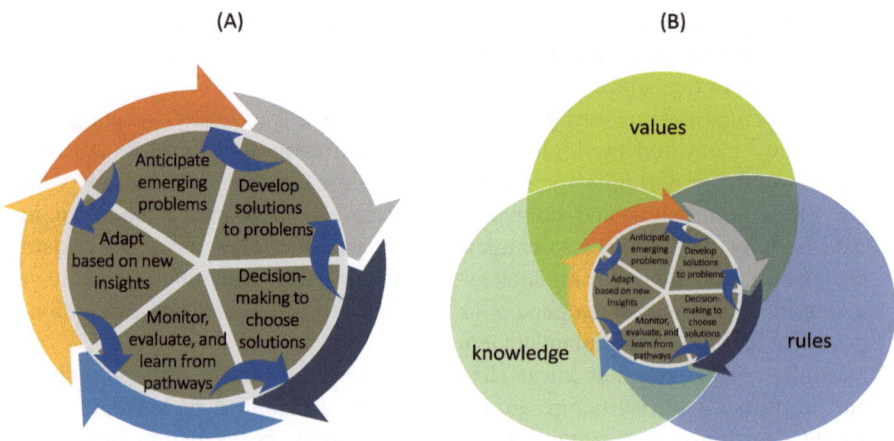

Fig. 2.1 Adaptive learning and decision-making cycle (**a**) operating and embedded within the societal systems of values, rules, and knowledge (**b**) that constrain and enable the types of decisions that can be legitimately, legally, and/or credibly made

relates to the sequence of steps to anticipate emerging problems, leading to the development of solutions.

Based on an adaptation pathways approach, whereby consideration of complexity is acknowledged by 'integrating incremental actions on proximate causes with the transformative aspects of societal change' (Wise et al. 2014). This relates to the sequence of steps: decisions for which outcomes are monitored, evaluated, and learned from, and solutions are adapted based on incremental and emerging insights.

> **Box 2.1 Shifting the Decision Contexts to Enable 'Resilience-Oriented' Decision-Making**
> The series of catastrophic disasters in Australia over the last decade—drought, bushfires, floods, and COVID—including the risk of cascading hazards (Kemter et al. 2021) and associated fast-growing economic costs (Deloitte Access Economics 2021), have highlighted the urgency of raising the priority of social and environmental considerations in investment and management decisions. High-performing and resilient communities and economies require human, social, and ecosystem capacities that enable effective coping and adaptability in the face of large-scale global changes. To achieve this requires more than just 'including' measures (i.e. monetising) of these non-market values in risk and business-case (cost–benefit) assessments. The ramifications of meaningfully factoring non-market (non-monetisable) factors such as social and natural capital into decision-making are significant and require fundamental changes to jurisdictional, market, and organisational rules (policies, regulations, standards, legislation) that incentivise, constrain, or enable more pluralistic and transdisciplinary ways for decision-makers to think about,

measure, and account for stakeholders' diverse values. This also requires broadening the kinds of knowledge currently prioritised or considered legitimate in decision processes (i.e. broaden from predominantly technocratic, managerial, and scientific knowledge to include traditional, relational, and experiential knowledge, and to promote the adoption of plural methodologies and heterodox approaches to scientific disciplines). Fundamental to the effectiveness of all of these is that all decision-making processes be required to inclusively and reflexively surface and deliberate about value priorities and value tensions so that priorities and objectives and risk-transfer and sharing arrangements amongst diverse actors and interests can be renegotiated to be more compatible with a radically transforming world (Buchtmann et al. 2022). As identified by Crosweller and Tschakert (2021, p. 203), this broadening of knowledge also needs to move away from a focus on individualistic responses and shift instead towards an *'ethic of compassion supported by a relational leadership framework that guides their resilience policy advice and decision-making'*.

2.4.2 Intervention 2: Build Competencies and Capacities That Promote Resilient Infrastructure

This intervention aims at enabling different ways of thinking about and acting to achieve resilient urban environments. These competencies and capacities are values- and relationship-focused and emphasise learning and expanding on adaptation skills such as systems and future thinking.

We argue that this is critical to support a new paradigm, and would enhance the skills, capacities, capabilities, and competencies of our formal and informal leaders, researchers, risk assessors, and decision-makers to meaningfully grapple with the complexities, uncertainties, and ambiguities of rapid and large-scale global change (Wise 2018). Skills are tools for change, while capacity is the ability, time, and resources to do something. Together skills and capacity are the capabilities or the set of resources or ways of functioning that an organisation needs to be able to command (Meharg 2022). When put into action, they relate to quality and consistency in action (competency). Therefore, competencies are a know-how or practice that people cultivate.

Meharg (2020) identified critical decision-maker or change agent competencies for effective systemic change as being good with people, willing to learn, and adaptive. Specific adaptation competencies include systems, future, strategic, and critical thinking; a normative perspective (values, principles, and goals); integration; and dealing with ambiguity. Critically, individuals do not need to have all of these competencies, but effective interventions need access to these competencies. These competencies align with the IPCC's desirable characteristics of decision-making under uncertainty (Jones et al. 2014), which include systems thinking; inclusive and compassionate stakeholder engagement; anticipatory social innovation and adaptive learning; and navigating power relationships.

To change the paradigm, decision-makers need to have access to the time and resources needed to make resilient infrastructure decisions (capacities). Specifically, infrastructure planners need

- Staffing time and competency in systems thinking that allows for.

  - Ongoing reflection and strategic conversations that allow assumptions to be challenged, emerging challenges and trends to be understood, and an awareness of blind spots in thinking. The practices that can be used to achieve this have been maturing in the field of futures thinking and scenario planning (Cork et al. 2023).
  - Ongoing collaborations across sectors and community groups that allow more and wider connection and involvement across sectors, and better policy integration across different levels of government.

- Systems-modelling competencies, drawing on the now-expansive urban systems science toolbox (Batten 2000; Groffman et al. 2017; Furtado et al. 2019). This includes capacity as well as funding and time to use tools like agent-based modelling, backcasting, scenario planning, Bayesian belief networks, systems optimisation, systems dynamics, integrated assessment modelling, or synchronous distributed collaboration to allow for.

  - Exploring multiple plausible future scenarios and their impact on infrastructure performance, their unintended consequences, etc.
  - Stress-testing the impacts of different decisions. This is the process of qualitatively evaluating policy in the context of multiple plausible future scenarios, attempting to avoid choices that could lead to significantly adverse impacts (van Asselt et al. 2014).

- Safe operating spaces that allow for experimentation and trialling of new options:

  - Competencies for managing risks in pilots that are run in a way that provides scalable solutions for cities.
  - Funding and staffing that help run pilots and experiments.

Importantly, planning agencies also require consistency of staffing that allows them to build a deep understanding and intuition over time of very complex issues. However, even with the necessary time and resources, resilient infrastructure decisions will require the cultivation of the change agent competencies, with a particular focus on

- **Interpersonal competencies** to enable effective working across domains, scales, and cultures. Combining normative competencies would enable the ability to undertake the necessary inclusive compassionate engagement required to support genuine, inclusive, ethical, and compassionate stakeholder engagement, collaboration, deliberation, and negotiation about complex and contested issues, preferences, and priorities. These are critical features of leadership and decision-making that reduce hazards, as identified by Crosweller and Tschakert (2021).

- **Learning competencies**, which require having the orientation and skills to learn individually and collectively. Given future uncertainties, it is unlikely that all decisions will be optimal, so learning is vital for any intervention to ensure that the activities are on track and achieve the intended aims. Such learning is in line with the probe–sense–respond framework of managing complexity (Snowden and Boone 2007) and can enable longitudinal assessment that builds a foundation of place-based knowledge, and potentially turns every decision into an experiment and pilot (Jain and Rohracher 2022).
- **Transdisciplinary competencies**, that is, learning through mixed-methods evaluation approaches; these mitigate the limitations of one approach and allow for multiple perspectives to be explored. Mixed methods can combine qualitative and quantitative data and analysis and inductive and deductive processes (Kasemir et al. 1999), and link or coordinate bottom-up and top-down approaches to analysis, funding, and governance.
- **Social learning competencies**, which develop a culture enabling respectful disagreement where deliberation and diversity are valued and encouraged. Combining interpersonal and learning competencies allow for inclusive, practice-based, and challenge-led initiatives (pilots), the application of a key method for navigating complexity, the development of alternative solutions that expand the options space, building decision-maker support, and the scaling of innovation from niche to mainstream (Sengers et al. 2019).
- **Systems thinking**, an **adaptation competency** that is fundamental to diagnosing problems and generating innovative solutions in highly dynamic and complex contexts where uncertainties and ambiguities are unavoidable.
- **Futures or anticipatory thinking**, which builds a future orientation and can be paired with adaptive sequences of learning initiatives to build shared understanding of the changing dynamics of systems and to evaluate novel interventions that underpin adaptability. Combining competencies such as systems, critical, and future thinking draws in tools like scenarios, backcasting, or forecasting, which can challenge status quo thinking and assumptions that the future will reflect an ongoing continuation of what has happened in the past. In practice, 'black swans', shocks, stresses, and fundamental shifts in the dynamics of society are the norm, rather than the exception (Feduzi et al. 2022), which is why experts are generally poor forecasters (Tetlock 2005).

Expanding, cultivating, and resourcing the development of decision-maker competencies and capacities across government departments, the private sector, and the community will enable

- **Democratisation of knowledge**: The recognition and meaningful adoption or integration of plural knowledge types and methodologies informed by the characteristics of the problems and the decision context (Midgley and Rajagopalan 2021).

- **Navigating power relationships**: Explicit recognition of prevailing power imbalances in terms of who sets and controls narratives and agendas, rules, and the modes of accumulation that are hampering innovation and change (i.e. doing things differently) for resilience and sustainability (Scoones 2016).
- **Diagnosis of root causes of problems and leverage points for sustainability:** Basing decision-making on understanding how urban systems behave and evolve (Webb et al. 2018), including an understanding of trade-offs and undesirable consequences such as inequality and biodiversity loss. This requires processes and competencies in systems thinking and mapping to understand and make explicit the fundamental causes of problems such as incentive structures, governance arrangements, and cultural ideologies, norms and practices, and the leverage points to address these (Abson et al. 2016; O'Connell et al. 2018; Walker 2020; Buchtmann et al. 2022).
- **Cross-sectoral policy design**: The intersection of infrastructure functions and services with multiple domains (i.e. interventions in one sector or asset class may simultaneously have implications for economic policy, land-use policy, national security, mobility, ecology, and health and well-being) (Keele and Coenen 2019). Policy, planning, and management processes and practices across these domains need to aspire to, and be supported in achieving, alignment and coordination. Digital technology can be used to enable such cross-sectoral and multi-level decision-making (governance) for urban resilience in action (Newton and Frantzeskaki 2021).
- **Change the sequence of decisions:** In Australia, infrastructure decisions in cities tend to be sequenced based on population projections. This results in limited engagement from key stakeholders in the early stages of planning, and, consequently, urban development that is not resilient. Instead, such problems can be better addressed by enabling multi-level and multi-sector collaboration in decision-making upfront.
- **Anticipatory governance**: Learning and adapting through mechanisms that rapidly relay feedback from policy and practice pilots to those with authority (formal and informal) to make a change. This also includes robust decision-making that allows complexity and future uncertainties to be factored into decision-making.
- **Building, through inclusive deliberation and negotiation, coalitions and networks for collective actions focused on leverage points**: Effective change cannot happen unless it is bipartisan, necessitating that individuals and groups have the competencies to build trust, partnerships, and networks based on an acknowledgement that there may be insurmountable differences in values, world views, and interests, but that it is still possible to arrive at joint agreements, and to develop and practice mechanisms for making the creation, ownership, and transfer of risks and rewards transparent and fair.

Unfortunately, these capacities and competencies are underdeveloped in the decision-making processes that currently govern cities, limiting paradigm shifts and resilient decision-making. The good news is that capacities and competencies can be cultivated and supported, as asserted by a growing body of literature on building such capacity (Webb et al. 2018; O'Connell et al. 2019; Colloff et al. 2021; Meharg 2022).

2.4.3 Intervention 3: Implement and Learn from Actions That Support Urban Resilience

This intervention aims at enabling systemic change based on each of the three interventions working synergistically through strategically coordinated pilots that catalyse innovation. Such pilots are underpinned by scaling mechanisms that identify, aggregate, and disseminate lessons to create positive tipping points in the thinking and practices of individuals and organisations across all relevant elements of urban infrastructure.

This type of experimentation, in its various forms, is an underdeveloped opportunity for learning to improve the resilience of cities (Heilmann 2008; Bulkeley and Castán Broto 2013; Brundiers and Eakin 2018). In the language of David Tàbara et al. (2018), pathways to urban resilience are *'progressive courses of action for achieving strategic objectives, or more broadly to attain transformative visions'*. In other words, by adopting a selection of the three game-changing interventions, a city may chart a pathway to urban resilience. These interventions need to be carried out at multiple governance scales (i.e. the local, state, and federal levels in Australia) with well-developed, supported, funded, and coordinated actions that are directed at innovation in the five transformation areas outlined below. By focusing efforts on these innovations, they can provide the enabling environments for a shift in paradigm and the build-up of competencies and capacities, amounting to an important tipping point in how Australia builds and sustains its infrastructure into the future.

We suggest that mission-directed innovation and investment can be framed around several complex, wicked urban challenges that, if tackled in strategic and coordinated ways (Heilmann 2008; Mayan et al. 2019; Janssen et al. 2021), could catalyse the fundamental drivers and positive tipping points required of resilient cities (Wolfram et al. 2016; Moglia et al. 2021). Those drivers are generally ones that either reduce pressure or increase capacity for adaptation. A portfolio to accomplish this (Moglia et al. 2018, 2021) could contain elements such as

- *Urban regeneration*: To reduce urban sprawl so as to mitigate a range of unintended consequences such as inadequate services, congestion, inequality, crime, loss of agricultural land, etc. Australia's pattern of urbanisation, which involves low-density urban development, is problematic, and the alternative requires injecting precinct-scale medium-density redevelopment—the 'missing middle'—as a model for re-urbanisation more reflective of cities in Europe, which deliver liveability more sustainably (Newton et al. 2020). To maximise the benefits of the more compact city, there is good value in combining it with increased walkability and providing enhanced access to nature.

- *Nature-based solutions*: To help reduce the risks of climate change, improve health and well-being outcomes, improve ecosystem health, and reduce inequality (Frantzeskaki et al. 2019). Importantly, providing more nature, such as more tree canopy, healthier waterways, and more permeable surfaces, is a relatively cheap way of achieving stormwater management objectives, and one of the only available ways to significantly reduce outdoor temperatures, thereby reducing the devastating impacts that Australian cities are likely to experience as a result of forecast future climate change-induced heatwaves (Kumar et al. 2024). Furthermore, more nature in cities tends to support health and well-being and allows for citizens to connect on a regular basis with nature.
- *Reduced dependence on private cars*: To facilitate an ecologically just mobility transition towards more efficient and sustainable modes of transport (including more equitable and better access to public and active travel), and away from private cars. This would help to free up valuable land and space, reduce productivity losses associated with congestion, improve health and well-being, reduce greenhouse gas emissions, and reduce public expenditure on car-based infrastructure and services such as roads, parking lots, signage, signalling, and traffic regulation enforcement. It is also a strategy that tries to 'do more with less' by increasing our reliance on alternative modes of mobility, such as telework, cycling, or walking, that require less investment and a smaller environmental footprint, but with better social-good outcomes. Importantly, these alternative modes of transport free up space for more nature in cities, as well as greener urban densification.
- *Affordable housing*: To provide cost-effective provision of this critical infrastructure, which will lead to reduced inequality, improved economic productivity, and improved health and well-being. Housing is perhaps the most important driver to address economic inequity and inequality in the Australian context, and in social systems higher levels of equality generally lead to higher levels of social capital, which in turn is strongly linked to the development of flexible and adaptable societies that can absorb unexpected shocks (Walker 2020). It follows that this is one of the key levers for improving urban resilience in Australian cities.
- *Investment in climate resilience*: To support an environment that enables the urgent and large-scale investments required to adapt and build the resilience of urban communities and infrastructure to meet a changing climate and growing disaster risks. This requires assembling people's assessment (scalable methodologies, tools, data platforms, and data) and engagement competencies to effectively diagnose the systemic causes of vulnerability. These can then be addressed through creating enabling mechanisms for governments, communities, and industry to drive the necessary changes confidently and collectively in thinking and practice needed to design and deliver resilience-building interventions and investments across local, national, and regional scales. One such mechanism is broad, place-based collaborations funded by the federal government, similar to the 'regional or cities deals' in Australia. However, such collaborations would benefit from mission-directed innovation policies and funding programmes tar-

geting climate-resilience outcomes (Infrastructure NSW, Infrastructure Australia 2022).

Importantly, a key aspect of such solutions is that, apart from housing, they require fewer financial investments and are more closely linked with urban land-use planning, retrofitting of existing infrastructure, and public policy to provide incentives for behaviour change. This brings it back into the domain of policy and legislation, for which particular challenges exist (Keele and Coenen 2019; Naderpajouh et al. 2019).

2.5 Summary: An Adaptive Pathway Towards Resilient Infrastructure

This chapter has described three pillars of actions that can support initiatives to build the capacity for resilience-based infrastructure investments:

- Shifting the paradigm of decision-making
- Building competencies and capacities for navigating a complex and uncertain future
- Developing and supporting a portfolio of resilience-enhancing actions

These interventions are synergistically linked in a way that activity in any one of them cannot be done without coordinating or collaborating with activities in the others at least one of the other two, and with benefits from one spilling across to others. It should also be noted that current planning of infrastructure in Australia is lacking in the consideration of uncertainty and complexity, which is a problem in the twenty-first century that looks to provide some considerable challenges in this respect.

Also, to implement these interventions requires a governance framework, champions, staffing resources, funding, and the building of competency for ongoing learning activity. Learning based on pilots and experimentation in all activity areas needs to be embedded within adaptive governance, which is the most fundamental mechanism of socio-ecological resilience.

Further, in this chapter, we have suggested five major opportunities for resilience-enabled innovations, but we note that there are many ways to achieve greater resilience, and that these actions need to be worked out in the local context.

We have also noted that large-scale adaptation of cities—and the large precincts, suburbs, growth areas, or other geographical units within them—to the types of challenges outlined in this chapter have tended to follow similar patterns of radical innovation and reorganisation, which are often triggered by significant events like pandemics, war, or disasters (Moglia et al. 2018). These experiences have shown that, for cities as places of human ingenuity and creativity, the key to resilience and adaptation is embedded in how innovation and adaptive decisions are enabled. To capitalise on such opportunities, it has been found that these successes are largely explained by the timeliness and execution of five iterative steps or phases of the adaptive learning and decision-making cycle, as outlined in this chapter.

It follows, therefore, that the key to ensuring cities' resilience and ability to adapt lies in people's and organisations' capacity, competencies, and enabling governance for the three transformative actions outlined in this chapter, and in the application of the five phases of ongoing adaptive learning and decision-making within a framework of missions-oriented innovation.

References

Abson DJ, Fischer J, Leventon J, Newig J, Schomerus T, Vilsmaier U, Von Wehrden H, Abernethy P, Ives CD, Jager NW, Lang DJ (2016) Leverage points for sustainability transformation. Ambio 46:30–39

Batten D (2000) Discovering artificial economics. Westview Press, Boulder, CO

Boschetti F, Gaffier C, Moglia M, Walke I, Price J (2017) Citizens' perception of the resilience of Australian cities. Sustain Sci 12:345–364

Brundiers K, Eakin HC (2018) Leveraging post-disaster windows of opportunities for change towards sustainability: a framework. Sustainability 10:1390

Buchtmann M, Wise R, O'Connell D, Crosweller M, Edwards J (2022) Reforming Australia's approach to hazards and disaster risk: national leadership, systems thinking, and inclusive conversations about vulnerability. Disaster Prev Manag 32:49

Bulkeley H, Castán Broto V (2013) Government by experiment? Global cities and the governing of climate change. Trans Inst Br Geogr 38:361–375

Colloff MJ, Gorddard R, Abel N, Locatelli B, Wyborn C, Butler JRA, Lavorel S, Van Kerkhoff L, Meharg S, Múnera-Roldán C, Bruley E, Fedele G, Wise RM, Dunlop M (2021) Adapting transformation and transforming adaptation to climate change using a pathways approach. Environ Sci Policy 124:163–174

Cork S, Alexandra C, Alvarez-Romero JG, Bennett EM, Berbés-Blázquez M, Bohensky E, Bok B, Costanza R, Hashimoto S, Hill R, Inayatullah S, Kok K, Kuiper JJ, Moglia M, Pereira L, Peterson G, Weeks R, Wyborn C (2023) Exploring alternative futures in the anthropocene. Annu Rev Environ Resour 48:25

Crona B, Folke C, Galaz V (2021) The Anthropocene reality of financial risk. One Earth 4:618–628

Crosweller M, Tschakert P (2021) Disaster management leadership and policy making: a critical examination of communitarian and individualistic understandings of resilience and vulnerability. Clim Pol 21:203–221

Cuthill M (2010) Strengthening the "social" in sustainable development: developing a conceptual framework for social sustainability in a rapid urban growth region in Australia. Sustain Dev 18:362–373

David Tàbara J, Frantzeskaki N, Hölscher K, Pedde S, Kok K, Lamperti F, Christensen JH, Jäger J, Berry P (2018) Positive tipping points in a rapidly warming world. Curr Opin Environ Sustain 31:120–129

Deloitte Access Economics (2021) Special report: update to the economic costs of natural disasters in Australia

Feduzi A, Runde J, Schwarz G (2022) Unknowns, black swans, and bounded rationality in public organizations. Public Adm Rev 82:958–963

Frantzeskaki N, McPhearson T, Collier MJ, Kendal D, Bulkeley H, Dumitru A, Walsh C, Noble K, Van Wyk E, Ordóñez C, Oke C, Pintér L (2019) Nature-based solutions for urban climate change adaptation: linking science, policy, and practice communities for evidence-based decision-making. Bioscience 69:455–466

Furtado BA, Fuentes MA, Tessone CJ (2019) Policy modeling and applications: state-of-the-art and perspectives. Complexity 2019:5041681

Gorddard R, Colloff MJ, Wise RM, Ware D, Dunlop M (2016) Values, rules and knowledge: adaptation as change in the decision context. Environ Sci Pol 57:60–69

Groffman PM, Cadenasso ML, Cavender-Bares J, Childers DL, Grimm NB, Grove JM, Hobbie SE, Hutyra LR, Darrel Jenerette G, McPhearson T, Pataki DE, Pickett STA, Pouyat RV, Rosi-Marshall E, Ruddell BL (2017) Moving towards a new urban systems science. Ecosystems 20:38–43

Heilmann S (2008) From local experiments to national policy: the origins of China's distinctive policy process. China J 59:1–30

Infrastructure Australia (2018) Infrastructure decision-making principles. Canberra

Infrastructure Australia (2021) A pathway to infrastructure resilience. Advisory paper 1: opportunities for systemic change, Canberra

Infrastructure NSW, Infrastructure Australia (2022) City deals. Australian Government, Canberra. https://www.infrastructure.gov.au/territories-regions-cities/cities/city-deals

Jain M, Rohracher H (2022) Assessing transformative change of infrastructures in urban area redevelopments. Cities 124:103573

Janssen MJ, Torrens J, Wesseling JH, Wanzenböck I (2021) The promises and premises of mission-oriented innovation policy—a reflection and ways forward. Sci Public Policy 48:438–444

Jones RN, Patwardhan, Cohen SJ, Dessai S, Lammel A, Lempert RJ, MMQ M, Von Storch H (2014) Foundations for decision making. Cambridge University Press, Cambridge and New York

Jorgenson S, Stephens JC (2022) Action research for energy system transformation. Educ Action Res 30:655–670

Kasemir B, Van Asselt MBA, Dürrenberger G, Jaeger CC (1999) Integrated assessment of sustainable development: multiple perspectives in interaction. Int J Environ Pollut 11:407–425

Keele S, Coenen L (2019) The role of public policy in critical infrastructure resilience. University of Melbourne and The Resilience Shift

Kemter M, Fischer M, Luna LV, Schönfeldt E, Vogel J, Banerjee A, Korup O, Thonicke K (2021) Cascading hazards in the aftermath of Australia's 2019/2020 Black Summer wildfires. Earth's Future 9

Klerkx L, Rose D (2020) Dealing with the game-changing technologies of Agriculture 4.0: How do we manage diversity and responsibility in food system transition pathways? Global Food Security 24: 100347

Kumar P, Debele SE, Khalili S, Halios CH, Sahani J, Aghamohammadi N, Andrade MdF, Athanassiadou M, Bhui K, Calvillo N, Cao S-J, Coulon F, Edmondson JL, Fletcher D, Dias de Freitas E, Guo H, Hort MC, Katti M, Kjeldsen TR, Lehmann S, Locosselli GM, Malham SK, Morawska L, Parajuli R, Rogers CDF, Yao R, Wang F, Wenk J, and Jones L (2024) Urban heat mitigation by green and blue infrastructure: Drivers, effectiveness, and future needs. The Innovation 5(2):100588

Mayan M, Pauchulo AL, Gillespie D, Misita D, Mejia T (2019) The promise of collective impact partnerships. Commun Dev J 55:515–532

Mazzucato M (2018) Mission-oriented research & innovation in the European Union. A problem-solving approach to fuel innovation-led growth. European Commission Directorate-General for Research and Innovation, Brussels

Mazzucato M (2019) Governing missions: governing missions in the European Union. European Commission

Meerow S, Newell JP (2016) Urban resilience for whom, what, when, where, and why? Urban Geogr 40:1–21

Meharg S (2020) Catalysing change agents through research for development. PhD, The Australian National University

Meharg S (2022) Critical change agent characteristics and competencies for ensuring systemic climate change interventions. Sustain Sci 18:1445–1457

Midgley G, Rajagopalan R (2021) Critical systems thinking, systemic intervention, and beyond. In: Metcalf GS, Kijima K, Deguchi H (eds) Handbook of systems sciences. Springer, Singapore

Moglia M, Cork SJ, Boschetti F, Cook S, Bohensky E, Muster T, Page D (2018) Urban transformation stories for the 21st century: insights from strategic conversations. Glob Environ Change 50:222–237

Moglia M, Cork S, Cook S, Muster T, Bohensky E (2019) The future of Sydney: scenarios to guide collaboration by the Sydney Common Planning Assumptions Group. CSIRO, Sydney

Moglia M, Frantzeskaki N, Newton P, Pineda-Pinto M, Witheridge J, Cook S, Glackin S (2021) Accelerating a green recovery of cities: lessons from a scoping review and a proposal for mission-oriented recovery towards post-pandemic urban resilience. Dev Built Environ 7:100052

Muiderman K, Zurek M, Vervoort J, Gupta A, Hasnain S, Driessen P (2022) The anticipatory governance of sustainability transformations: hybrid approaches and dominant perspectives. Glob Environ Chang 73:102452

Naderpajouh N, Matinheikki J, Hills M (2019) The role of legislation in critical infrastructure resilience. RMIT University and The Resilience Shift

National Infrastructure Commission (2020) Anticipate, react, recover. Resilient Infrastructure Systems. Report prepared by the UK National Infrastructure Commission

Newton PW, Doherty P (2014) The challenges to urban sustainability and resilience. Resilient sustainable cities: a future. Routledge

Newton P, Frantzeskaki N (2021) Creating a national urban research and development platform for advancing urban experimentation. Sustainability 13:530

Newton P, Glackin S, Garner S, Witheridge J (2020) Beyond small lot subdivision: pathways for municipality-initiated resident supported precinct scale residential infill regeneration in greyfield suburbs. Urban Policy Res 38:338–356

O'Connell D, Wise RM, Doerr V, Grigg N, Williams R, Meharg S, Dunlop M, Meyers J, Edwards J, Osuchowski M, Crosweller M (2018) Approach, methods and results for co-producing a systems understanding of disaster. Technical report supporting the development of the australian vulnerability profile. CSIRO, Canberra

O'Connell D, Maru Y, Grigg N, Walker B, Abel N, Wise R, Cowie A, Butler J, Stone-Jovicich S, Stafford-Smith M, Ruhweza A, Belay M, Duron G, Pearson L, Meharg S (2019) Resilience, adaptation pathways and transformation approach. A guide for designing, implementing and assessing interventions for sustainable futures (version 2). CSIRO, Canberra

OPSI (2022) "Mission-Oriented Innovation." Retrieved 12th October 2024, 2024, from https://oecd-opsi.org/workareas/mission-oriented-innovation/

Patterson J, Schulz K, Vervoort J, Van Der Hel S, Widerberg O, Adler C, Hurlbert M, Anderton K, Sethi M, Barau A (2017) Exploring the governance and politics of transformations towards sustainability. Environ Innov Soc Trans 24:1–16

Pereira P, Baró F (2022) Greening the city: thriving for biodiversity and sustainability. Sci Total Environ 817:153032

Raymond CM, Frantzeskaki N, Kabisch N, Berry P, Breil M, Nita MR, Geneletti D, Calfapietra C (2017) A framework for assessing and implementing the co-benefits of nature-based solutions in urban areas. Environ Sci Policy 77:15–24

Resilience Shift (2022) The global hub for resilience best practice. Lloyd's Register Foundation, London

Schipper ELF (2020) Maladaptation: when adaptation to climate change goes very wrong. One Earth 3(4):409–414. https://doi.org/10.1016/j.oneear.2020.09.014

Schweizer P (2019) Governance of systemic risks for disaster prevention and mitigation. Contributing paper to the UNDRR global assessment report on disaster risk reduction. Institute for Advanced Sustainability Studies, Potsdam, Germany

Schweizer P, Renn O (2019) Governance of systemic risks for disaster prevention and mitigation. Contributing paper to the UNDRR global assessment report on disaster risk reduction. Institute for Advanced Sustainability Studies, Potsdam, Germany

Scoones I (2016) The politics of sustainability and development. Annu Rev Environ Resour 41:293–319

Sengers F, Wieczorek AJ, Raven R (2019) Experimenting for sustainability transitions: a systematic literature review. Technol Forecast Soc Chang 145:153–164

Smith A, Stirling A (2010) The politics of social-ecological resilience and sustainable sociotechnical transitions. Ecol Soc 15

Smith MS, Horrocks L, Harvey A, Hamilton C (2011) Rethinking adaptation for a 4°C world. Philos Trans R Soc A Math Phys Eng Sci 369:196–216

Snowden DJ, Boone ME (2007) A leader's framework for decision making. Harv Bus Rev 69–76

Steffen W, Broadgate W, Deutsch L, Gaffney O, Ludwig C (2015) The trajectory of the anthropocene: the great acceleration. Anthropocene Rev 2:81–98

Stirling A (2014) Transforming power: social science and the politics of energy choices. Energy Res Soc Sci 1:83–95

Stirling AC, Scoones I (2009) From risk assessment to knowledge mapping: science, precaution, and participation in disease ecology. Ecol Soc 14

Tetlock PE (2005) Expert political judgment. Princeton University Press, Princeton

Van Asselt MBA, Van't Klooster SA, Veenman SA (2014) Coping with policy in foresight. J Futur Stud 19:53–76

Voß JP, Newig J, Kastens B, Monstadt J, Nölting B (2007) Steering for sustainable development: a typology of problems and strategies with respect to ambivalence, uncertainty and distributed power. J Environ Policy Plan 9:193–212

Walker B (2020) Finding resilience: change and uncertainty in nature and society. CSIRO Publishing, Canberra

Webb R, Bai X, Smith MS, Costanza R, Griggs D, Moglia M, Neuman M, Newman P, Newton P, Norman B, Ryan C, Schandl H, Steffen W, Tapper N, Thomson G (2018) Sustainable urban systems: co-design and framing for transformation. Ambio 47:57–77

Wise RM (2018) Key capabilities for long-term development strategies in the face of unprecedented and uncertain large-scale global change. World Resources Institute. https://www.wri.org/climate/expert-perspective/key-capabilities-transformational-long-term-development-strategies

Wise RM, Fazey I, Stafford Smith M, Park SE, Eakin HC, Archer Van Garderen ERM, Campbell B (2014) Reconceptualising adaptation to climate change as part of pathways of change and response. Glob Environ Chang 28:325–336

Wise RM, Marinopoulos J, O'Connell D, Mesic N, Tieman G, Gorddard R, Chan J, Flett D, Lee A, Meharg S, Helfgott A (2022) Enabling resilience investment guidance (version 1). Report prepared by CSIRO, Value Advisory Partners, University of Adelaide. CSIRO, Canberra

Wolfram M, Frantzeskaki N, Maschmeyer S (2016) Cities, systems and sustainability: status and perspectives of research on urban transformations. Curr Opin Environ Sustain 22:18–25

Magnus Moglia is an Associate Professor of Urban Sustainability at the Centre for Urban Transitions, Swinburne University of Technology, Australia. He has a PhD in environmental management and complex systems science, from the Australian National University's Crawford School for Public Policy, as well as degrees in engineering, physics, and mathematics from the Royal Institute of Technology in Sweden. From 2001 to 2020, he was a researcher at the CSIRO, Australia's National Science Agency, holding the positions of Principal Research Scientist and Research Team Leader in Urban Systems Science. With a background in both physical and social sciences and extensive experience in participatory and foresight methods, he is a transdisciplinary researcher, focusing on how solutions for sustainability can be equitably promoted and implemented in cities and regions. His research topics include water management, climate adaptation, freight and transport decarbonisation strategies, sustainable agriculture, circular economy, urban regeneration, working-from-home practices, and infrastructure resilience. Much of his research has been in the international context, including in Vietnam, Laos, and Kiribati. He has published more than 70 peer-reviewed journal articles, as well as numerous book chapters and industry reports. He is currently the lead Chief Investigator for projects on circular economy, freight decarbonisation, intermodal rail, and sustainable aquaculture.

Russell M. Wise is a Principal Sustainability Economist at the CSIRO, Australia's national research agency. Russ is passionate about working with diverse groups of people across industry, government, research, and community to make sense of and respond to global changes in climate, ecosystems, and socio-economic development. Over the past 15 years, Russ has pursued his passion by leading and creating transdisciplinary teams and large-scale research programs focused on building stakeholders' capacities to understand and respond to these unprecedented changes. Russ has delivered this work in Australia, Indonesia, Papua New Guinea, and South Africa.

Seona Meharg is an interdisciplinary social scientist at the CSIRO, Australia's national research agency. She specialises in impact evaluation and catalysing the competencies required for climate adaptation. She collaboratively designs, leads, and delivers projects with governments and communities, both within Australia and across the Asia-Pacific, often in situations of high ambiguity and contested contexts, such as climate adaptation, disaster risk reduction, and sustainable development. Seona has an eclectic background in ecology, communication, environmental law, and project management, exploring how change can be catalysed through projects, exploring the theory and practice of implementation, including capacity considerations for adaptation research.

Chapter 3
Transitions to Sustainable Urban Water Systems

Mojtaba Moravej, Beata Sochacka, Steven Kenway, Peter Newton, Cassady Swinbourne, and Ka Leung Lam

Abstract Radical changes are needed in metropolitan-scale strategic planning to better integrate land use, transport planning, and urban water planning, as well as new models for water-sensitive urban design at building and precinct scales that deliver liveability and ecosystem benefits. This is a mission-scale challenge. Transition pathways involve combinations of new technology, innovative urban design, enabling policies and regulations, novel planning processes and urban development, and demand-side changes in consumers' attitudes regarding urban lifestyles related to water and energy use. The chapter draws on 10 years of applied research undertaken collaboratively with government and industry to illustrate how integrated plans and designs can be established and tested. Examples of good design spanning the architectural and technological realms supported by quantified performance analysis and institutional change across the entire water cycle including natural and anthropogenic systems are provided. They address urban water transitions that need to be accelerated across scales, including site, precinct, and city to achieve more sustainable water-sensitive urban regions.

Keywords Urban water metabolism · System performance · Built environment · Water-sensitive urban design (WSUD) · Integrated urban water management

M. Moravej (✉) · B. Sochacka · S. Kenway · C. Swinbourne
University of Queensland, St Lucia, QLD, Australia
e-mail: m.moravej@uq.edu.au

P. Newton
Swinbourne University of Technology, Hawthorn, VIC, Australia

K. L. Lam
Duke Kunshan University, Kunshan, Jiangsu, China

© The Author(s) 2025 57
N. Frantzeskaki et al. (eds.), *Future Cities Making*, Theory and Practice of
Urban Sustainability Transitions, https://doi.org/10.1007/978-981-97-7671-9_3

3.1   Grand Challenges in Urban Water Management

Water is a vital resource for the effective functioning of cities, the health of their citizens, economic development, and overall prosperity. Cities suffer when there is not enough water, or too much, or it is too polluted. There are also inequality issues in accessing water services that hinder sustainable growth and environmental justice. Addressing these issues is a grand challenge.

The global population continues to grow and urbanise (United Nations 2018). Since the start of the twenty-first century, Australia's cities have seen their fastest growth in the nation's history, centred mainly on its major cities (Newton et al. 2022b). Climate change is another major global externality. The United Nations (https://www.unwater.org/water-facts/water-and-climate-change) wrote, 'Climate change is primarily a water crisis. We feel its impacts through worsening floods, rising sea levels, shrinking ice fields, wildfires and droughts'. All of these elements influence our cities, where our populations are most heavily concentrated, and which are often located beside or close to water. Risk to Australia's cities and regions from climate change is high (Australian Academy of Science 2021).

The intensity and frequency of rainfall are altering due to climate change. It is becoming harder to predict the components of the urban water cycle and more challenging to provide urban water services. A warmer atmosphere, as a result of global warming, can hold more water, leading to more intense precipitation. This, in turn, often causes crop damage and soil erosion, threating food security, and flooding, risking lives, infrastructure, and livelihoods. At the same time, warmer air has a higher dew point, making condensation in the atmosphere less likely, which leads to more and longer dry spells. Globally, the extent of urban areas exposed to floods and droughts is projected to at least double by 2030 (Güneralp et al. 2015). Australia has experienced several episodes of extensive and extreme drought, megafire, and flooding events in recent years. Globally, this will amplify existing patterns of migration towards cities, which might not be prepared to house and integrate them (Zaveri et al. 2021).

Sustainably providing essential water services (a potable water supply, safe disposal and treatment of wastewater, and flood protection) is becoming more challenging for urban water managers. Cities' water demand is increasing as urban populations grow, but the available water supply is less predictable due to climate change. This means that competition for water resources for urban, agricultural, industrial, and environmental uses is becoming fierce as supporting catchments reach their carrying capacity. Urban water managers often rely on energy-intensive technologies (such as desalination) to meet the increasing water demand. For example, in the 1960s, 88% and 12% of Perth's urban water resources were sourced from dams and groundwater, but in 2020, 48% came from seawater desalination, 40% from groundwater, and 10% from dams (Byrne et al. 2020). Similar trends can be seen in other cities globally, shifting a water problem to an energy problem that exacerbates greenhouse gas emissions (Lam et al. 2017). Flood protection is another major issue. One in five people around the world live in areas directly exposed to a 1-in-100-year flood risk (McDermott 2022).

Urban water services are traditionally provided through large centralised systems. These source water from catchments, treat and distribute it to the end users, collect and treat wastewater and stormwater, and safely convey them to the receiving water bodies. There are four issues associated with the continuation of this approach. First, the urban water infrastructure, especially in developed countries, is ageing. Climate change, urbanisation, city growth, and ageing infrastructure have led to a widening gap between actual and required rates of investment for expanding urban water infrastructure. Second, single-purpose, centralised systems are at odds with the principles of multi-functionality, redundancy, diversity, and multi-scale connectivity required for urban resilience (Ahern 2011). This has made urban areas inefficient and vulnerable to future disturbances. Third, centralisation relies on a linear 'take–use–dispose' approach, which is widely criticised for being unsustainable and wasteful, and for missing opportunities for local capture, recycling, and reuse of resources (Kenway et al. 2011; Renouf et al. 2018). Fourth, there is an overconcentration of infrastructure associated with the hydraulic mission legacy that produced a system that exceeds its technical optimal degree of centralisation. For example, in a case study in Switzerland, Eggimann et al. (2015) showed that the optimal degree of centralisation for wastewater infrastructure is 13% lower than the actual degree of centralisation. This example is also valid for Australian cities, where the level of centralisation of water infrastructure is above 90% and urban sprawl has created large, inefficient water infrastructures.

Meeting emerging urban water management objectives is another grand challenge for centralised urban water systems. The three conventional objectives of (i) provision of safe drinking water, (ii) handling wastewater for public health, (iii) and protecting residents against flooding are now being joined by emerging objectives of recognising water for environment, ecosystems, recreation, biodiversity, improved aesthetics, urban heat mitigation, and liveability. Meeting these emerging objectives calls for bringing water onto the urban landscape instead of burying it in underground pipe networks. This requires a paradigm shift in urban water planning and management practices: from an unsustainable (wasteful) linear and siloed system to an integrated system involving collaborative and communicative design processes connecting urban water engineers with other urban professionals such as designers, architects, planners, and environmental and social scientists (McEvoy et al. 2018; Moravej and Leardini 2022; Moravej et al. 2022b; van de Ven et al. 2016).

Another motive for a paradigm shift in urban water management stems from the argument that planning and management in the water sector operate on a fundamentally incorrect assumption of stationarity: that the future is predictable using actuarial records and expectations involving 'management' interventions. However, it has been argued that 'stationarity is dead' (Haddad and Moravej 2015; Milly et al. 2008; Moravej 2016) due to deep uncertainties caused by changes in climate, societies, and institutions (Urich and Rauch 2014). The conventional engineering paradigm of predictability and control has led to compartmentalisation of water supply, sewerage, and stormwater services (Brown 2012; Pahl-Wostl et al. 2009). The compartmentalisation is both physical (i.e. separate urban water infrastructure for different water services) and institutional (in terms of institutions' jurisdiction and operation) (Brown 2012; Pahl-Wostl et al. 2009). Moving away from this paradigm requires removing barriers at both the individual level (e.g. skills and knowledge) and the organisational level (e.g. organisational and political inertia). This is a substantial technical, educational, societal, and institutional challenge.

The complex nature of these challenges, the threats to established urban infrastructure, and the large number of people affected demonstrate the global relevance and urgency of urban water management issues. This chapter focuses on integrated planning and design for water-sensitive urban development, drawing on 10 years of applied industry-oriented research undertaken collaboratively with government, supported by quantified performance analysis across the entire water cycle, including natural and anthropogenic systems (CRC for Water Sensitive Cities 2022; Sochacka et al. 2021a). It showcases frontier multi-disciplinary research that supports an accelerated transition towards sustainable urban water systems.

3.1.1 Changing Paradigms for Water-Sensitive Cities

As socio-political drivers and urban water management objectives are becoming more numerous and complex, there has been an emergence of water-themed city visions such as water-sensitive cities, water-wise cities, and sponge cities. We use an urban water management transition framework developed by Brown et al. (2009) to characterise the sequence of socio-technical and political economy transitions towards sustainable urban water management. In this context, the objective of water-sensitive cities represents a hydro-social contract that values water security, public health, flood control, environmental protection, liveability, amenity, and long-term economic sustainability that encompasses environmental justice and intergenerational equity under climate change.

The paradigm shift for enabling water-sensitive cities is fundamental. It shares a worldview that water systems can contribute to broader productivity and welfare gains for cities beyond essential water services. It calls for more integrated approaches that manage water holistically in both a physical sense (i.e. infrastructure as one connected system) and in institutional arrangements (i.e. shared responsibility). Changing the focus from urban water infrastructure within the urban landscape to the urban landscape itself as a complex integrated system enables consideration of innovative solutions to deliver multiple objectives and outcomes (Table 3.1).

Increased references to liveability in urban water management serve as a good example of the water sector's growing ambition to tackle problems that go beyond its traditional mandate. In Australia, the contribution of water management to liveability is most typically defined as an amenity associated with water used for greenspace irrigation and water-sensitive urban design features, primarily for treating and retaining stormwater (Sochacka et al. 2021b). These design features in turn

Table 3.1 Paradigm shift: major differences in conventional and desired urban water management vision and attributes

Aspects	Conventional	Desired
Objectives	Supply potable water, collect and treat wastewater, control flood and overland flows	Recognise multiple functions of water in cities including recreation, amenity, aesthetic values, heat mitigation, biodiversity, transport, energy in addition to the conventional objectives
System boundary	Water infrastructure in isolation	Urban systems as a whole
Ontology	Reductionist and anthropocentric, prediction and control	Complex system recognising humans as part of the environmental context and not independent
Perception of problems	Tame problems: Simple, well-defined and stationary, problems are independent of each other	Wicked problems: Complex, problems are dependent on each other, non-stationarity is recognised
Perception of solutions	A unique optimal solution exists for each problem	No optimal solution, each solution might be more attractive from one point of view but they alter the system so they might create problems elsewhere
Management approach	Compartmentalisation and optimisation of single components of the water cycle, economic-focused	Integrated urban water management including water-related aspects of the urban system such as water-energy-heat nexus, value-focused based on wide social, environmental, and economic values
Expertise	Narrow, technical, and siloed disciplines	Interdisciplinary, collaborative
Resilience	Resist change and recover quickly from disturbances	Adapt and synergise with change, avoid conflict with the environment, safe to fail
Mode of service delivery	Centralised, optimised for technical and economic performance of the infrastructure, single mode, underground pipes	Diverse, flexible solutions (technical, social, economic, ecological, etc.), functioning at multiple scales, bring water onto the urban landscape
Model	Linear, take–use–dispose approach	Circular, multiple uses of resources via cascading, demand minimisation, recycling and reuse at multiple scales (appliances, households, precincts, city, etc.)
Connection with the supporting environment	Highly reliant, sensitive to external disturbances such as climatic variabilities	Reduced pressure on the environment by improving self-sufficiency, adaptive to external disturbances
Governance	Water managed by government on behalf of communities	Shared responsibility, co-management of water between government, business and communities
Risk	Regulated and controlled by the government	Shared risk and diversified

Source: adapted and expanded from Keath and Brown (2009) and Franco-Torres et al. (2021)

contribute to urban cooling and health benefits associated with the recreational use of vegetated public spaces. Systematic approaches to water-related liveability also include appreciation for the role water services have traditionally played in supporting cities' liveability: by providing households with water and sewage disposal, protecting cities from flooding, and building infrastructure (e.g. reservoirs) that can also be used for recreation. There is a recognition that delivering the whole range of liveability outcomes requires a rethinking of urban water governance, as many of the potential liveability benefits of better urban water management currently lie outside the scope of water utilities' responsibilities. After all, liveability has conventionally belonged in the domain of urban planning and the design of the public realm. This overlap of competencies has been recognised as one of the barriers to realising the vision of water-sensitive cities, where insufficient integration between water and land-use planning and opportunities for cross-sectoral collaboration limit the benefits multi-functional infrastructure could deliver. At the same time, this highlights significant opportunity for a better coordination of efforts in this area inspiring a move from isolated water-sensitive urban design features to water-sensitive urban developments, where liveability and water objectives are pursued simultaneously from the early planning phase (see Chesterfield et al. 2021).

3.1.2 Emergence of a Resilience Framework with Specific Implications for Water

Resilience is often referenced in urban water strategies. This has emerged from a general shift in the water sector from building infrastructure to be fail-safe, towards building a system that is 'safe-to-fail'. This shift was motivated by a realisation that building highly efficient or resistant infrastructure systems eroded the long-term resilience of the urban water cycle and its ability to deliver key services (Holling 1996). With rapid urbanisation and the increased severity of floods, droughts, and heatwaves due to climate change, it is unlikely that traditional grey infrastructure will be able to resist anticipated stresses (Marks et al. 2022). To prevent catastrophic failures, urban water systems should be designed to be flexible, with built-in adaptive capacity and redundancy to maintain function across a range of temporal and spatial scales.

The need to consider multiple spatial and temporal scales, especially larger scales, results in considerable uncertainty about the level of resilience required of our cities. This uncertainty, paired with the large upfront price tag of investing in large-scale infrastructure upgrades, means that governments are sometimes reluctant to invest in mitigation measures at the required level and scale. Currently, only 3% of Australia's disaster funding targets preparation and long-term resilience, with the overwhelming majority being spent on response and recovery measures. However, with the 2022 floods in Queensland alone costing an estimated

AU\$7.7 billion (Deloitte 2022), it is clear that the price of not investing in future-proofing our cities is substantial.

Resilience is generally built by increasing the strength of advantageous feedback loops, which can be achieved by increasing the circularity in the urban water system. Building diversity of water sources and multi-functionality into urban infrastructure helps integrate flexibility and adaptability into a previously linear system. However, there are trade-offs within resilience that need to be better understood. Australia needs to build an urban water system that is simultaneously resilient against droughts, floods, and heatwaves, which may occur in quick succession. The protection against a particular threat should not compromise a city's resilience against another threat; for example, water-related cooling strategies to combat heatwaves should be managed so as not to degrade water supply security. A thorough understanding of how particular water-sensitive infrastructure or designs affect the entire water cycle is required to prevent problem-shifting and ensure the resilience of the whole urban water system.

3.2   Current Frontiers to Deliver Transitions

There are four interconnected frontier pathways for delivering sustainable transitions in urban water management: (i) technology, (ii) individual behaviour, (iii) innovative urban design, and (iv) institutional processes and governance (Newton et al. 2019). Each of these can be applied to a range of scales (Table 3.2). It is important to note that progression in all four innovative frontiers at a range of scales is needed.

Technological pathways involve the development of more efficient technologies that can disturb poor performing systems and replace them with those that offer superior performance. They are often key for local harvesting (e.g. rainwater tanks),

Table 3.2 Scales of action in urban water management

Scale	Example actions
Site	Create site-specific urban design and water servicing solutions to minimise the impacts of urban sites upstream (water supply network) and downstream (receiving environments)
Precinct	Integrate and scale the site-scale actions to create multi-functional water and design solutions, suitable for the local context. Consider communal measures connecting multiple sites and their integration into the city-wide water infrastructure
City/town	Define a vision for the water management of the city. Align the physical assets (infrastructure), institutions, and urban systems to achieve the vision. Total water cycle or integrated urban water management applied at this scale
River catchment	Minimise the urban water footprint to alleviate the competition between cities and agriculture (and other uses). Ensure the quality and quantity of water meet all needs (including those of the environment)

Source: adapted from CRC for Water Sensitive Cities (2022)

reusing (e.g. on-site greywater reuse), retaining and detaining stormwater (e.g. on-site storage), increasing efficiency (e.g. efficient appliances, recirculating showers), and supporting local greenery.

Individual behaviour pathways encapsulate the societal aspects by encouraging water-sensitive practices. This includes abandoning highly consumptive behaviours (e.g. long showers) or those that negatively affect the urban environment. Housing preferences (Iftekhar et al. 2022), gardening practices, vehicle dependency for transport, littering and waste disposal, and pet ownership are some examples of individual behaviours that affect the quantity and quality of urban water flows (Müller et al. 2020).

Innovations in urban design and planning are more influential than advances in technologies (Moravej et al. 2022b). However, the most effective options are those that integrate the two. Given the forecast rapid urban development projected for the next three decades, development and application of processes, tools, and platforms for systematic scenario creation, urban design optioneering, and performance quantification and testing is crucial.

Finally, new institutional frontiers aim to overcome contemporary resistance and barriers to technical and social innovations by creating new governance structures and processes for urban development. New governance recognises that (i) there are multiple stakeholders (actors) involved, (ii) the process should allow for their participation and engagement, (iii) multi-scale governance structures should correspond with a nested biophysical scale, (iv) local solutions can provide valuable opportunities but processes for incentivising them are needed, and (v) adaptive governance principles are needed to create a responsive institutional system that can accommodate the dynamic and complex nature of urban environments.

The programme of research undertaken by the CRC for Water Sensitive Cities, which provided the basis for the research reported in the following sections, explored challenges and opportunities involved in implementing water-sensitive developments at the site and precinct scale (IRP4 2021; Sochacka et al. 2021a). The project responded specifically to the problem of 'infill development', the currently most popular redevelopment practice, in which established residential properties are subdivided into two or more smaller allotments for medium-density dwellings, but often at the expense of useable greenspace (see the chapter by Newton and Glackin in this volume for a more detailed explanation of current 'greyfield redevelopment' practices). Such infill redevelopment increases imperviousness and has significant adverse impacts on urban hydrology and the amenity of urban areas. Thus, developing guidelines for densification that both improve liveability outcomes and reduce water-related impacts presents significant opportunity for creating water-sensitive cities through higher-density urban redevelopment. The project responded to this need by developing alternative designs for dwelling typologies and the public realm (London et al. 2020b) and assessing them using the Infill Performance Evaluation Framework (Moravej et al. 2020; Renouf et al. 2020a) in two case studies in South Australia (Renouf et al. 2020b) and Western Australia (London et al. 2020a). The framework is also used in a case study in Brisbane (Moravej et al. 2022a).

3.2.1 Site Scale

The individual site represents a scale at which most urban redevelopment decisions are made. Redevelopment sites often involve idle, abandoned, underutilised, or ageing commercial, industrial, or residential properties. Examples are abandoned sites in ports and harbours in coastal cities occupying high-value waterfronts and abandoned manufacturing sites from a previous industrial era in large metropolitan areas, known as 'brownfields' (Newton et al. 2022a), and ageing (but occupied) residential properties where land value, rather than buildings, is the main economic asset, known as 'greyfields' (Newton et al. 2020). Under compact city strategies, brownfields and greyfields represent priority targets for infill development, in an attempt to reduce 'greenfield' sprawl (Newton 2010). Sustainable redevelopment of greyfield sites requires a collaborative approach to urban design that involves collaboration between local and state governments, developers, architects, urban and landscape planners, water, transport, and energy engineers, and local resident communities. There are multiple objectives: quality of living spaces, cross-ventilation, passive cooling and urban greening for alleviating urban heat, liveability, affordability, meeting market demand, and providing community benefit and sustainability more broadly.

A first stage involved the development of an infill performance evaluation framework (Renouf et al. 2020a) for design typology analysis (Moravej and Leardini 2022) and systematic evaluation of multi-disciplinary decisions at site- and precinct-scale development (Fig. 3.1). The framework has four major steps that streamline the processes of scenario and system boundary definition, understanding the relevant input parameters, quantifying performance via modelling, and performance reporting.

The Site-scale Urban Water Mass Balance Assessment (SUWMBA) model (Moravej et al. 2020, 2021) was developed to integrate the factors affecting water performance of site-scale redevelopment projects, considering architectural design, on-site water service technologies, and environmental conditions. The model was used in an iterative, collaborative urban design process to create a new set of design technology configurations specific to three Australian urban environmental conditions in Brisbane (sub-tropical), Melbourne (temperate), and Adelaide (semi-arid). The results identified design typologies and required shifts in urban design (London et al. 2020b) and urban planning practices (Matthew and Moravej 2022; Moravej et al. 2022c) to achieve water-sensitive goals. Barriers to support the innovative design and uptake of on-site technologies were identified as lack of regulatory incentives, economic factors, market preferences (Iftekhar et al. 2022), unwillingness to pay, and lack of knowledge about sustainable residential development and its effectiveness (Göçmen 2013).

A metamodeling study (Moravej et al. 2022b) comparing urban water flows under two different design technology transition pathways before and after redevelopment showed that current urban design and planning regulations could result in increased population density, stormwater discharge, and water demand to 198%,

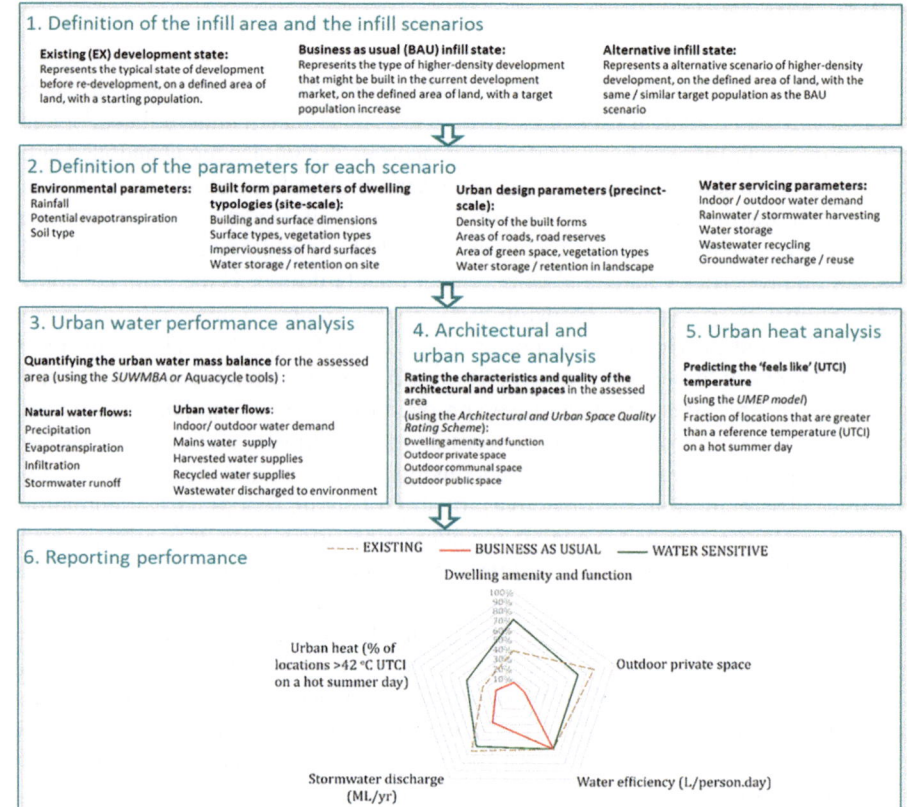

Fig. 3.1 Components of infill evaluation framework. (Source: adapted from Renouf et al. 2020a)

144%, and 185% of conditions prior to infill redevelopment. Furthermore, such designs would also reduce healthy green spaces and infiltration to 47% and 66% of pre-infill conditions (Fig. 3.2). An innovative design employing more efficient use of vertical and horizontal spaces on the site and amalgamating properties instead of subdividing individual lots showed superior population density and water performance. This was further enhanced when on-site water-sensitive technologies such as efficient appliances and water fixtures, local harvesting of water (e.g. via rainwater and stormwater tanks), water reuse (e.g. greywater recycling and cascading), and introduction of on-site green infrastructure (e.g. green roofs and permeable pavement) were considered. It was shown that these initiatives could support a higher population density, up to 241% compared to before infill redevelopment, with lower impacts on urban water flows.

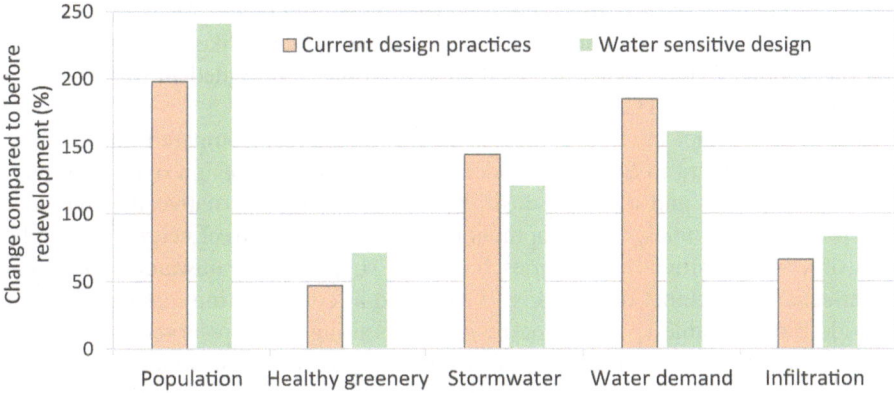

Fig. 3.2 Comparisons between two site-scale design technology development pathways. Note: evapotranspiration was used as a proxy of healthy greenery. (Source: data is plotted from Moravej et al. 2022b)

3.2.2 Precinct Scale

Mid-scale urban planning at the neighbourhood or precinct level offers significant opportunities for water-sensitive city experimentation, despite a number of challenges associated with both urban and water planning at that scale. The precinct is not a common planning scale for water service providers, who typically serve much larger urban areas. Nor is it a planning scale for stormwater drainage, where hydrological scales of catchments are more typical. While state governments have instruments for precinct-scale planning (precinct structure plans) used for master-planned greenfield developments and large brownfield redevelopments, they are much less common in established (greyfield) urban areas with significant (re)development potential (Newton et al. 2022b). Precinct-scale planning is more challenging in established residential areas, where coordinated redevelopment (e.g. urban renewal) is obstructed by fragmented land ownership among residents and legacy infrastructures associated with water and roads.

Despite these challenges, the precinct scale offers unique opportunities for integrated water management. It is a scale that is large enough for hybrid infrastructures with a wider range of decentralised supply systems (e.g. wastewater recycling, stormwater storage) and enables implementation of larger water-sensitive urban design features such as wetlands. Uniquely, it allows an appreciation of the combined effect of increased imperviousness on natural water flows and development of more systematic measures for retaining perviousness in private settings (e.g. permeable driveways) as well as the public realm (e.g. street design and parkland planning). Given the ageing infrastructure and housing stock in middle (greyfield) suburbs, coupled with targets for increasing urban density in most Australian cities, more opportunities for coordinated precinct-scale redevelopment are likely in such areas. The concept of urban regeneration of greyfields—at the precinct scale—promotes the ideas that such redevelopment could cater for both improved liveability and more sustainable outcomes, and that water is likely to play a major role in achieving these synergies (a detailed outline is provided in this volume's chapter by Newton and Glackin).

The first precinct case study presented in this chapter compared current low-density development in Salisbury (Adelaide, SA) with two design options: typical infill development and a water-sensitive, higher-density alternative that included new housing typologies, redevelopment of greenspace and street verges, and inclusion of water-sensitive technologies for water supply (e.g. rainwater tanks). The water-sensitive precinct design (Fig. 3.3) aimed at consolidating outdoor space to provide more useable space to the residents, limiting space reserved for vehicles while replacing it with multi-use permeable spaces, retaining mature trees, and ensuring an adaptability of spaces to multiple uses. This was achieved through compact built-form typologies to retain greenspace area and useability, narrower street design with green verges composed of public and private land and converting some of the underutilised residential and commercial lots to pocket parks. A number of lots were identified as potential sites for future redevelopment, and the higher-density dwelling options for these lots were explored. These included dwelling typologies for both single-occupancy dwellings on smaller lots and multi-dwelling buildings (e.g. apartments). Water-sensitive typologies determined not only the

Buildings	Trees	Green space outside the development site
Permeable surface - car	Garden	
Permeable surface - car park	Green deck/terrace	
Permeable surface - paved/people	Non-permeable surface - road	

Fig. 3.3 An example of water-sensitive design for the Salisbury case study. (Source: adapted from Renouf et al. (2020b) and London et al. (2020b), https://watersensitivecities.org.au/content/project-irp4/)

buildings' envelope but also outdoor areas that could be used for tree planting, given the space required for deep roots and potential spaces for water storage (e.g. rainwater tanks). Water technologies covered both small systems for single dwellings (rainwater tanks) and larger schemes for the whole precinct (wastewater recycling and stormwater recycling).

The designs were then compared based on their modelled performance in three key areas: (i) water performance informed by the water mass balance of all water flows within the precinct, (ii) architectural and urban space quality, based on expert appraisal using a scoring system against a number of key criteria, and (iii) urban heat performance based on a thermal comfort model (Fig. 3.4). Evaluation confirmed the adverse effects of business-as-usual infill development on urban water performance, liveability, and thermal comfort. By comparison, water-sensitive design was able to accommodate more residents than conventional infill development, while retaining or improving natural hydrology and creating more liveable built environments. For example, the analysis showed that 42% and 50% of rainfall volume was converted to runoff under the low-density and business-as-usual infill development scenarios, respectively. By comparison, water-sensitive design reduced the volume of runoff to 39% of rainfall through a combination of greenspace and rainwater tanks.

Further evidence of benefits associated with water-sensitive urban design was provided by another case study undertaken in Perth (Western Australia), which was evaluated using the same methodology and three development scenarios. The analysis confirmed that water-sensitive designs contributed to better retention of

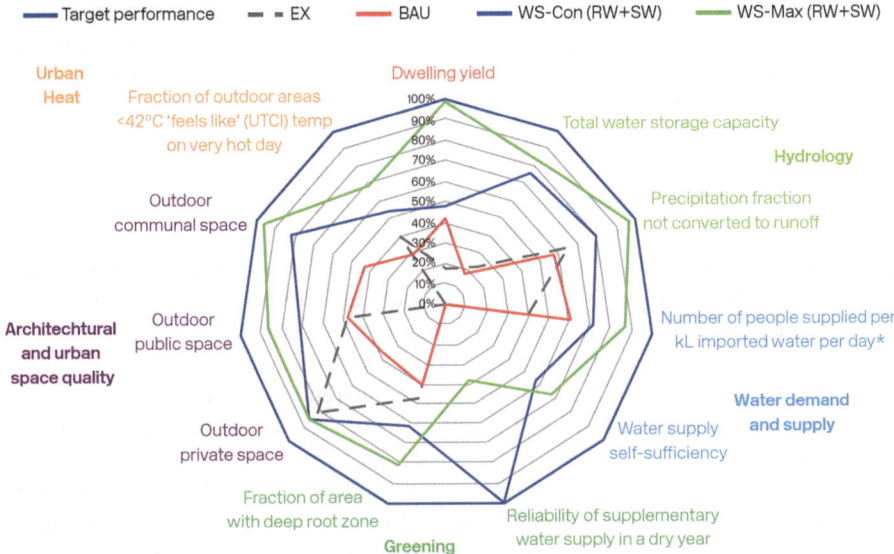

Fig. 3.4 An example of multi-criteria assessment for the Salisbury case study. *EX* existing conditions, *BAU* business-as-usual development, *WS* water-sensitive, *Con* conservative scenario, *RW* rainwater harvesting, *SW* stormwater harvesting. (Source: adapted from Sochacka et al. 2021a)

stormwater than conventional (business-as-usual) infill development and improved water efficiency. Since both the business-as-usual infill development and water-sensitive development scenarios assumed an increase in population, the total water demand in both of these scenarios was higher than that under the low-density scenario. However, water-sensitive designs achieved higher water-use efficiency per capita. Per capita potable water demand, which was entirely sourced from a centralised water supply system in the (historical original) low-density development scenario, was 325 L/person/day. Increasing residential density in the business-as-usual infill development scenario reduced this to 161 L/person/day, mostly due to reduced garden space area on the residential lots. Water-sensitive development further reduced this performance indicator to 86–174 L/person/day depending on the combination of decartelised water service technologies used. The water-sensitive development scenario also retained a larger area of greenspace without the need to change the irrigation patterns (London et al. 2020a).

The benefits generated by water-sensitive design extended beyond water performance. In both case studies, water-sensitive designs, compared to other development options, improved the architectural quality of indoor and outdoor spaces by improving the provision of gardens, vegetation and trees, solar access, and cross-ventilation of the outdoor private space. The interventions in the public realm also improved important features of the greenspace accessible to all residents of the precinct: greenspace availability and diversity, access and connectivity, multi-functionality, and outlook to vegetation were improved in the water-sensitive design option. Consideration was also given to the effects redevelopment would have on the urban heat island effect. Typically, densification is associated with adverse effects on urban thermal comfort, which, in the context of climate change and increasing numbers of heatwaves, is likely to further reduce the liveability appeal of higher-density neighbourhoods to residents. The water-sensitive design, however, was shown to improve thermal comfort in the Adelaide case study. On a typical hot summer day, 77% of the case study area in the existing low-density precinct was exposed to temperatures above 42 °C, which was reduced to 68% of the area in the water-sensitive alternative and could be reduced even further to 59% should water-sensitive design be adopted across the entire precinct.

Uniquely, the project brought together innovation in architectural design with quantitative urban water performance and thermal comfort analysis to define the level of ambition for water-sensitive densification. But while the findings related to how the particular designs can inspire and inform similar innovation in other locations, the project also provided some interesting insights into important elements of enabling environments that are key to delivering water-sensitive regeneration on a precinct scale.

First, the selection of case studies was coordinated with case study proponents who exercised a level of responsibility and influence over the precinct development. In both case studies, these were organisations that already had both expertise and responsibilities spanning the planning of both water and land use. In the Adelaide case study, the local council (City of Salisbury) was involved in establishing a purple pipe scheme to supply recycled stormwater to commercial and residential customers. In the Perth case study, the proponent was the Western Australian government's state land development agency, which plans and develops large greyfield and brownfield sites as urban renewal projects that champion sustainability performance. Both of the case study proponents had experience in delivery of relatively more innovative and sustainable forms of urban development: specifically, a system of wetlands to capture, filter, and recharge a local aquifer, which was then used to augment water supply; and shared community rainwater harvesting systems.

Second, both of the case study sites had environmental challenges that required unconventional approaches and innovations. In the Perth case study, groundwater that other residential developments in the area used for augmentation of supply was contaminated due to previous industrial land use. The developer was also concerned about the possible wastewater discharge volumes requiring fees for sewerage system upgrades. This created a desire to consider the reuse of wastewater to meet water demand. Proponents of the Adelaide case study, in contrast, were concerned about the low demand for the purple pipe stormwater recycling scheme. Thus, they were interested in non-potable water uses (i.e. irrigation) that could provide liveability benefits for the residents (such as contributing to the reduction of ambient temperatures during heatwaves).

Finally, the timing of the case studies was opportune, as it coincided with major planning reforms in both states. These reforms focused on changes to building codes, as they related to medium-density housing, creating opportunities for a review of standard industry practices that would enable better urban redevelopment and increased societal acceptance of densification targets.

3.2.3 City/Region Scale

Urban water management visions are defined at city scale. Visions often require a combination of top-down and bottom-up initiatives directed at operational scales (precinct and site) and are coupled with transition pathways to determine the extent of centralisation versus decentralisation and how social benefits are balanced against economic returns. Creating staged objectives for operationalising the envisioned pathways is also important when translating high-level visions to tangible actions that are meaningful in operational, spatial, and organisational contexts (Newton and Rogers 2020).

Benchmarking tools and indicators such as the Water Sensitive Cities Index (Rogers et al. 2020) and City Blueprints (van Leeuwen et al. 2012) are proving useful for establishing current states (or baselines) as well as for monitoring the progression towards sustainable urban water management aspirations. They are particularly valuable in stakeholder engagement for creating a common understanding of the current situation, setting future directions for collaboration, and building partnerships to transform built environments. One example involved synergies among water and carbon goals in cities. The water sector can contribute to a low-carbon transition pathway through multi-disciplinary collaboration. The building of partnerships between water utilities and other sectors is an important enabling factor for fostering wider sustainable opportunities in cities (Lam and van der Hoek 2020). Yarra Valley Water (one of the water retailers in Melbourne) provides an Australian example of an anaerobic digestion facility for co-digestion of sewage sludge and food waste to generate biogas for energy recovery. The opportunity was supported by insights from the partnership with the East Bay Municipal Utility District (EBMUD) in California, which is the first net-positive wastewater treatment plant in the USA.

In the urban water sector, some utilities are more advanced in setting sustainability goals and reporting their performance routinely, including greenhouse gas emissions (Lam et al. 2022). It is critical for utilities to quantify their baseline performance in order to set any improvement targets. Globally, many utilities or national water statistical reporting agencies have established performance indicator frameworks and are reporting utilities' performance results annually (Lam et al. 2017). Some leading examples include the Bureau of Meteorology's National Performance Report in Australia and the World Bank's International Benchmarking Network (IBNET). These reporting frameworks usually cover a range of environmental indicators, including greenhouse gas emissions, biosolid disposal, and wastewater treatment levels. Inter-utility reporting provides a means for benchmarking performance required for progressing towards a sustainable transition.

Performance quantification is essential for systematically comparing, prioritising, and selecting among the wide range of innovative urban designs and technology options. Although best-performing designs are a combination of all four transition pathways (innovative design, technology, water-sensitive behaviour, and adaptive governance), architectural design is up to three times more influential than on-site water service technologies; this emphasises the role architects and urban planners are required to play in addressing urban water management issues.

3.3   Conclusions

Over 50 years ago Abel Wolman published a seminal study driven by then-urgent crises of water and energy shortages and threats to clean air in the USA. His conclusion was that that there was no shortage of water; instead, there was a need for long-term thinking and planning (Wolman 1965). In the twenty-first century, with cities facing multiple water challenges in the future, the need for long-term thinking and coordinated multi-disciplinary approaches is clear.

Urban planning and design is an essential multi-disciplinary collaborative ingredient for meeting these challenges. There is also a clear need for a 'multi-scale' approach for transitioning to sustainable urban water systems, as well as a 'multi-agency' approach. The urban water sector alone cannot achieve the vision of water-secure, liveable, resilient, and water-sensitive cities. How to achieve this collective vision and accelerate the implementation of the multiple solutions that we know exist remains a major transitioning challenge. A principal challenge for city-scale strategic planning is effective collaboration between agencies responsible for long-term land use and water planning (among others). Objectives and processes that have operated in the past are no longer fit for purpose, as climate change is requiring urban forms and structures that mitigate generation of greenhouse gases (primarily transport, but also buildings) and adapt to urban heating (where nature-based services for greening require increased supplies of water). Planning for these transitions is now, urgent given the time scale involved in realising necessary urban change.

Acknowledgements Mojtaba Moravej acknowledges the funding of the Cooperative Research Centre for Water Sensitive Cities (CRCWSC) through the IRP4 and Tools and Products (TAP) projects. Prof. Steven Kenway was the project leader of Water Sensitive Outcomes for Infill Developments (IRP4) project, and acknowledges the funding of ARC DECRA (DE160101322) and the University of Queensland (UQ) Amplify funding schemes. Prof. Peter Newton was a member of the CRC for Water Sensitive Cities Advisory Committee from 2013 to 2021, with specific focus on water-sensitive urbanism and regenerative infill redevelopment.

References

Ahern J (2011) From fail-safe to safe-to-fail: sustainability and resilience in the new urban world. Landsc Urban Plan 100(4):341–343

Australian Academy of Science (2021) The risks to Australia of a 3°C warmer world. Australian Academy of Science, Canberra

Brown R (2012) Transitioning to the water sensitive city: the socio-technical challenge. In: Howe C, Mitchell C (eds) Water sensitive cities. IWA Publishing, London, pp 29–42

Brown RR, Keath N, Wong TH (2009) Urban water management in cities: historical, current and future regimes. Water Sci Technol 59(5):847–855

Byrne J, Taylor M, Wheeler T, Breadsell JK (2020) WGV: quantifying mains water savings in a medium density infill residential development. Sustainability 12(16):6483

Chesterfield C, Tawfik S, Malekpour S, Murphy C, Bertram N, Furlong C (2021) Practising integrated urban and water planning: framework and principles. Cooperative Research Centre for Water Sensitive Cities, Melbourne

CRC for Water Sensitive Cities (2022) Urban WaterGuide—a guide for building sustainable and resilient cities. CRC for Water Sensitive Cities, Canberra

Deloitte (2022) The social, financial, and economic costs of the 2022 South East Queensland Rainfall and Flooding Event. Queensland Reconstruction Authority, Brisbane

Eggimann S, Truffer B, Maurer M (2015) To connect or not to connect? Modelling the optimal degree of centralisation for wastewater infrastructures. Water Res 84:218–231

Franco-Torres M, Rogers BC, Harder R (2021) Articulating the new urban water paradigm. Crit Rev Environ Sci Technol 51(23):2777–2823

Göçmen ZA (2013) Barriers to successful implementation of conservation subdivision design: a closer look at land use regulations and subdivision permitting process. Landsc Urban Plan 110:123–133

Güneralp B, Güneralp İ, Liu Y (2015) Changing global patterns of urban exposure to flood and drought hazards. Glob Environ Chang 31:217–225

Haddad OB, Moravej M (2015) Discussion of "Trend, independence, stationarity, and homogeneity tests on maximum rainfall series of standard durations recorded in Turkey" by Tefaruk Haktanir and Hatice Citakoglu. J Hydrol Eng 20(10):07015016

Holling CS (1996) Engineering resilience versus ecological resilience. In: National Academy of Engineering (ed) (ed) Engineering within ecological constraints. National Academies Press, Washington, DC, pp 31–43

Iftekhar MS, Polyakov M, Rogers A (2022) Social preferences for water sensitive housing features in Australia. Ecol Econ 195:107386

IRP4 (2021) Water sensitive outcomes for infill developments. https://watersensitivecities.org.au/content/project-irp4/

Keath NA, Brown RR (2009) Extreme events: being prepared for the pitfalls with progressing sustainable urban water management. Water Sci Technol 59(7):1271–1280

Kenway S, Gregory A, McMahon J (2011) Urban water mass balance analysis. J Ind Ecol 15(5):693–706

Lam KL, van der Hoek JP (2020) Low-carbon urban water systems: opportunities beyond water and wastewater utilities? Environ Sci Technol 54(23):14854–14861

Lam KL, Kenway SJ, Lant PA (2017) Energy use for water provision in cities. J Clean Prod 143:699–709

Lam KL, Liu G, Motelica-Wagenaar AM, van der Hoek JP (2022) Toward carbon-neutral water systems: insights from global cities. Engineering 14(7):77–85

London G, Bertram N, Renouf M, Kenway S, Sainsbury O, Todorovic T, Byrne J, Pype M, Sochacka B, Surendran S, Moravej M (2020a) Knutsford case study final report: water sensitive outcomes for infill development. Cooperative Research Centre for Water Sensitive Cities, Melbourne

London G, Bertram N, Sainsbury O, Todorovic T (2020b) Infill typologies catalogue. Cooperative Research Centre for Water Sensitive Cities, Melbourne

Marks NK, Hosseiny H, Bill VP, Ahn KL, Crimmins MC, Kremer P, Smith VB (2022) Spatial integration of urban runoff modeling, heat, and social vulnerability for blue-green infrastructure planning and management. J Water Resour Plan Manag 148(11)

Matthew P, Moravej M (2022) Summary of building codes in Brisbane: site setbacks, height limits and site cover restrictions. The School of Architecture, Brisbane

McDermott TKJ (2022) Global exposure to flood risk and poverty. Nat Commun 13(1):3529

McEvoy S, van de Ven FH, Blind MW, Slinger JH (2018) Planning support tools and their effects in participatory urban adaptation workshops. J Environ Manag 207:319–333

Milly PCD, Betancourt J, Falkenmark M, Hirsch RM, Kundzewicz ZW, Lettenmaier DP, Stouffer RJ (2008) Stationarity is dead: whither water management? Science 319(5863):573–574

Moravej M (2016) Investigating climate change using AK stationarity test in the Lake Urmia basin. Int J Hydrol Sci Technol 6(4):382–407

Moravej M, Leardini P (2022) Design typological analysis for urban water management: why quantification is needed and how t can be done? In: 1st International e-conference on green and safe cities 2022 (IeGRESAFE). UiTM Perak, Malaysia, pp 699–717

Moravej M, Renouf M, Lam KL, Kenway S (2020) User manual for site-scale urban water mass balance assessment (SUWMBA) tool V2. Cooperative Research Centre for Water Sensitive Cities, Melbourne

Moravej M, Renouf MA, Lam KL, Kenway SJ, Urich C (2021) Site-scale urban water mass balance assessment (SUWMBA) to quantify water performance of urban design-technology-environment configurations. Water Res 188:116477

Moravej M, Leardini P, Kenway S (2022a) Collaborative water sensitive design and urban performance analysis. School of Architecture, University of Queensland

Moravej M, Renouf M, Kenway S, Urich C (2022b) What roles do architectural design and water servicing technologies play in the water performance of residential infill? Water Res 213(4):118109

Moravej M, Sochacka B, Leardini P, Matthew P (2022c) Housing typologies in Greenslopes, Brisbane (draft). School of Architecture, University of Queensland, Brisbane

Müller A, Österlund H, Marsalek J, Viklander M (2020) The pollution conveyed by urban runoff: a review of sources. Sci Total Environ 709:136125

Newton P (2010) Beyond greenfields and greyfields: the challenge of regenerating Australia's greyfield suburbs. Built Environ 36(1):81–104

Newton PW, Rogers BC (2020) Transforming built environments: towards carbon neutral and blue-green cities. Sustainability 12(11):4745

Newton P, Prasad D, Sproul A, White S (2019) Decarbonising the built environment: charting the transition. Springer, Singapore

Newton P, Glackin S, Witheridge J, Garner L (2020) Beyond small lot subdivision: towards municipality-initiated and resident-supported precinct scale medium density residential infill regeneration in greyfield suburbs. Urban Policy Res 38(4):338–356

Newton PW, Newman PW, Glackin S, Thomson G (2022a) Greening the greyfields. Springer, Singapore

Newton PW, Newman PW, Glackin S, Thomson G (2022b) Greening the greyfields: new models for regenerating the middle suburbs of low-density cities. Palgrave Macmillan, Singapore

Pahl-Wostl C, Sendzimir J, Jeffrey P (2009) Resources management in transition. Ecol Soc 14(1):46

Renouf MA, Kenway SJ, Lam KL, Weber T, Roux E, Serrao-Neumann S, Choy DL, Morgan EA (2018) Understanding urban water performance at the city-region scale using an urban water metabolism evaluation framework. Water Res 137:395–406

Renouf M, Kenway S, Bertram N, London G, Todorovic T, Sainsbury O, Nice K, Moravej M, Sochacka B (2020a) Water sensitive outcomes for infill development: infill performance evaluation framework. Cooperative Research Centre for Water Sensitive Cities, Melbourne

Renouf MA, Kenway SJ, Bertram N, London G, Sainsbury O, Todorovic T, Nice K, Surendran S, Moravej M (2020b) Salisbury case study final report: water sensitive outcomes for infill development. Cooperative Research Centre for Water Sensitive Cities, Melbourne

Rogers BC, Dunn G, Hammer K, Novalia W, de Haan FJ, Brown L, Brown RR, Lloyd S, Urich C, Wong THF, Chesterfield C (2020) Water Sensitive Cities Index: a diagnostic tool to assess water sensitivity and guide management actions. Water Res 186:116411

Sochacka B, Kenway S, Bertram N, London G, Renouf M, Sainsbury O, Surendran S, Morave M, Nice K, Todorovic T, Tarakemehzadeh N, Martin D (2021a) Water sensitive outcomes for infill development: final report. Cooperative Research Centre for Water Sensitive Cities, Melbourne

Sochacka BA, Kenway SJ, Renouf MA (2021b) Liveability and its interpretation in urban water management: systematic literature review. Cities 113:103154

United Nations (2018) World urbanization prospects: the 2018 revision, key facts. United Nations, Department of Economic and Social Affairs, Population Division, New York

Urich C, Rauch W (2014) Exploring critical pathways for urban water management to identify robust strategies under deep uncertainties. Water Res 66:374–389

van de Ven FH, Snep RP, Koole S, Brolsma R, van der Brugge R, Spijker J, Vergroesen T (2016) Adaptation planning support toolbox: measurable performance information based tools for co-creation of resilient, ecosystem-based urban plans with urban designers, decision-makers and stakeholders. Environ Sci Pol 66:427–436

van Leeuwen CJ, Frijns J, van Wezel A, van de Ven FHM (2012) City blueprints: 24 indicators to assess the sustainability of the urban water cycle. Water Resour Manag 26(8):2177–2197

Wolman A (1965) The metabolism of cities. Sci Am 213(3):178–193

Zaveri E, Russ J, Khan A, Damania R, Jägerskog A (2021) Ebb and flow, vol 1. Water, migration, and development. World Bank Publications, Washington, DC

Mojtaba Moravej is a post-doctoral research fellow at The University of Queensland. His research focuses on model development, simulation, and optimisation of urban water systems, with a particular emphasis on Water Sensitive Urban Design and Sustainable Urban Water Management, to achieve sustainable, resilient, and liveable cities. He has developed new integrated modelling tools to support collaborative urban design processes and optimise co-design options. He has authored 19 peer-reviewed research articles, two book chapters, and seven research and development reports for industry. His work was showcased in the State Library of Queensland's "Purpose Built: Architecture for a Better Tomorrow" public exhibition in 2023. He has also won multiple awards for teaching, including Australian Awards for University Teaching (AAUT), Citations for Outstanding Contributions to Student Learning category in 2023.

Beata Sochacka is a PhD researcher at the Australian Centre for Water and Environmental Biotechnology, University of Queensland, Australia, investigating measurable aspects of the relationship between urban water management and liveability, with the aim of supporting the water sector in demonstrating its contribution to the creation of better urban environments. Beata has interdisciplinary background in both social science and water management and experience in community engagement for sustainability across different geographic contexts. Her areas of expertise and research interests include social aspects of urban water management, integrated approaches to resources management, urban greening, and conceptual underpinnings of urban sustainability frameworks.

Steven Kenway is a water leader with senior experience in research, industry, and government, gained through roles with the University of Queensland, CSIRO, Brisbane Water, Sydney Water, and private consulting. His work addresses urban water security, the water-energy nexus, and circular economies to achieve key goals such as (i) a net zero carbon water cycle, (ii) hybrid, decentralised, and integrated systems, and (iii) sustainable urban design and planning. He has authored over 60 Scopus-listed articles in high-quality journals, 20 books or major CRC public reports, 10 book chapters, and over 67 conference articles (a total of over 200 publications).

Peter Newton is an Emeritus Professor in the Centre for Urban Transitions at Swinburne University of Technology in Melbourne, Australia. From 2007 to 2021, he held the position of Research Professor, following 15 years as Chief Research Scientist with the Commonwealth Scientific and Industrial Research Organisation (CSIRO). He has had a distinguished career in built-environment research and was elected a Fellow of the Academy of Social Sciences in Australia in 2014. His principal fields of research have focused on the technology of planning, sustainability science, and urban transitions. He has published over 25 books, including *Greening the Greyfields* and *Migration and Urban Transitions in Australia* in 2022.

Cassady Swinbourne is a PhD student researching within the Water-Energy-Carbon Group at the Australian Centre for Water and Environmental Biotechnology at the University of Queensland. Her research focuses on the co-benefits and trade-offs between various components of the urban water cycle, with a particular focus on how these interactions manifest in water-sensitive urban-design infrastructure and during periods of droughts and extreme rainfall. Her work is informed by systems and resilience theory, and her background in mathematics, which she studied for her Bachelor of Science at the University of Tasmania.

Ka Leung Lam is an Assistant Professor of Environmental Science at Duke Kunshan University, China. Prior to joining DKU, he was a Marie Skłodowska Curie Fellow at Delft University of Technology, the Netherlands. His research focuses on how we can sustainably transition our approaches to water management: reducing greenhouse gas emissions, reducing environmental footprints, enabling resource recovery, and improving resource efficiency. He has published approximately 20 peer-reviewed articles related to sustainability.

Chapter 4
Accelerating Climate Innovation in Cities

Alessandro Ossola and Brenda Lin

Abstract The rate at which climate action is currently achieved is often outpaced by the increasing speed of climate impacts in many urbanised regions globally. Actions related to coping with climate extremes, improving adaptive responses, and ultimately transforming cities for climate resilience have had variable success and have often failed to be implemented at scale. Climate stresses and impacts are accelerating; current climate action is not, at least not fast enough.

Here we propose a novel conceptual framework based on the concept of 'climate innovation': the pervasive, strategic application of new and yet-to-be-born ideas, knowledge, and technology that can significantly accelerate the mitigation of climate change impacts and realise climate adaptation in human societies.

We identify a typology of climate innovations—incremental, sustaining, radical, and disruptive—highlighting possible foci and examples of innovations. We further discuss how climate innovation and adoption curves, as well as innovation series, could be idealised and implemented to fast-track climate change mitigation and adaptation strategies at multiple scales.

This framework can hopefully help societies and decision-makers to better contextualise how meaningful climate action can be envisioned, prioritised, and implemented—and discontinued when it fails to accelerate climate innovations and meet climate goals in increasingly accelerated timeframes.

Keywords Climate threats · Sustainability transitions · Climate business and finance · Climate disruption

A. Ossola (✉)
University of California, Davis, Davis, CA, USA

University of Melbourne, Parkville, VIC, Australia
e-mail: aossola@ucdavis.edu

B. Lin
CSIRO Environment, Dutton Park, QLD, Australia

© The Author(s) 2025
N. Frantzeskaki et al. (eds.), *Future Cities Making*, Theory and Practice of
Urban Sustainability Transitions, https://doi.org/10.1007/978-981-97-7671-9_4

4.1 The Increased Urgency for Climate Action

The increased focus on cities over the last decades has revealed a large push by local and regional governments to bring about activities, ideas, and transformative changes to help cities address the impacts of climate change (Pancost 2016; Ürge-Vorsatz et al. 2018). However, cities are complex and dynamic systems, and inter-disciplinary approaches are increasingly required to provide insights that take the context of the city into account (Lin et al. 2021). Even for cities that have the governance and financial capacity to best respond to climate pressures (Fig. 4.1)—like some in the Global North—urbanisation and a fast-paced climate change pathway may erode their ability to promptly respond in an equitable and sustainable matter. Because of this, climate change may amplify social injustices or intensify communities' inequalities and adaptive capacities between and within cities (Schlosberg and Collins 2014).

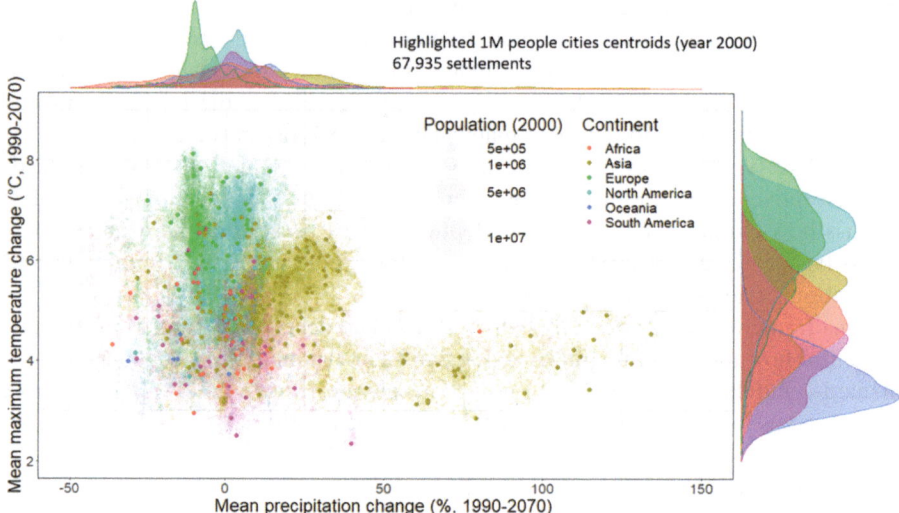

Fig. 4.1 Visualisation of climatic changes (1990–2070) predicted for 67,935 urban settlements globally. Mean maximum temperature change (°C) and percent precipitation change (%) are calculated for each settlement from a 1960–1990 climate baseline map (30 arc-s resolution) and the HadGEM2-ES model projected to 2060–2080 by using the Representative Concentration Pathway 8.5 (Hijmans et al. 2005; WorldClim v 1.4). Urban settlement points are provided by the Global Rural-Urban Mapping Project, Version 1 (GRUMP v1 2011), created by the Center for International Earth Science Information Network (CIESIN), Columbia University, the International Food Policy Research Institute (IFPRI), the World Bank, and the Centro Internacional de Agricultura Tropical (CIAT) (http://sedac.ciesin.columbia.edu/data/dataset/grump-v1-settlement-points). Population estimates refer to the baseline year 2000. Cities with populations exceeding one million people have their centroids highlighted in bold for clarity. Settlements are colour-coded by continent and frequency plots on the secondary axes represent the frequency distribution of settlements for each in continent in relation to mean maximum temperature change and mean precipitation change, respectively

The variable speed of change processes and the urgency of current challenges (i.e. climate change and rapid population growth) mean that we need proactive and rapid change, and we will thus require a significant acceleration in the realm of climate adaptation and mitigation advances, particularly in cities that are foci of climate impacts. These impacts are readily apparent and occurring at a faster rate than scientists have predicted. However, advances in climate action based on current norms and practice take time to build (Sparkman et al. 2021). Collaboration and cooperation between cities and across disciplines, governments, and industries also take substantial trust and time to create. Interdisciplinary approaches that consider the natural and built environment, social dynamics, interactions with stakeholders, and an assessment of community vulnerability at a local level have been proposed to sustain climate action (Bai et al. 2018). Multiple solutions in flexible and adaptable pathways have been proposed to bring forth plans to protect cities and their citizenry (Wise et al. 2014; Buurman and Babovic 2016).

Yet, the rate at which climate action is currently achieved is often outpaced by the increasing speed of climate impacts in many urbanised regions globally. Actions related to coping with climate extremes, improving adaptive responses, and ultimately transforming cities for climate resilience have had variable success and have often failed to be implemented at scale.

Here we propose a new conceptual framework based on the concept of 'climate innovation': *the pervasive, strategic application of new and yet-to-be-born ideas, knowledge, and technology that can significantly accelerate the mitigation of climate change impacts and realise climate adaptation in human societies*. This framework can hopefully help communities and decision-makers to better contextualise how meaningful climate action can be envisioned, prioritised, and implemented— and discontinued when it fails to accelerate climate innovations and meet climate goals in suitable timeframes ('climate-action sunsetting').

4.2 Climate Actions Lag Climate Impacts

Current and popular approaches for climate action in urban areas include (i) measures for coping with acute stresses, such as those from weather extremes, (ii) actions for improving responses to periodic and recurrent climate shocks, and (iii) plans that seek to promote urban transformations to deal with the chronic stresses brought about by climate change in the short and medium term (Fig. 4.2).

Many cities globally are increasingly relying on short-term strategies, plans, and actions aimed at *coping* with weather and climate extremes (e.g. Larsen 2015). These immediate and short-term actions targeting extreme, acute stresses—while important to increase resilience and adaptability of urban systems—offer tactical but limited strategic pathways to support cities for a different climate change future. Among those, warning systems for approaching cyclones, large-scale flooding events, and bushfires are becoming commonplace in many urban areas, in both the Global North and South (O'Connell et al. 2020). Because of the large impact these

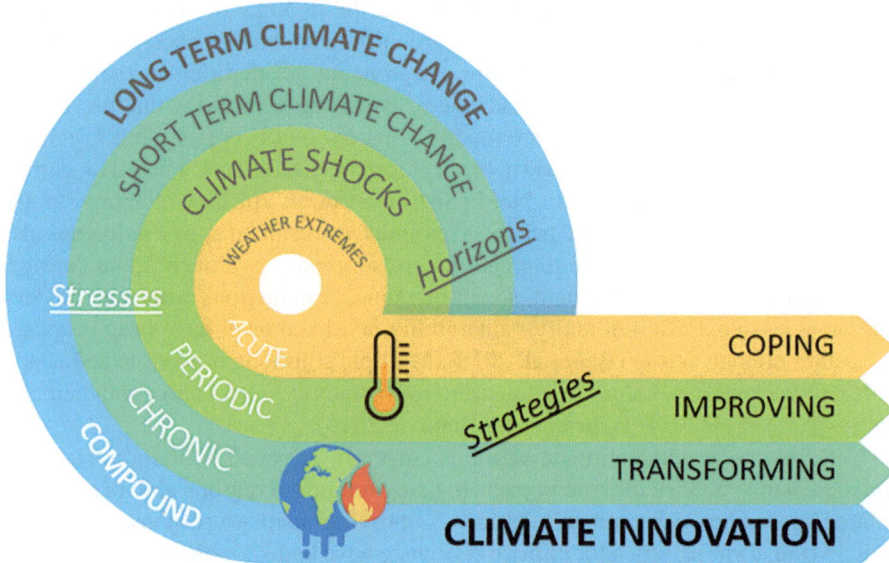

Fig. 4.2 Climate innovation is a holistic strategy that moves beyond traditional approaches aimed at coping with weather extremes, improving human and infrastructure resilience to climate shocks, and transforming urban systems for adapting to climate change in the short term. Climate innovation aims at accelerating climate action and outcomes in the long term through an integrated series of new and yet-to-be-born technologies, actions, policies, programmes, and interventions that can significantly amplify their impacts over time (see also Figs. 4.3 and 4.4)

extreme, acute events have on cities and their communities, governments and community groups focus strongly on coping strategies aimed at survival and the reduction of lost infrastructure and life. For example, increased flooding events due to prolonged storm systems have increased the use of levees in cities to protect important infrastructure. As weather and climate extremes are the short-term stresses mostly felt by urban dwellers, coping strategies targeting short and acute impacts are those most likely to affect the public perception and ability to take meaningful action for personal and community protection.

More sophisticated climate actions and strategies aim at *improving* urban responses to periodic climate shocks and stresses such as the Millennial drought or seasonal heatwave or coldwave events. The effect of global circulation changes on large-scale climate patterns such as El Niño-Southern Oscillation (ENSO) and the Indian Ocean Dipole (IOD) influence the frequency and duration of rainfall distribution patterns, which can have long-lasting effects on cities in terms of too much and not enough water to sustain the population and environmental landscapes in and around cities. While long-term periods of high rainfall can lead to a greater threat of flooding from sudden extreme storms, prolonged periods of low rainfall have led to cities opening up desalinisation plants and recycling water to better manage uncertainties in highly variable water supplies.

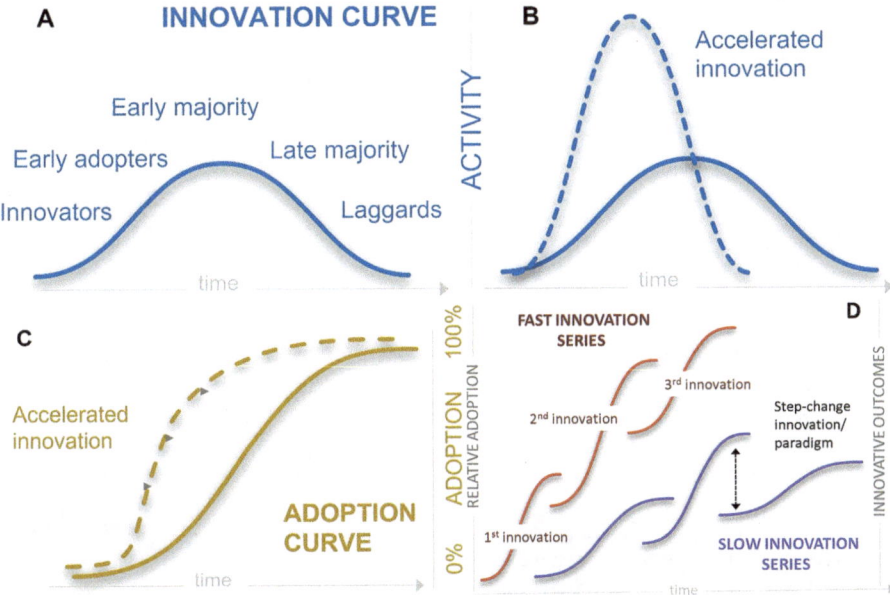

Fig. 4.3 Representation of climate innovation (**a, b**) and adoption (**c**) curves following Everett Roger's work on the diffusion of innovations (1962) and its modifications. Representation of fast and slow theoretical climate innovation series (**d**)

In the last decade a new approach based on *transforming* cities (also known as urban transformation) has been increasingly adopted to foster climate change resilience and adaptive outcomes (Rosenzweig et al. 2018). These actions recognise and try to address chronic climate stresses, particularly those brought about by climate change in the short to medium term, but with the realisation that many of these trends will continue into the future. One of the biggest issues in many of the world's cities is the continuing trend towards increasing temperatures along with increased urbanisation and urban heat islands. Urban transformation is largely geared towards reshaping and changing existing urban areas and fabrics to retrofit climate sustainability elements and incorporate new climate policies and actions (Egerer et al. 2021). The broader scope and forward-looking goals of this transformative approach have stimulated recent scholarship and practices that use transdisciplinary approaches to navigate the complexity of urban and climate responses in the short and medium term (Lin et al. 2021). The hope is that the current global agreements on net zero and other mitigation pathways will begin to ameliorate some of these chronic trends. Urban transformation initiatives, such as those incorporating wide-scale nature-based solutions and urban forests, while promising in some areas, are often very slow to be implemented at scale (e.g. trees take decades to grow to maturity) or have intrinsic limits in their implementation (e.g. tree canopy targets of greater than 30% are often impossible to achieve due to lack of space for tree plantings and climate limitations).

Fig. 4.4 Definitions and examples of a possible typology of climate innovations based on technology newness and likely climate outcomes and impacts. The definitions of climate innovation, technology newness, and climate outcomes are relative concepts and depend upon local characteristics of urban settings, as well as social and economic milieus. Icons are in the free public domain from UXWing and Wikimedia Commons

Approaches for coping, improving, and transforming cities to survive weather and climate issues have undeniably set many urban areas onto positive pathways for climate action. Despite these advances, several issues can limit the value and efficacy of these measures, particularly when dealing with the many climate stresses and impacts that will arise with climate change in the long term. First, many of the examples discussed above are often one-off tactical actions, rather than concerted strategies placed into a synergistic framework. Many of these climate actions are themselves a product of ideas, policies, technologies, plans, and frameworks envisioned even before climate change was known to be an issue. At the same time, as the population increasingly experiences so-called 'climate fatigue', it becomes increasingly hard to convince urban dwellers to implement old solutions (e.g. water-use restrictions during drought) that have been proven numerous times to bring often limited or intangible outcomes. Further, the lack of integration, progression, and flexibility in climate outcomes arising from these actions can significantly reduce their overall impact.

Second, these climate actions are often difficult to translate and scale to other urban contexts, and their uptake might be limited due to low buy-in, poor resources, and high costs. The implementation of multiple and integrated climate solutions remains limited to progressive and well-funded urban areas globally and is far from being mainstream practice. Ultimately, current climate actions are adopted at a speed that rarely matches the rate of climate change expected in many urban areas (Fig. 4.1). Climate policies driven by decision- and policymakers often stall for years—if not decades—before they can generate often delayed and less tangible impacts. Older climate strategies and tactics are mostly based on the assumption that climate change is solely an issue to be dealt with, rather than a *once-in-a-civilization* opportunity to further society and the environment. Too many of our current climate actions are trite, unimaginative imitations of attempts devised elsewhere, rather than a coherent set of locally based, scientifically grounded strategies able to win the war on climate change. The proof of this is that most urban areas globally are trying just to survive—not thrive—under climate change.

4.3 Accelerating Climate Innovation in Cities

The current *age of accelerations* has led to massive shifts in technology, globalisation, and progress that are argued to fundamentally reshape the complex ways cities and societies function (Friedman 2017). Despite some limited advances, these accelerated waves of innovation are far from providing the solutions that futurists have imagined would already be invented to save humanity from the impacts of climate change. Climate stresses and impacts are accelerating; current climate action is not, at least not fast enough.

Here we propose a framework of ideas based on the concept of climate innovation that could complement and amplify current climate action approaches across urban areas (Fig. 4.2) while significantly accelerating climate change adaptation and mitigation strategies in the long term. To do so, we revisit the sociological theory of the diffusion of innovations by Everett Rogers (1962) by applying a modern 'climate lens'. Here we define climate innovation as 'the pervasive, strategic application of new and yet-to-be-born ideas, knowledge and technology that can significantly accelerate the mitigation of climate change impacts and realize climate adaptation in human societies'. In doing so we recognise some caveats and assumptions. We propose climate innovation to be a variable concept with no universal value and applicability; what represents a climate innovation in one city might not be a true innovation in another city. Thus, climate innovation may occur in different ways and trajectories, depending not only on the rate of change of climatic variables (Fig. 4.1), but also on city-based environmental and social variables, among others. Climate innovation can be (i) directly intended to create disruptive outcomes for climate-human systems (e.g. planning a climate-positive city) or (ii) an innovation in fields other than those related to climate (e.g. energy) that can indirectly benefit climate adaptation and mitigation (e.g. LED lights). Climate innovation is intended here as involving and encouraging not only new and emerging technologies, products, and policies, but also those likely to come into existence in the near future. Ultimately, climate innovation can be intended as both an additive and a subtractive process. For instance, the creation of a new disruptive climate policy as well as the removal of an outdated climate policy ('sunsetting') could lead to progressive climate outcomes.

Compared to other approaches for climate action (e.g. coping, Fig. 4.2), climate innovation engenders a more positive, forward-looking connotation that is underpinned by the intrinsic human ability to create disruptive and original ideas that can lead to a step change in practice. In addition to the 'climate innovators' themselves, 'early adopters' are those climate actors with the drive, capital, skills, technology, and capacity to adopt and implement new climate actions before an 'early majority' can act (Fig. 4.3a). As seen when analysing innovation curves in other fields, delays shown by 'late adopters' and 'laggards' are often triggered by economic reasons and policy constraints rather than a true lack of drive and capacity to pursue innovative climate outcomes. The rate at which climate actors adopt and mainstream climate innovation can dictate the shape and rate of change in innovation and adoption curves and the timing with which a climate innovation will fully penetrate a particular climate-human system (Fig. 4.3).

Due to the intrinsic temporal nature of climate and other innovations, innovation curves can be rather flat and extended over time (Fig. 4.3b) or compressed, as for 'accelerated innovation'. Similarly, adoption curves can be extended or compressed in a temporal dimension depending on the rate of adoption in a local urban context. For example, innovators and early adopters were the first to make use of the LED light bulb, a relatively recent incremental innovation that has achieved some energy saving and reductions in greenhouse gas emission (Fig. 4.3a). Late-majority and laggard actors embraced this innovation later, largely because of policies aimed at discontinuing the use of traditional low-efficiency incandescent light bulbs and changes in market forces. Compared to the traditional incandescent light bulb, the LED light bulb can be considered an 'accelerated innovation' in many developed countries, where almost full adoption was reached in years, and not decades, as for the traditional light bulb (Fig. 4.3b, c). In contrast, in many urban areas in the Global South that are still not fully connected to the electric grid, the LED light still represents an innovation with indirect benefits for the climate system that is yet to be realised and scaled to size.

Because of the accelerated nature of climate change impacts in many urban areas globally, the selection and prioritisation of 'accelerated climate innovation' could shorten the timeframe over which meaningful climate mitigation and adaptation efforts could be mainstreamed and scaled to an appropriate size. Despite being technically and economically feasible, some climate innovations might be simply too slow to fully penetrate a climate-human system to make a tangible impact soon enough to make a significant difference. For instance, some recent urban forest policies that aim to increase urban tree canopy cover to more than 30% might not match the rate of projected changes in the local climate system when drought-tolerant (but slow-growing) tree species are prioritised (Ossola and Lin 2021). However, effective policies, incentives, and regulations could be implemented to mainstream climate innovations in particular contexts (e.g. phase-outs of old technologies such as incandescent light bulbs).

Multiple climate innovations are often sequential and require a step change in established technological, social, economic, or environmental paradigms to be effective (Fig. 4.3d). For instance, technological advances in energy storage and small batteries enabled the creation of incremental innovations such as E-bikes and electric scooters, which have had a variable impact on greenhouse emissions from the personal mobility sector. Further advances in information technology and mobile apps allowed the creation of new personal mobility-as-a-service (MaaS) companies, some of which were able to locally disrupt transportation markets, while others quickly perished.

When strategising innovation for climate action, climate actors could not only prioritise accelerated climate innovations that could match predicted shifts in the local climate, but also plan for innovation series and pathways (i.e. multiple innovations) that could deliver outcomes more quickly. Faster innovation series could be achieved by shortening and accelerating adoption curves (Fig. 4.3b, c), allowing the flow of concurrent innovations (Fig. 4.3d), and integrating complementary types of innovations (Fig. 4.4). Slow climate innovation series could, in contrast, lead to plateau outcomes and hamper climate action and progress towards meaningful adaptation and mitigation goals. While it might be difficult to prioritise yet-to-be-born technologies and innovations, flexible approaches that could accommodate parallel and complementary climate innovation trajectories and pathways could mitigate the risk of unanticipated (or lack of) innovation, as could 'climate action sunsetting' when needed. At the same time, new innovation could emerge by realigning human and ecological systems with each other—a tenet that has been fundamental to many indigenous knowledge systems for thousands of years—or by looking at older solutions now forgotten (Ossola et al. 2018).

As noted earlier regarding the local nature of single climate innovations, innovation series that enable certain climate outcomes in a particular urban area do not guarantee that the same outcomes could be achieved at other locations with different climates and socio-economic contexts.

4.4   A Climate Innovation Typology for Cities

Innovation theories from economics, business development, and social science have been used in numerous attempts to classify the many types of innovations that have fuelled economic and social cycles of human societies. However, a typology of climate innovations for cities has not yet been established. Here we envision and codify a possible typology along two axes related to technology newness and potential climate impacts and outcomes (Fig. 4.4). We acknowledge that this typification is far from complete, as it only represents one of the many possible ways a climate innovation landscape can be structured and analysed. Our effort aims to initiate—rather than resolve—the effort to establish a larger climate innovation typology and taxonomy for cities. As discussed earlier, a climate innovation could be included in a particular typology in some cities but not in others, depending on the local context, climate, and socio-eco-technological milieu.

Incremental climate innovations are actions characterised by low technology newness, which can lead to low or moderate climate impacts and outcomes (Fig. 4.4). These innovations rely on existing ones, but they can be implemented with a low risk of failure or unanticipated outcomes. Some incremental innovations, such as the installation of urban solar reflectors (see, for instance, https://www.meer.org), cooling materials, or simple white roofs, are often small in nature but can be scaled up to achieve compound outcomes across urban landscapes.

Sustaining climate innovations also have low technology newness, but they can lead to greater climate impacts and outcomes. For instance, some urban plans and policies, such as million tree planting programmes in arid cities or the adoption of green energy options such as solar panels, aim at creating a step change in the uptake of current climate actions without a significant departure from climate strategies or adaptation and mitigation plans.

Radical climate innovations have higher technology newness than do incremental and sustaining innovations, but can lead to limited climate impacts and outcomes, particularly when not scaled up. Smart city technologies might allow some urban areas to measure, adjust, and optimise processes and fluxes of energy and materials, and to respond to climate stresses in a quasi-real-time fashion (Obringer and Nateghi 2021). Some of these innovations might require significant time and economic resources to mature and scale up to determine larger climate impacts.

Disruptive climate innovations can be envisioned at the higher cutting-edge end of the technology spectrum as well as have the potential to determine high-impact climate benefits and outcomes. These innovations, most of which have to still come to fruition, can not only minimise but also reverse climate impacts and stresses. For instance, the advent of circular cities, the creation of new carbon-absorbing materials and concrete, as well as floating cities could give way to the rise of carbon-negative cities able to progressively offset and reverse human emissions from urban areas. The Green Riyad Project aims to plant over seven million trees across a desert city alongside new technical and governance enablers to sustain its growth and viability (https://www.rcrc.gov.sa/en/projects/green-riyadh-project). While this project is anticipated to spur significant climate innovations, outcomes for a range of industries (e.g. healthcare, IT, finance, etc.), and a \$19 billion USD return on investment (ROI), its risk of failure and costs remain high.

4.5 A Call for Innovative Climate Action

Reframing climate action through an 'innovation lens' has the potential to accelerate climate change adaptation and mitigation efforts and bring them up to speed with the climate impacts that are mushrooming across cities and towns globally. This brief contribution aims to spur a dialogue among climate innovators, decision-makers, and researchers to step up tactical efforts towards well-concerted, dynamic strategies that can accelerate action while pursuing dynamic pathways to climate action.

The move towards a more climate-innovative approach may seem confronting to current governance systems, where failure is not often accepted by constituents. Failure that has led to a loss of taxpayer money has especially been seen as a difficult political pill to swallow. However, innovation, and especially fast innovation, requires the public to allow government to test trial a number of options quickly while evaluating the potential to scale up. Teaming up with scientists and stakeholders to trial these new ideas will be essential to co-create solutions that parties accept because they see the value in testing the new innovation, even with the potential for failure.

It remains to be seen which actors could take on board the charge of fostering and coordinating climate innovators and innovations, although this role could be well suited to a plethora of international, governmental, non-governmental, and non-profit organisations (such as the Intergovernmental Panel on Climate Change or the United Nations Environment Programme).

References

Bai X, Dawson RJ, Ürge-Vorsatz D, Delgado GC, Barau AS, Dhakal S, Dodman D, Leonardsen L, Masson-Delmotte V, Roberts DC (2018) Six research priorities for cities and climate change. Nature 555(7694):23–25. https://www.nature.com/articles/d41586-018-02409-z

Buurman J, Babovic V (2016) Adaptation pathways and real options analysis: an approach to deep uncertainty in climate change adaptation policies. Polic Soc 35:137–150

Egerer M, Haase D, McPherson T, Frantzeskaki N, Andersson E, Nagendra H, Ossola A (2021) Urban change as an untapped opportunity for climate adaptation. NPJ Urban Sustain 1:22

Friedman TL (2017) Thank you for being late: an optimist's guide to thriving in the age of accelerations (version 2.0). Picador USA

Hijmans RJ, Cameron SE, Parra JL, Jones PG, and Jarvis A (2005) Very high resolution interpolated climate surfaces for global land areas. Int. J. Climatol 25:1965–78. https://doi.org/10.1002/joc.1276

Larsen L (2015) Urban climate and adaptation strategies. Front Ecol Environ 13:486–492

Lin BB, Ossola A, Alberti M, Andersson E, Bai X, Dobbs C, Elmqvist T, Evans KL, Frantzeskaki N, Fuller RA, Gaston KJ (2021) Integrating solutions to adapt cities for climate change. Lancet Planet Health 5:479–e486

O'Connell D, Grigg N, Hayman D, Bohensky E, Measham T, Wise R, Maru Y, Dunlop M, Patterson S, Vaidya S, Williams R (2020) Disaster-resilient and adaptive to change—narratives to support coordinated practice and collective action in Queensland. CSIRO, Canberra

Obringer R, Nateghi R (2021) What makes a city "smart" in the Anthropocene? A critical review of smart cities under climate change. Sustain Cities Soc 75:103278

Ossola A, Lin B (2021) Making nature-based solutions "climate-ready" for the 50°C world. Environ Sci Policy 123:151–159

Ossola A, Egerer M, Lin B, Rook G, Setälä H (2018) Lost food narratives can grow human health in cities. Front Ecol Environ 16:560–562

Pancost RD (2016) Cities lead on climate change. Nat Geosci 9:264

Rogers EM (1962) Diffusion of innovations. The Free Press, New York

Rosenzweig C, Solecki W, Romero-Lankao P, Mehrotra S, Dhakal S, Ibrahim S (2018) Pathways
    to urban transformation. In: Rosenzweig C, Solecki W, Romero-Lankao P, Mehrotra S, Dhakal
    S, Ibrahim SA (eds) Climate change and cities: second assessment report of the urban climate
    change research network. Cambridge University Press, Cambridge, pp 3–26
Schlosberg D, Collins LB (2014) From environmental to climate justice: climate change and the
    discourse of environmental justice. Wiley Interdiscip Rev Clim Chang 5:359–374
Sparkman G, Howe L, Walton G (2021) How social norms are often a barrier to addressing climate
    change but can be part of the solution. Behav Public Policy 5:528–555
Ürge-Vorsatz D, Rosenzweig C, Dawson RJ, Rodriguez RS, Bai X, Barau AS, Seto KC, Dhakal S
    (2018) Locking in positive climate responses in cities. Nat Clim Chang 8:174
Wise RM, Fazey I, Smith MS, Park SE, Eakin H, Van Garderen EA, Campbell B (2014)
    Reconceptualising adaptation to climate change as part of pathways of change and response.
    Glob Environ Chang 28:325–336

Alessandro Ossola is an urban ecologist and environmental scientist, and Assistant Professor at
the University of California, Davis, USA. He is honorary faculty at the University of Melbourne,
Australia, a 2022 FFAR New Innovator Awardee, and a former NASEM Associate with
US-EPA. His research encompasses urban ecology, climate change science, urban sustainability,
and governance. Over the years, he has advised several government agencies in various countries
on urban nature-based solutions for climate adaptation. He is interested in applied, co-produced
research that bridges environmental management, ecological design, and science communication.

Brenda Lin 's research examines how vegetation cover is changing in our cities and how components of natural systems can be reintegrated to provide ecosystem services that optimise environmental, human, and social well-being. She is a Senior Scientist within CSIRO's Sustainable,
Liveable, and Resilient Cities Program and a former postdoc fellow with the Earth Institute at
Columbia University. She undertook an AAAS Science and Technology Policy Fellowship with
the US-EPA, translating scientific research for policy application.

Part II
Low-Carbon and Circular Cities

Chapter 5
Delivering Sustainable, Resilient, and Low-Carbon-Built Environments

Bao-Jie He and Deo Prasad

Abstract The current climate emergency is now a matter of urgent attention, as evidenced by the growing number of scientific reports on the subject and media generally. In response, there is a global movement within the built environment sector to explore how net-zero carbon targets can best be met for this sector, globally, by 2030. Many businesses, governments, and other organisations have released their commitments with benchmarks, targets, and pathways to achieve net-zero carbon-built environments. Such commitments have mainly prioritised carbon pollution metrics, while urban liveability indicators related to cooling, comfort, health, and well-being of communities have often not been well integrated into built environment decarbonisation goals. This chapter aims to present pathways for achieving climate-linked goals for resilient planning and design systems for built environments. It analyses global guides, plans, and actions designed to lead to sustainable, resilient, and low-carbon-built environments and presents a robust, comprehensive, and integrative discussion of the findings of the CRC for low-carbon living to inform strategic decision-making for built environments in Australia.

Keywords Low-carbon-built environment · Decarbonisation · Climate resilience · Targets and pathways · Techniques and strategies

B.-J. He
Centre for Climate-Resilient and Low-Carbon Cities, School of Architecture and Urban Planning, Chongqing University, Chongqing, China

Institute for Smart Cities, Chongqing University, Chongqing, China

Key Laboratory of New Technology for Construction of Cities in Mountain Area, Ministry of Education, Chongqing University, Chongqing, China

Network for Education and Research on Peace and Sustainability (NERPS), Hiroshima University, Hiroshima, Japan

School of Built Environment, University of New South Wales, Kensington, NSW, Australia

D. Prasad (✉)
School of Built Environment, University of New South Wales, Kensington, NSW, Australia
e-mail: d.prasad@unsw.edu.au

© The Author(s) 2025
N. Frantzeskaki et al. (eds.), *Future Cities Making*, Theory and Practice of Urban Sustainability Transitions, https://doi.org/10.1007/978-981-97-7671-9_5

5.1   Introduction

Climate change is a global challenge that has arisen largely due to the release of polluting gases from burning fossil fuel (e.g. coal, oil, and gas). Average global temperature in 2022 on Earth has risen by 1.0–1.1 °C above the pre-industrial level, already causing weather extremes such as heatwaves, droughts, flooding, hurricanes, and wildfires (UNEP 2020). For instance, heatwaves, where temperatures remain high from a few to dozens of days, are one of the most lethal climate-related disasters. A record-breaking heatwave that hit the European continent in July 2022 resulted in a 16% increase in mortality, involving about 53,000 excess deaths (European Union 2022a). Apart from threats to humans, natural ecosystems are also under significant risk. For example, in Australia's 2019–2020 bushfire season, nearly three billion animals were killed and, with ongoing climate change, many endangered species are likely to be driven to extinction (WWF 2020). What is worse, well-developed scientific models have projected that warming will be a major trend in the coming decades and that future climate-related disasters and hazards will be even more frequent and severe (Freychet et al. 2022; IPCC 2023). Importantly, climate change is now recognised as a threat-multiplier globally and locally, intensifying the full spectrum of extreme events: heatwaves, storm systems and their associated flooding, damage from hail, snow, and wind, megafires, and droughts.

Addressing climate change is thus an urgent task not only because it is crucial to the sustainability of the present generation, but because it is critical for future liveability and prosperity. A direct pathway to dealing with ever-changing climate is mitigation through reducing the emission of greenhouse gases, including carbon dioxide and methane. However, there is also a need to concentrate on disaster risk reduction and climate change adaptation, to protect people from climate-related disasters and risks. The reasons are twofold. First, because successful decarbonisation is a significant change in the way humans on the earth construct and live, it cannot be achieved in the short term. Second, accompanying the efforts to fully achieve decarbonisation, climate keeps changing, and associated impacts on people (especially vulnerable and disadvantaged groups) remain intractable (Rosenzweig et al. 2018).

5.1.1   International Agendas on Built Environment Infrastructure Decarbonisation

Decarbonisation represents a direct response to decelerating climate change and limiting warming below 1.5 °C, beyond which lives, livelihoods, and economies will be locked into more serious risks (IPCC 2018). A necessary target to achieve this is to curb global carbon emission intensity to less than 50% of the 2019 emission level by 2030 (Ritchie and Roser 2022). Many nations have now recognised the importance of shifting their societies and economies onto a path of decarbonisation and have committed to diverse versions of frameworks for carbon neutrality.

Cities are considered to be a major opportunity for carbon-neutral action as they constitute the main settlement form for over half of the world's population, as well as being economic growth engines. However, cities also account for about 70% of global greenhouse gas emissions (United Nations 2022a). Different sectors of the economy, such as energy, land, industry, and infrastructure (e.g. buildings and transportation), have formulated independent targets and identified implementable pathways for reducing demand for buildings and materials with high embodied and operating energy, converting their energy supply to carbon-free electricity and fuels (such as green hydrogen), and increasing carbon sequestration and storage to produce carbon-neutral cities (Quigley 2019).

The construction and operation of buildings and other forms of infrastructure account for 34% of global energy use and represent 37% of global operational energy and process-related CO_2 emissions (UNEP 2021). The building and construction sector has responded by identifying quick, deep, and cost-effective greenhouse gas mitigation targets (Table 5.1) (Huovila et al. 2009). For instance, the Global Alliance for Building and Construction (GABC) launched a global roadmap to achieve a common vision of a zero-carbon building and construction sector. In this vision, new buildings are expected to meet net-zero operation-ready codes and policies by 2030, and most new buildings are expected to achieve the net-zero whole-of-life carbon target by 2050. A target of net-zero operational carbon emissions among most existing buildings by 2050 is expected to be achieved by actions such as renovation, repair, refurbishment, and retrofits (GlobalABC 2020). (Appendix A contains a definition of decarbonisation terms used in this chapter.)

5.1.2 Baseline Challenges in Decarbonising the Built Environment

Along with extensive carbon emissions and the associated climate change, cities are facing many mega-challenges, including urbanisation, population increase and ageing, environmental deterioration, economic growth, and biodiversity loss. City decision-makers need to understand how and where these challenges in population, economy, environment, and well-being intersect with the existing and emerging risks and threats generated by climate change. An improper decision in urban planning and design can have knock-on effects for many other issues, such as automobile dependence, heat-island effects, air pollution, and urban flooding.

Low-density and dispersed urban growth, which is a typical development pattern of Australian cities, leads to cities that are highly dependent on automobiles and high consumers of resources (such as basic raw concrete and asphalt materials for building longer roads, pipes, and wires) and emissions (such as pollutant exhausts, waste heat, and CO_2) (Thomson et al. 2019). Another problem with this type of built

Table 5.1 Roadmaps of selected global and national net-zero carbon-built environment initiatives (WGBC 2020; Architecture 2030 2021; IEA 2021a; LETI 2021; RIBA 2021; UNEP 2021)

Organization	2020	2025	2030	2040	2050
Global Alliance for Buildings and Construction, World			Net-zero operational ready for new buildings		Net-zero for most new and existing buildings
				Net-zero for new buildings in some countries	Net-zero for most new buildings
World Green Building Council, World			Net-zero for all new buildings		Net-zero for all buildings
			40% reduction in embodied carbon for new buildings, infrastructures, and renovations		Net-zero for all buildings and infrastructure
Royal Institute of British Architects, UK	Adopt 2025 targets, but aim for 2030 targets	≈40–50% reduction	60% reduction with offsets to net-zero		
	Adopt 2025 targets, but aim for 2030 targets	≈30% reduction in embodied carbon for new domestic and office buildings	40% reduction in embodied carbon for new domestic and office buildings (and NTE built targets). Offsets remaining carbon emissions		
London Energy Transformation Initiative, UK	10% of all new buildings designed to net-zero	All new buildings designed to net-zero	Net-zero operational for all new buildings		Net-zero for all buildings
	≈30–40% reduction in embodied carbon in all new buildings		≈65% reduction in embodied carbon in all new buildings		
Architecture 2030 Challenge, USA	80% reduction for all new buildings and major renovations	90% reduction for all new buildings and major renovations	Net-zero for all new buildings		

(continued)

Table 5.1 (continued)

Organization	2020	2025	2030	2040	2050
	40% reduction for all buildings, infrastructure, and associated materials	45% reduction for all buildings, infrastructure, and associated materials	65% reduction for all buildings, infrastructure, and associated materials	Zero embodied carbon for all buildings, infrastructure, and associated materials	
International Energy Agency, World			All new buildings are zero carbon ready. 20% of existing buildings retrofitted to be zero carbon ready	50% of existing buildings retrofitted to zero carbon ready levels	More than 85% of buildings are zero carbon ready
			40% reduction per square metre of new floor area		95% reduction in embodied carbon due to net-zero carbon emissions in other linked sections

environment is that concrete and asphalt pavements often have strong solar radiation absorption and heat storage capacity, causing local temperature increases called urban heat islands (Roth 2012). Furthermore, hard surfaces often prevent cities from draining excessive water, thereby placing their populations at risk of urban flooding, especially during the heavy precipitation that is increasingly frequent in cities as climate change accelerates (Wang et al. 2022). Low-density and dispersed precincts are also criticised as the causes of undesirable economic and social performance, particularly in comparison with other development patterns (Kjaersgaard et al. 2019).

Other improper built environment designs leading to intensification of heat-island effects include reduced urban greenery and water bodies, increased building height and density and city footprints, reduced ventilation performance, and increased anthropogenic heat release associated with air-conditioner and vehicle operations (Roth 2012). In developed cities, the urban heat island has already become a typical phenomenon of local climate change. For instance, Australia is a highly urbanised nation, and cities are generally 5 °C hotter than their surrounding suburban areas (Palin 2017).

Geographical location can also amplify heat-island effects. Within the Greater Sydney Region, the inland suburbs, due to their local microclimate and lack of a sea breeze, can be 12 °C hotter than the coastal suburbs (Santamouris et al. 2017). Under climate change and ongoing urban development, both heatwaves and heat islands increase people's exposure to extreme heat challenges (He et al.

2021a). Extreme heat reduces outdoor activity and productivity, obstructs transport operation, drives air pollution (e.g. ozone), increases energy and water use, and causes anti-social behaviour and biodiversity loss (Santamouris and Kolokotsa 2016).

5.1.3   Overcoming Inertia in Achieving Decarbonisation of the Built Environment

Factors slowing built environment decarbonisation include population increase, climate change, and people's increasingly high demands for comfort, health, and well-being and their preference for car-based travel in the absence of convenient alternatives. Population increase leads to growth in building footprints, implying the use of large amounts of materials (embodied carbon) and the adoption of mechanical heating, ventilating, and cooling for regulating indoor and outdoor environment temperatures (operational carbon).

Climate change reshapes the carbon emission pattern of the building and construction sector, mostly leading to an upward trend in energy demand. Temperature increases and associated natural disasters from extreme weather events increase the likelihood of material and structural damage, or even building collapse and the reduction of service life. This adds to embodied carbon emissions for building maintenance, repair, and refurbishment, or for new construction (Prasad et al. 2023).

Outdoor climates shape indoor climates through heat transfer and penetration (Path 1 in Fig. 5.1). Deterioration of indoor climate quality (e.g. temperature) is a driver for the extensive adoption of HVAC systems for cooling purposes, resulting in a significant increase in electricity use and a concomitant increase in operational carbon emissions, as the majority of electricity grids worldwide currently depend mainly on fossil fuels. It should be noted that air-conditioning systems have been a primary adaptive approach for rapid post-1960s population and city growth in hot tropical and arid regions (Dick and Rimmer 1999; He et al. 2021b). The increasing use of HVAC systems also boosts outdoor extreme heat (Path 2 in Fig. 5.1). For example, in the Phoenix metropolitan area, waste heat emitted from air-conditioning systems at night elevated the average air temperature at 2 m by more than 1 °C for some urban locations (Salamanca et al. 2014).

These drivers shape the potential for passive design techniques and strategies for aspects related to site space and form, building type, envelope properties, ventilation performance, lighting, renewable energy use, and other passive technologies such as surface coating and vertical greening. Overall, addressing climate change is a task relevant not only to decarbonisation, but also to building urban resilience and adaptation capacity. This chapter aims to describe how to develop sustainable, resilient, and low-carbon-built environments.

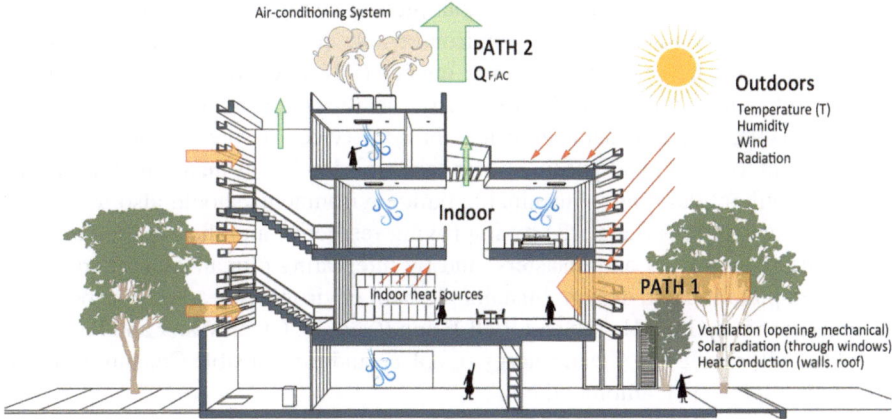

Fig. 5.1 The indoor–outdoor heat transfer and release, generating a negative loop of indoor over-heating and outdoor extreme heat. (Source: Author)

5.2 International Visions for Sustainable, Resilient, and Low-Carbon-Built Environments

The development of sustainable, resilient, and low-carbon-built environments for cities requires a foundational framework for coping with climate-related challenges, in the context of increasing population and urban and economic growth. Low-carbon, resilient cities typically attempt a win-win solution with mitigation through reducing greenhouse gas emissions and the achievement of urban resilience by reducing disaster risk and encouraging climate change adaptation (Rosenzweig and Solecki 2018). Low-carbon, resilient cities nest within sustainable cities, where a broader set of environmental indicators, in addition to pollution and carbon emissions, include energy and water use, water quality, energy mix, waste and recycling rates, green space ratios, primary forests, and agricultural land loss. Further, key socio-economic indicators include social equity, accessibility, density, and variety (de Jong et al. 2015). Many urban philosophies, concepts, and movements, such as healthy cities, garden cities, green cities, sustainable cities, low-carbon cities, and eco-cities, embrace elements of low-carbon cities. A number of more comprehensive international frameworks have also emerged and are identified below.

5.2.1 Sustainable Development Goals

The Sustainable Development Goals (SDGs) chart a blueprint for all sectors to achieve a better and more sustainable future with a broad scope, future time horizon, and transformational perspective (UN SDG 2015). The SDGs have also elaborated contributions from the building sector that present significant challenges and valuable opportunities for improving environmental, economic, and social benefits whilst addressing the problem of climate change (UN SDG 2015). SDG Sustainable Cities and Communities (Goal 11) explicitly specifies the need to provide adequate, safe, inclusive, and affordable housing, basic services, transport systems, waste management systems, and public spaces, and to adopt local materials for sustainable and resilient buildings. Sustainable built environments should also respond to SDG Climate Actions (Goal 13) by improving resilience and the capacity to adapt to climate-related risks and disasters, and by integrating climate change measures into urban planning and design. Sustainable built environments are also expected to accomplish SDG Good Health and Well-being (Goal 3), Clean Water and Sanitation (Goal 6), Affordable and Clean Energy (Goal 7), and Responsible Consumption and Production (Goal 12), among others.

5.2.2 UNEP's Three-Pillar Pathway

From the broad perspective of urban sustainability, cities are under the constraints of economic change, resource scarcity, social change, environmental deterioration, and climate change. To address such challenges, the United Nations Environmental Programme (UNEP) has established a three-pillar pathway for creating resource-efficient cities: clean, resource-efficient, and green and healthy cities (Fig. 5.2). Regarding resource-efficiency improvement, the UNEP prioritises measures to achieve a circular economy and 3R (reduce, reuse, and recycle) principles, promotes

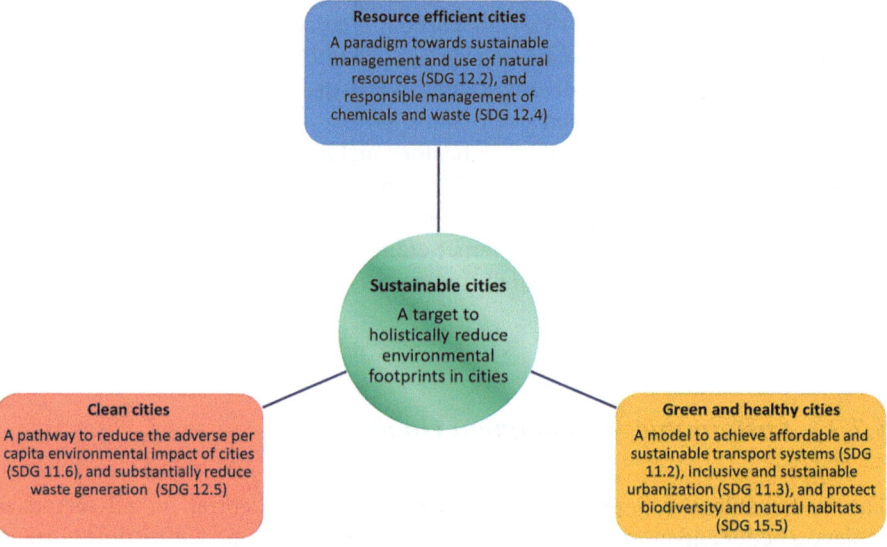

Fig. 5.2 Pathways of sustainable, resilient, and low-carbon cities defined by the UNEP (2023)

lifecycle analysis of material and energy use, and recommends the adoption of smart technologies. Suggestions about sustainable consumption and production, payment for pollution and waste, and accountability mechanisms are solutions for making cities cleaner. Land-use planning, mobility management, and socio-economic equity improvement are solutions leading towards green and healthy cities (UNEP 2023).

5.2.3   Carbon Neutral Cities Alliance

The Carbon Neutral Cities Alliance presents a community of leading global cities dedicated to carbon neutrality between 2030 and 2050 (Carbon Neutral Cities Alliance 2022). Table 5.2 demonstrates carbon-neutral targets and actions established by some leading cities. Among these cities, Copenhagen, Denmark (66% of

Table 5.2 Leading cities working on carbon-neutral cities

No.	City	Targets and actions
1	Copenhagen, Denmark	Copenhagen aims to be the first carbon-neutral capital in 2025. Its Climate Plan 2025 consists of four pillars: energy consumption, energy production, mobility, and city administration
2	Glasgow, UK	Glasgow is committed to becoming a Net Zero Carbon City by 2030. The city council declared a climate and ecological emergency and produced 61 recommendations on how the city could respond to the emergency
3	Helsinki, Finland	Helsinki aims to achieve the carbon-neutral target by 2030. This goal will be achieved by reducing the greenhouse-gas emissions in Helsinki by 80%
4	New York City, USA	New York City is committed to reducing its greenhouse gas emissions by 80% by mid-century and is investing $20 billion to adapt neighbourhoods to climate change risks, such as flooding, heat, and sea level rise
5	Stockholm, Sweden	The vision of a climate-smart Stockholm forms the basis of a strategy for a fossil fuel-free Stockholm by 2040
6	Sydney, Australia	Across the local government area, the City of Sydney has set targets for 50% renewables by 2030, 70% reduction in 2006 levels of greenhouse gas emissions levels by 2030, and net-zero emissions by 2050
7	Toronto, Canada	By 2050, 100% of vehicles in Toronto will use low-carbon energy; 75% of trips under 5 km will be walked or cycled
8	Vancouver, Canada	The Renewable City Strategy is the city's long-term plan to shift building and transportation energy use in the entire city to 100% renewables before 2050
9	Washington, DC, USA	In December 2017, Washington, DC pledged to become carbon-neutral by 2050. Washington, DC has developed Clean Energy DC, a roadmap to cut greenhouse gas emissions by 50% by 2032 through reductions in energy use and increased use of renewable energy
10	Yokohama, Japan	Yokohama City is committed to addressing urban problems such as a hyper-ageing society and reduction of CO_2 emissions, along with the revitalisation of the city's economy

Source: Carbon Neutral Cities Alliance (2022)

carbon emissions from energy and 34% from transport), has a target of carbon neutrality by 2025 (Carbon Neutral Cities Alliance 2022). It presents a four-pillar pathway including initiatives addressing energy consumption, energy production, mobility, and city administration. The City of Toronto aspires to reduce greenhouse gas emissions by 80% by 2050, based on 1990 levels. All new buildings should be near-zero greenhouse gas emissions by 2030, and all existing buildings should be retrofitted to gain a 40% improvement in energy performance by 2050. All vehicles must be powered by low-carbon energy by 2050, at which time 75% of trips within 5 km should be walked or cycled (Carbon Neutral Cities Alliance 2022). The key principles include social equality improvement, affordability improvement, poverty reduction, public health improvement, and infrastructure-resilience improvement (Carbon Neutral Cities Alliance 2022).

5.2.4   Race to Zero by C40

The Race to Zero, launched by the C40, Local Governments for Sustainability, the Global Covenant of Mayors, Carbon Disclosure Project, United Cities and Local Governments, the World Resources Institute, and the World Wildlife Fund, is an emerging campaign to rally leadership and support from businesses, cities, regions, and investors for resilient, healthy, and low-carbon cities (United Nations 2022b). As a district from the decarbonisation coalition, which aims to limit warming below 1.5 °C, the Race to Zero also scales up a series of actions in job opportunities and quality, working skills and productivity, public service availability and accessibility, environmental improvement, new partnerships, and new investments. To help alleviate inequality, the Race to Zero groups cover workers, business and academia, women, communities of colour, indigenous groups, and vulnerable people relevant to ethnicity, origin, gender, age, or social conditions (United Nations 2022b).

5.2.5   NetZeroCities by the European Union

The European Union launched the NetZeroCities project to deliver 100 climate-neutral and smart cities by 2030; such cities were expected to serve as successful experimentation and innovation hubs that be followed by all other cities in Europe by 2050. During the process of driving sustainable urban transformation by achieving climate neutrality before 2030, the NetZeroCities project offers their citizens the co-benefits of cleaner air, safer transport, and less congestion and noise. The European Union has developed a guidebook for net-zero cities that prioritises six themes for reaching zero emissions: stationary energy, energy generation, mobility and transport emissions, green industry, circular economy, and nature-based solutions (European Union 2022b).

5.2.6 The International Energy Agency

The International Energy Agency suggests the integration of smart technologies into net-zero-emission cities by identifying the opportunities in the built environment, district heating and cooling, energy communities, streetlighting, mobility and transport, EV charging infrastructure, and management of municipal services (IEA 2021b). Beyond net-zero carbon emissions by 2050, the International Energy Agency is looking for opportunities for co-benefits such as stable and affordable energy supplies, universal energy access, and robust economic growth. For instance, transitions towards clean energy must consider the social and economic impacts on individuals and communities, especially in the area of clean-energy jobs. Meanwhile, the International Energy Agency anticipates that all benefits (e.g. air quality, public health, energy access) from energy transitions towards new clean energy will be shared by all (Bouckaert et al. 2021).

5.2.7 The Cool Coalition

UNEP's Cool Coalition launched the guide *Beating the Heat: A Sustainable Cooling Handbook for Cities* to systematically address extreme heat challenges caused by heatwaves and heat-island effects. The guide identifies the international awareness of the importance of cooling cities and communities for healthy, safe, and comfortable living environments and the need for actions to constrain the warming trend. The pathways for sustainable urban cooling consist of reducing heat at the urban scale, reducing cooling needs in buildings, and providing cooling needs in buildings efficiently. Furthermore, this guide offers planners an encyclopaedia of proven options, with 80 supporting case studies and examples, to help cool cities (Campbell et al. 2021). For instance, the city of Ljubljana, Slovenia, renovated major roads to be walking- and cycling-friendly, planted trees, and modernised the public transport system for cooling purposes. The programme also generated co-benefits related to the reduction of air and noise pollution and transport carbon emissions (Campbell et al. 2021).

Overall, the vision of sustainable, resilient, and low-carbon cities has already been highlighted and advocated by international leading and pioneering organisations through the release of plans, roadmaps, projects, and guidelines. To work in practice, such international visions should be implemented at the national, state, and city levels. However, the achievement of sustainable, resilient, and low-carbon cities is challenging. For instance, actions to support the SDGs have been taking place in many nations and cities, but the required speed or scale has been lagging expectations for progress. Key reasons why existing guides and plans cannot be implemented are the multiple linkages among different urban challenges (e.g. energy, water, waste, transport, air quality, biodiversity, housing, and health) and the lack of a far deeper investigation into the development of targets, benchmarks, strategies, and policies needed to achieve sustainable, resilient, and low-carbon cities. The Cooperative Research Centre for Low Carbon Living (CRCLCL) is a leading research and innovation hub in Australia dedicated to driving the built environment sector of Australia to be sustainable, resilient, and low-carbon. The next section introduces the CRCLCL's visions for sustainable, resilient, and low-carbon-built environments. These visions focus on how to transition global concepts, goals, objectives, and performance indicators into a national and local context, drawing primarily on the outputs from a 7-year body of applied research.

5.3 The CRCLCL Visions for Sustainable, Resilient, and Low-Carbon-Built Environments

The CRCLCL is committed to low-carbon-built environments, in which the achievements of sustainable and resilient buildings, homes, communities, and precincts are highlighted. The CRCLCL has identified three parallel pathways for research and action, including integrated building systems, low-carbon precincts, and engaged communities (Newton et al. 2019). The CRCLCL, based on a collaboration of Australian industries, governments, and university researchers, has undertaken more than 100 research projects and developed practical ways to decarbonise the built environment. A series of national, state, and city plans, roadmaps, projects, and guidelines for sustainable, resilient, and low-carbon-built environments, including decarbonising the current built environment, have been formulated, and some have been integrated into government, business, and community practice (Fig. 5.3).

5.3.1 Race to Net Zero Carbon: A Climate Emergency Guide for New and Existing Buildings in Australia

The CRCLCL proposed a net-zero whole-of-life carbon target for the Australian built environment (buildings, precincts, and infrastructure) by 2050, and the interim targets of net-zero operational carbon by 2030 and net-zero embodied carbon by 2040 for all new buildings and major renovations (Prasad et al. 2021). The Race to

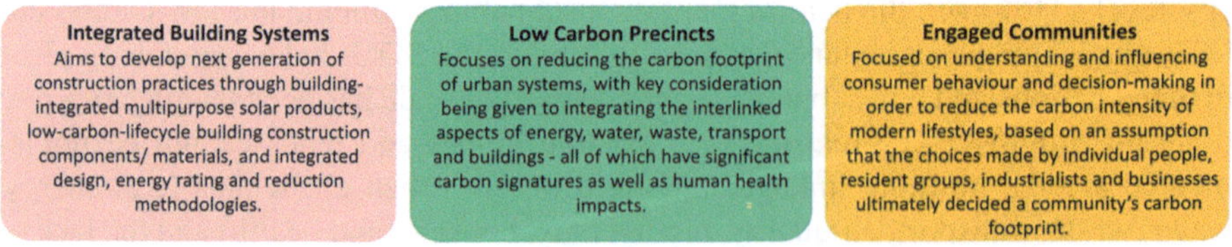

Integrated Building Systems	Low Carbon Precincts	Engaged Communities
Aims to develop next generation of construction practices through building-integrated multipurpose solar products, low-carbon-lifecycle building construction components/ materials, and integrated design, energy rating and reduction methodologies.	Focuses on reducing the carbon footprint of urban systems, with key consideration being given to integrating the interlinked aspects of energy, water, waste, transport and buildings - all of which have significant carbon signatures as well as human health impacts.	Focused on understanding and influencing consumer behaviour and decision-making in order to reduce the carbon intensity of modern lifestyles, based on an assumption that the choices made by individual people, resident groups, industrialists and businesses ultimately decided a community's carbon footprint.

Fig. 5.3 CRCLCL's three pathways towards sustainable, resilient, and low-carbon-built environments. (Source: CRCLCL 2019)

Net Zero Carbon guide starts by defining the key variables affecting operational and embodied carbon performance. Climate, building classification and conditions, building area measurement methods, and building design and its systems are key variables affecting operational carbon. Building classification, functional unit area definition, lifecycle inventory calculation method, overall embodied carbon calculation method, scope of building included, and country of material origin are the key variables affecting embodied carbon. The advantage of this guide is to define operational and embodied carbon benchmarks, measurement methods, current performance, and strategies and techniques. For instance, the achievement of net-zero operational carbon depends on energy efficiency, clean energy generation, and carbon offset, and the achievement of net-zero embodied carbon follows the strategies of no build, build less, build smarter, and maximise the efficiency of supply chain and procurement methods. This guide is a partner document to an accompanying book that discusses design strategies and systems in depth and provides exemplars from around the world. It also gives policy snapshots from various countries and develops benchmarks and targets for delivering on net-zero-carbon buildings globally (Prasad et al. 2023).

5.3.2 Series Guide to Low-Carbon Buildings

To guide the decarbonisation of buildings, the CRCLCL developed a series guide to low-carbon buildings that summarises best practice in various phases of the building lifecycle—construction, retrofit, and operation—for a range of building types in the residential and commercial sectors. The series guides are *Guide to Low Carbon Residential Buildings—New Build; Guide to Low Carbon Residential Buildings—Retrofit; Guide to Low Carbon Households; Guide to Low Carbon Commercial Buildings—New Build; and Guide to Low Carbon Commercial Buildings—Retrofit*. These guides contain important advice for stakeholders (e.g. architects and building designers, contractors and drafters, state and local government planning agencies, private developers, and owner-builders) in the building and construction sector to improve indoor environmental quality, while reducing energy use and carbon emissions. For instance, the *Guide to Low Carbon Residential Buildings—New Build* offers advice on how to reduce a building's carbon footprint in all phases of planning, design, construction, and system selection. It also showcases strategies and techniques for embodied emission reduction, waste reduction, and low-carbon transport (Byrne et al. 2019).

5.3.3 Guide to Low-Carbon Precincts and Landscape

The CRCLCL touches on low-carbon neighbourhood developments by developing the *Guide to Low Carbon Precincts*. This guide frames comprehensive solutions for councils and developers on the generation of strategic planning decisions when implementing low-carbon neighbourhoods. The guide defines actions that prioritise public transport, designing with nature, optimising urban structure, promoting precinct-scale energy systems, and integrating water and waste systems. Suggestions for addressing barriers (e.g. scale, regulation, collaboration, physical limitation, vision, and investment) are provided as aspects of human processing. An attractive element of this guide is that it offers a checklist of principles and themes that can standardise, and thus facilitate, the achievement of sustainable and low-carbon precincts (Newton 2019; Thomson et al. 2019).

The CRCLCL also promotes low-carbon landscapes through the *Guide to Low Carbon Landscapes*, which addresses landscapes relevant to residential retrofit, small to medium enterprises, and commercial development and precincts. It is a supplement aiming to reduce carbon emissions and enhance opportunities for carbon sequestration in the planning and design of built environments (Kjaersgaard et al. 2019).

5.3.4   Guide for Urban Cooling Strategies

The CRCLCL developed the *Guide for Urban Cooling Strategies* to offer practical guidance for built environment professionals and regulatory agencies on how to optimise development projects to moderate urban microclimates and alleviate heat-island effects in major urban centres across different climates in Australia (Osmond and Sharifi 2017). This guide provides a 3D matrix of cooling performance by different strategies (e.g. cool paving, cool envelope, green envelope, tree canopy, evaporative cooling, and shading structures) suitable for the public realm and tailored to urban form, climate type, and the nature of the intervention. The matrix supports urban planners and designers in making proper decisions about cooling-strategy selection. *The Guide for Urban Cooling Strategies* is a good reference for subsequent cooling guides, decision support tools, and a mitigation performance index system. For instance, the Cooling Sydney Strategy was developed to support the Sydney strategic plan for 2050 of 'living with our climate' by offering mitigation strategies for urban overheating (Ding et al. 2019a). The microclimate and heat-island mitigation decision support tool was developed to bridge the gap between research on urban microclimates and its practical application, particularly through identifying the cooling performance of different cooling interventions in areas such as energy use, heat island intensity, and thermal comfort (Ding et al. 2019b).

5.4   Framework for Sustainable, Resilient, and Low-Carbon-Built Environments

Following the CRCLCL outcomes, a robust and integrative framework for urban sustainability was developed to inform strategic decision-making for liveable, economically productive, socially inclusive, and environmentally sustainable cities

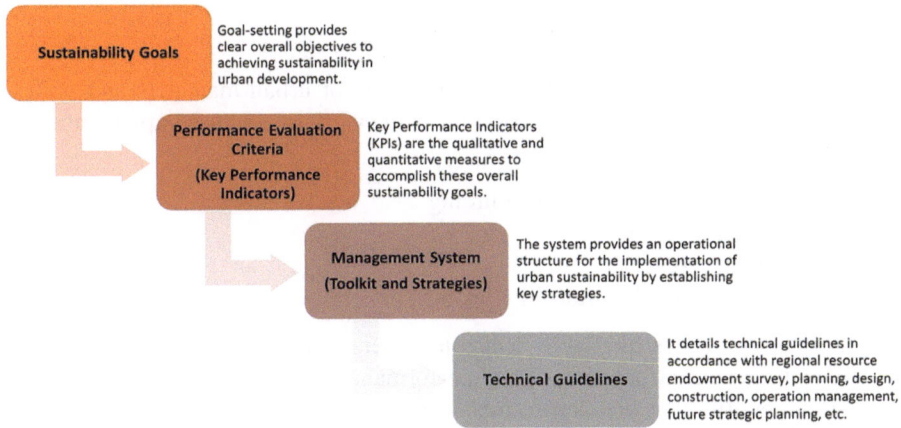

Fig. 5.4 Diagram of a robust and integrative sustainable development framework. (Source: Prasad et al. 2017)

(Prasad et al. 2017). The framework clarifies goals, objectives, and principles, as well as guidelines and techniques to support implementation and performance assessment (Fig. 5.4). The framework is designed to be applicable to different contexts and jurisdictions.

5.4.1 Objectives and Performance Measurement

The framework outlines nine objectives for sustainable, resilient, and low-carbon-built environments, including (i) safe and affordable cities, (ii) transportation and accessibility, (iii) land-use efficiency, (iv) cultural and natural heritage, (v) city disaster resilience, (vi) healthy eco-environment and climate mitigation, (vii) safe and sustainable public spaces, (viii) resource efficiency, and (ix) city management and policy. A quantification of sustainable goals can help cities and governments to measure the implementation. To support this, the framework provides a set of key primary performance indicators; secondary indicators are also applicable if needed. For instance, total city population, population density, annual population change, areal size of informal settlements as a percent of city area, urban population living in slums or informal settlements, and housing affordability are adopted to determine if a city is safe and affordable. Among these indicators, house affordability is further defined by the number of total houses available in the lowest price quartile of the local market for new housing, whether a minimum of 5% of a project total is offered to accredit not-for-profit housing providers for affordable rental housing, and a housing security rate of 90% (Prasad et al. 2017).

5.4.2 Five Strategies

The management system provides five strategies for urban managers and policy-makers on how to practically transform the framework into actual implementation:

1. Understanding the development context
2. Goal-setting and institutional resourcing
3. Implementation pathway
4. Monitoring progress
5. Lessons and knowledge transfer (Fig. 5.5)

The first strategy is to enable decision-makers at the regional or city scale to understand the current baseline of key performance indicators and identify implementation opportunities and constraints. Data collection and processing are needed for the pre-assessment. Based on this, the next strategy involves identifying the key areas and associated targets and goals for sustainability implementation, developing a collaborative team, and setting governance and financing options to alleviate the barriers to implementation. Guided workshops, peer-to-peer learning programmes, and other stakeholder consultation options support the identification of key areas.

The third strategy is the implementation pathway, which includes informing government officials about evidence-based decisions. City governments should be exposed to a broad scope of sustainability options and the most potentially effective solutions. It is of particular importance to provide supporting data and information for data modelling and scenario analysis to quantify the potential benefits of the selected solutions.

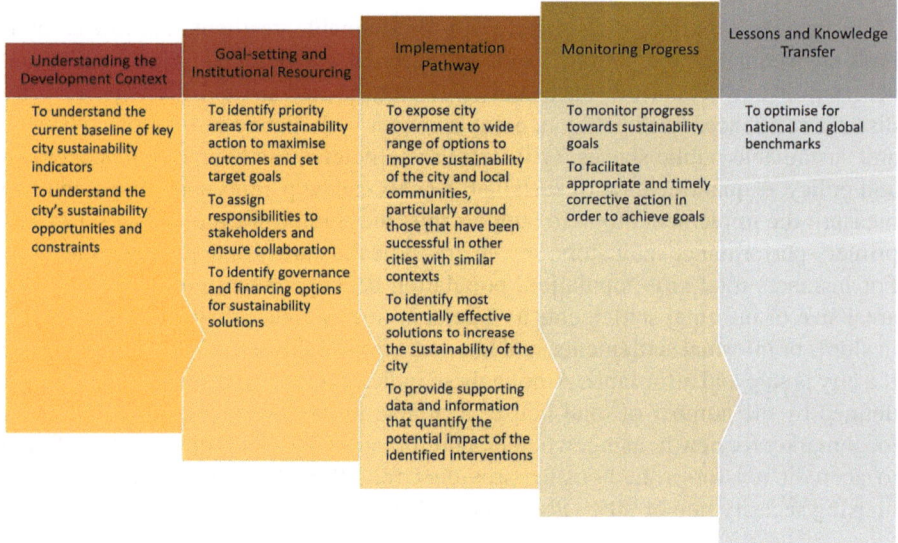

Fig. 5.5 Operational structure for the implementation of urban sustainability. (Source: Prasad et al. 2017)

The fourth strategy is to monitor progress towards sustainability goals and make timely and proper corrections for better implementation, and the fifth strategy is to optimise national and global benchmarks based on an effectiveness comparison. Overall, the sequential and progressive strategies enable different cities and communities to adapt the framework in different contexts and implement it with better performance.

5.4.3 Best-Practice Guidance

The framework further provides built environment professionals and governmental agencies with techniques, strategies, and best practices for practical achievement of sustainable, resilient and low-carbon cities. The techniques, strategies, and best practices are well developed, robust, and reliable, their sources developed from the literature, stakeholder meetings and workshops, expert advice, and research projects. Moreover, the technical guides scientifically follow the seven aspects inherent in the lifecycle process of urban development, including (i) resource endowment survey (project areas), (ii) planning and design, (iii) product and materials selection, (iv) building and construction, (v) operation and management, (vi) future strategic planning, and (vii) city performance assessment tools (Table 5.3).

Table 5.3 Technical guidelines for achieving sustainable, resilient, and low-carbon-built environments

Lifecycle process	Strategies and technologies	Best practice
Resource endowment survey	(i) Develop indicators that prepare for a realistic projection of urban land needs, management of different land-use patterns, and reduction of additional land-related resources (ii) Develop a vegetation plan that provides clear guidelines on a statutory basis and conditions (iii) Identify advanced tools that help in assessing climatic features of the project area, such as solar access or access to cooling breezes, and other microclimate matters in order of priority	Microclimate and Urban Heat Island Mitigation Decision Support Tool, Australia
Planning and design	(i) Develop a long-term strategic plan enabling a sustainable transition of non-urban land to urban land (ii) Develop and implement an 'integrated urban design' to make cities more cohesive, lively, and sustainable (iii) Embody 'localness', encourage diversity of land use, and protect and enhance eco- and cultural infrastructure. (iv) Develop plans, policies, and designs	Sponge cities, Changde, China Sydney Green Grid, Australia Hammarby Sjöstad, Sweden Greater Sydney Regional Plan, Australia

(continued)

Table 5.3 (continued)

Lifecycle process	Strategies and technologies	Best practice
Product and materials selection	(i) Reduce the amount of waste and environmental footprint of buildings (ii) Promote the implementation of locally sourced and recycled content sustainable materials and construction solutions (iii) Encourage advanced sustainable building materials (iv) Adopt sustainable, smart building materials and products that conserve energy and promote human well-being	Sustainable building construction, Auroville, India
Building and construction	(i) Design and construct buildings that minimise the impact on climate and buildings that adapt to the changing climate (ii) Design and construct carbon-zero, carbon-neutral, zero-energy, or zero-emission buildings using sustainable and intelligent technologies	Josh's House, Perth, Australia Singapore zero-energy building
Operation and management	(i) Encourage innovative tools, indicators, and technologies to manage safe places and to support informed decision-making on safety and vulnerability (ii) Encourage sustainable and low-carbon transport systems to manage traffic, air quality, and accessibility (iii) Undertake urban planning initiatives that promote and protect local, natural, marine, and other protected environments (iv) Develop intelligent energy initiatives to reduce energy demand and use and improve urban climates (v) Adopt and promote sustainable and smart technologies to manage and improve quality (vi) Implement policies, programmes, and sustainable and smart technologies to manage waste, energy demand and use, water, and sanitation	Smart Energy Management in Kashiwa-no-Ha Smart city, Japan Joint Emergency Management, Nanning City, China Rainwater storage below buildings, Rotterdam, The Netherlands
Future strategic planning	(i) Create new innovative, dynamic, and integrated policy and strategic frameworks (ii) Encourage strategic plans, policies, and programmes through open processes and creative participatory methods to foster the development of globally competitive, innovative, cultural, creative, and connected cities	U-City model, Korea
City performance assessment tools	(i) Develop performance assessment models or tools based on indicators grounded on a suitable conceptual framework that explains and prioritises relationships within and between criteria and allows the systematic assessment of performance (ii) Collect and process information on an ongoing basis and provide conclusions on policies or actions taken	Siemens' City Performance Tool Sustainable Systems Integration Model (SSIM)

Source: Prasad et al. (2017)

Regarding decarbonisation, the framework suggested planning and designing carbon-positive or carbon-neutral precincts through maximising sustainable building materials (materials with low embodied energy), providing wind and solar electricity to the grid, introducing energy-efficiency technologies for buildings, considering electricity and gas (biogas) needs and transport systems (fuel cells for vehicles), installing water recycling and storage systems, reusing and recycling household waste, providing district heating and cooling, ensuring sustainable food production (community and rooftop gardens), and integrating sustainable architecture (e.g. building integrated solar PV systems, solar roofs, green roofs and walls, and green spaces).

At the building scale, net-zero operational carbon buildings ought to produce as much energy as they use over the course of a year, thereby reducing the use of non-renewable energy in the building sector. Regarding products and materials selection, the guideline recommends low-carbon materials and recycled materials for reducing embodied carbon, and cool and permeable materials to regulate environmental performance (e.g. to mitigate heat islands and urban flooding).

5.5 Conclusions

Addressing climate change and creating inclusive, safe, resilient, and sustainable cities are crucial problems for now and the future. Decarbonisation of the built environments has been widely recognised and implemented to mitigate climate change, while many other improvements to the quality of the built environment (e.g. comfort, health, safety, water, and air quality) are still unresolved. A brief review of the vision of sustainable, resilient, and low-carbon-built environments indicates that action has been mainly advocated by leading and pioneering international organisations with the release of plans, roadmaps, projects, and guidelines. There is a gap between the implementation of these planning initiatives and the insights learned from efforts to achieve sustainable, resilient, and low-carbon-built environments at national, state, and city scales.

Therefore, to avoid taking a doubtful path by focusing only on climate change mitigation, this chapter advocates that climate change mitigation, adaptation, and urban sustainability should be holistically considered in urban development. Following the conclusions from the CRCLCL regarding integrated building systems, low-carbon precincts, and engaged communities, this chapter reports on a robust, comprehensive, and integrative sustainable development framework for sustainable, resilient, and low-carbon-built environments, with sustainability goals, key performance indicators, implementation strategies, and technical guidelines. The next steps are (i) looking for pilot precincts for testing the guidelines' performance and providing feedback for revisions, improvement, and optimisation, and (ii) engaging governments, the private sector, and communities in accelerating change for sustainable, resilient, and low-carbon-built environments. Overall, we expect this chapter to point towards better delivery of low-carbon cities and buildings, while addressing climate-related and emerging urban challenges.

Appendix: Definition of Decarbonisation Terms Used in This Chapter

Terms	Definition
Carbon neutral	A carbon neutral building refers to one whose design, construction, and operation do not contribute to GHG emissions that cause climate change
Zero energy	A zero-energy building, also known as a net-zero operational carbon building, is one with net-zero energy consumption, meaning the total amount of energy used by the building on an annual basis is equal to the amount of renewable energy created on the site
Net-zero carbon	A net-zero carbon building, also known as a zero-carbon building, is one that is highly energy efficient and fully powered from on-site and/or off-site renewable energy sources
Embodied carbon	Embodied carbon refers to the total of all direct and indirect GHG emissions arising from the production, transportation, and processing of materials used in the construction of buildings
Low carbon	A low-carbon building refers to a building which is specifically engineered with GHG reduction in mind. It is generally characterised using integrated passive design strategies, high-performance building envelopes, and energy-efficient heating, ventilation, and air-conditioning systems
Operational carbon	Operational carbon refers to the total direct (scope 1) and/or indirect (scope 2) GHG emissions from all operational energy consumed during the use stage of the building lifecycle. It includes both: • Regulated loads, for example, heating, cooling, ventilation, and lighting and • Unregulated/plug loads, for example, ICT equipment, cooking, and refrigeration appliances
Whole life cycle carbon	The carbon emissions resulting from the production of materials, construction, and the use of a building over its entire life, including its demolition and disposal

Source: New York State (2020), Prasad et al. (2023), and US DOE (2015)

References

Architecture 2030 (2021) Architecture 2030. https://architecture2030.org

Bouckaert S, Pales AF, McGlade C, Remme U, Wanner B, Varro L, D'Ambrosio D, Spencer T, Abergel T, Arsalane Y, Bains P, Bermudez Menendez JM, Connelly E, Crow D, Dasgupta A, Delmastro C, Goodson T, Gouy A, Hugues P, Lee L, Levi P, Mandova H, Millot A, Olejarnik P, Paoli L, Papadimoulis F, Papapanagiotou S, Pavan F, Petropoulos A, Pośpiech R, Staas L, Tattini J, Teter J, Tonolo G, Vass T, Wetzel D (2021) Net zero by 2050: a roadmap for the global energy sector. International Energy Agency, Paris. https://www.iea.org/reports/net-zero-by-2050

Byrne J, Taylor M, Ambrose M, Berry S, Sproul AB (2019) Guide to low carbon residential buildings—new build. Low Carbon Living CRC. https://apo.org.au/node/239826

Campbell I, Sachar S, Meisel J, Nanavatty R (2021) Beating the heat: a sustainable cooling handbook for cities. United Nations Environment Programme. https://www.unep.org/resources/report/beating-heat-sustainable-cooling-handbook-cities

Carbon Neutral Cities Alliance (2022) Mobilizing transformative climate action in cities. https://carbonneutralcities.org/

CRCLCL (2019) Cooperative Research Centre for Low Carbon Living. https://www.lowcarbon-livingcrc.com.au/research+

de Jong M, Joss S, Schraven D, Zhan C, Weijnen M (2015) Sustainable—smart—resilient—low carbon—eco—knowledge cities: making sense of a multitude of concepts promoting sustainable urbanization. J Clean Prod 109:25–38

Dick H, Rimmer P (1999) Privatising climate: first world cities in third world settings. In: East West perspectives on 21st century urban development. Ashgate, Aldershot

Ding L, He B, Craft W, Petersen H, Osmond P, Santamouris M, Prasad D, Bartesaghi Koc C, Derksema C, Midlam N (2019a) Cooling Sydney strategy: planning for Sydney 2050. Cooperative Research Centre for Low Carbon Living

Ding L, Petersen H, Craft W (2019b) Microclimate and urban heat Island mitigation decision-support tool (project short report). CRC for Low Carbon Living. https://apo.org.au/node/246691

European Union (2022a) Excess mortality hits +16%, highest 2022 value so far. Excess Mortality in July 2022. https://ec.europa.eu/eurostat/web/products-eurostat-news/-/ddn-20220916-20220911

European Union (2022b) NetZeroCities Pilot Cities Programme Guidebook. NZC Consortium coordinated by EIT Climate-KIC. https://netzerocities.eu/wp-content/uploads/2022/2006/Pilot-Cities-Guidebook.pdf

Freychet N, Hegerl GC, Lord NS, Lo YTE, Mitchell D, Collins M (2022) Robust increase in population exposure to heat stress with increasing global warming. Environ Res Lett 17(6):064049

GlobalABC (2020) GlobalABC regional roadmap for buildings and construction in Africa: towards a zero-emission, efficient and resilient buildings and construction sector. IEA, Paris. https://globalabc.org/roadmaps-buildings-and-construction

He B-J, Wang J, Liu H, Ulpiani G (2021a) Localized synergies between heat waves and urban heat islands: Implications on human thermal comfort and urban heat management. Environ Res 193:110584

He B-J, Zhao D, Xiong K, Qi J, Ulpiani G, Pignatta G, Prasad D, Jones P (2021b) A framework for addressing urban heat challenges and associated adaptive behavior by the public and the issue of willingness to pay for heat resilient infrastructure in Chongqing, China. Sustain Cities Soc 75:103361

Huovila P, Ala-Juusela M, Melchert L, Pouffary S, Cheng C-C, Urge-Vorsatz D, Koeppel S, Svenningsen N, Graham P (2009) Buildings and climate change: summary for decision-makers. United Nations Environment Programme: Sustainable Buildings & Climate Initiative. https://www.uncclearn.org/wp-content/uploads/library/unep207.pdf

IEA (2021a) Net zero by 2050: a roadmap for the global energy sector. International Energy Agency, https://www.iea.org/reports/net-zero-by-2050, pp 3–24

IEA (2021b) Empowering cities for a net zero future: unlocking resilient, smart, sustainable urban energy systems. International Energy Agency. https://iea.blob.core.windows.net/assets/4d5c939d-939c937-490b-bb953-932c930d923f932cf933d/G920EmpoweringCitiesforaNetZeroFuture.pdf

IPCC (2018) Summary for policymakers. In: Masson-Delmotte V, Zhai P, Pörtner H-O, Roberts D, Skea J, Shukla PR, Pirani A, Moufouma-Okia W, Péan C, Pidcock R, Connors S, Matthews JBR, Chen Y, Zhou X, Gomis MI, Lonnoy E, Maycock T, Tignor M, Waterfield T (eds) Global warming of 1.5°C. An IPCC special report on the impacts of global warming of 1.5°C above pre-industrial levels and related global greenhouse gas emission pathways, in the context of strengthening the global response to the threat of climate change, sustainable development, and efforts to eradicate poverty. Cambridge University Press, Cambridge and New York, pp 3–24

IPCC (2023) Synthesis report of the IPCC sixth assessment report (AR6): summary for policymakers. Intergovernmental Panel on Climate Change, 2023. https://report.ipcc.ch/ar6syr/pdf/IPCC_AR6_SYR_SPM.pdf

Kjaersgaard SP, Evans CB, Harris MS (2019) Guide to low carbon landscapes. CRC for Low Carbon Living

LETI (2021) LETI climate emergency retrofit guide. London

New York State (2020) Carbon neutral buildings. https://www.nyserda.ny.gov/All-Programs/Carbon-Neutral-Buildings/Carbon-Neutral-Buildings-State-Fair#:~:text=What%20is%20a%20carbon%20neutral,meet%20the%20State's%20climate%20goals

Newton P (2019) The performance of urban precincts: towards integrated assessment. In: Newton P, Prasad D, Sproul A, White S (eds) Decarbonising the built environment: charting the transition. Springer, Singapore, pp 357–384. https://doi.org/10.1007/978-981-13-7940-6_19

Newton P, Prasad D, Sproul A, White S (2019) Decarbonising the built environment: charting the transition. Springer, Singapore. https://link.springer.com/book/10.1007/978-981-13-7940-6

Osmond P, Sharifi E (2017) Guide to urban cooling strategies. Low Carbon Living CRC

Palin M (2017) Urban Island heat effect: Rising temperatures in Aussie cities could create death traps. Australia News. https://www.news.com.au/technology/environment/climate-change/urban-island-heat-effect-rising-temperatures-in-aussie-cities-could-create-death-traps/news-story/0b035c4707ea4708f4781e4732ee4700df4704fa4546bf

Prasad D, Ding L, Yenneti K, Fan H, Craft W, Sanchez AX, Li X, Amold P, Bouhmad K, Earley R, Wang Y, Wu J, Li F, You N, Fox J, Aragon JD (2017) Guidelines for sustainable cities and communities (China). Cooperative Research Centre (CRC) for Low Carbon Living, United Nations Environment Programme, Jia Cui Environment Promotive Center 2(9):805–814. https://apo.org.au/node/185056

Prasad D, Kuru A, Oldfield P, Ding L, Dave M, Noller C, He B (2021) Race to net zero carbon: a climate emergency guide for new and existing buildings in Australia. Low Carbon Institute. https://www.unsw.edu.au/content/dam/pdfs/unsw-adobe-websites/arts-design-architecture/built-environment/net-zero-guide/2023-2001-2013-Net-Zero-guide-online-version.pdf

Prasad D, Kuru A, Oldfield P, Ding L, Dave M, Noller C, He B (2023) The climate emergency and the built environment. In: Delivering on the climate emergency: towards a net zero carbon built environment. Springer, pp 1–27. https://doi.org/10.1007/978-981-19-6371-1

Quigley EV (2019) Five deep decarbonization strategies. Clean Energy Transition Institute. https://www.cleanenergytransition.org/post/five-deep-decarbonization-strategies

RIBA (2021) RIBA 2030 challenge. Version 2. London

Ritchie H, Roser M (2022) CO_2 and greenhouse gas emissions. Our World in Data. https://ourworldindata.org/co2-and-other-greenhouse-gas-emissions

Rosenzweig C, Solecki W (2018) Action pathways for transforming cities. Nat Clim Chang 8(9):756–759

Rosenzweig C, Solecki W, Romero-Lankao P, Mehrotra S, Dhakal S, Ibrahim SA (2018) Pathways to urban transformation. In: Rosenzweig C, Romero-Lankao P, Mehrotra S, Dhakal S, Ali Ibrahim S, Solecki WD (eds) Climate change and cities: second assessment report of the Urban Climate Change Research Network. Cambridge University Press, Cambridge, pp 3–26. https://doi.org/10.1017/9781316563878.008

Roth M (2012) Urban heat islands. In: Handbook of environmental fluid dynamics, vol 2. CRC Press, pp 162–181. https://doi.org/10.1201/b13691-15/

Salamanca F, Georgescu M, Mahalov A, Moustaoui M, Wang M (2014) Anthropogenic heating of the urban environment due to air conditioning. J Geophys Res Atmos 119(10):5949–5965

Santamouris M, Kolokotsa D (2016) Urban climate mitigation techniques. Routledge, London

Santamouris M, Haddad S, Fiorito F, Osmond P, Ding L, Prasad D, Zhai X, Wang R (2017) Urban heat Island and overheating characteristics in Sydney, Australia. An analysis of multiyear measurements. Sustain For 9(5). https://doi.org/10.3390/su9050712

Thomson G, Newton P, Newman P, Byrne J (2019) Guide to low carbon precincts

UN SDG (2015) The 17 goals. Department of Economic and Social Affairs: Sustainable Development, United Nations. https://sdgs.un.org/goals

UNEP (2020) Facts about the climate emergency. United Nations Environment Programme. https://www.unep.org/facts-about-climate-emergency

UNEP (2021) 2021 global status report for buildings and construction: towards a zero-emission, efficient and resilient buildings and construction sector. Nairobi

UNEP (2023) Sustainable cities. Home regions Asia and the Pacific Regional initiatives supporting resource efficiency. https://www.unep.org/regions/asia-and-pacific/regional-initiatives/supporting-resource-efficiency/sustainable-cities

United Nations (2022a) Cities and climate change. Our World in Data. https://www.unep.org/explore-topics/resource-efficiency/what-we-do/cities/cities-and-climate-change

United Nations (2022b) Race to zero campaign. United Nations Framework Convention on Climate Change. https://unfccc.int/climate-action/race-to-zero-campaign

US DOE (2015) A common definition for zero energy buildings. US Department of Energy. https://www.energy.gov/sites/prod/files/2015/2009/f2026/A%2020Common%2020Definition%2020 20for%2020Zero%2020Energy%2020Buildings.pdf

Wang J, Meng Q, Zou Y, Qi Q, Tan K, Santamouris M, He B-J (2022) Performance synergism of pervious pavement on stormwater management and urban heat Island mitigation: a review of its benefits, key parameters, and co-benefits approach. Water Res 221:118755

WGBC (2020) Whole life carbon vision. World Green Building Council. https://www.worldgbc.org/advancing-net-zero/whole-life-carbon-vision

WWF (2020) Australia's 2019–2020 bushfires: the wildlife toll: interim report. New WWF report: 3 billion animals impacted by Australia's bushfire crisis. https://www.wwf.org.au/news/news/2020/2023-billion-animals-impacted-by-australia-bushfire-crisis/

Bao-Jie He is a (Full) Professor of Climate Change and Sustainable Built Environment at the School of Architecture and Urban Planning, Chongqing University, China. Bao-Jie has published more than 150 peer-reviewed journal papers (h-index: 52). Bao-Jie is a Global Highly Cited Researcher (Clarivate). Bao-Jie serves as Topic Editor-in-Chief, Leading Guest Editor, Associate Editor, Section Editor, Editorial Board Member, Conference Chair, Sessional Chair, and Scientific Committee in various reputable international journals and conferences. Bao-Jie received the Green Talents Award in Germany in 2021 and was ranked as a World's Top 2% Scientist in 2021, 2022, 2023, and 2024.

Deo Prasad is a Scientia Professor in the field of sustainable built environments at the University of New South Wales in Sydney, Australia. His expertise covers sustainable, low carbon, smart, resilient, and regenerative buildings and cities. He has published over 300 refereed publications including ten books. The last of his books on Climate Emergency: Towards a Net Zero Carbon Built Environment (Palgrave) was released early in 2023. He is currently the CEO of the NSW Decarbonisation Innovation Hub and previously served as the CEO of the Co-operative Research Centre for Low Carbon Living. Deo has received acknowledgement of his contributions from all levels of government in Australia, including an Order of Australia, a Fellowship of the Australian Academy of Technological Sciences and Engineering, Fellow of the Australian Institute of Architects, NSW Government's Green Globe Award, and the Global Impact Award as well as the National Leadership in Sustainability Prize from the Australian Institute of Architects. He is currently leading the commercialisation of proof-of-concept technologies and systems by creating an industry-government-research collaborative community at scale.

Chapter 6
Transitioning to a Circular Economy: Understanding the Circular Economy Ecosystem in Victoria, Australia

Usha Iyer-Raniga, Oanh Thi-Kieu Ho, and Akvan Gajanayake

Abstract Australia's journey towards a circular economy is in its initial stage. Demand for resources is putting pressure on Australia's urban environments, which house over 75% of the country's population and account for over 80% of national GDP. Until 2018, recyclable waste was largely exported to other countries for processing. This has now changed and has led to national and state-based discussions on more effective reprocessing of waste resources.

This chapter aims to present findings based on research conducted to understand the current circular economy ecosystem in the state of Victoria (Australia) to support its plans for waste reduction and transition to a circular economy. A mixed-methods approach was taken, comprising desktop research, interviews with key actors, and a survey targeting a wide range of businesses. It was found that there was no consistent or systemic understanding of the concept of the circular economy; rather, the narrative revolved around waste management and recycling. For transitioning to a circular economy, a systematic shift is needed, supported by a clear policy directive, financial outlay, technical know-how, education, awareness, engagement, and collaboration across traditional isolated sectors.

6.1 Introduction: Cities and Circular Economy

Cities occupy just 3% of Earth's land but account for 60–80% of energy consumption and 75% of carbon emissions (UN Environment 2022). Rapid and often unplanned urbanisation has led to increased pressure on urban environments. As a result of the high concentration of people, infrastructure, housing, and economic activity, cities are particularly vulnerable to climate change and natural disasters. According to the OECD (2022a, b), cities produce an estimated 50% of global

U. Iyer-Raniga (✉) · O. T.-K. Ho · A. Gajanayake
School of Property, Construction and Project Management, RMIT University, Melbourne, VIC, Australia
e-mail: usha.iyer-raniga@rmit.edu.au

© The Author(s) 2025   117
N. Frantzeskaki et al. (eds.), *Future Cities Making*, Theory and Practice of Urban Sustainability Transitions, https://doi.org/10.1007/978-981-97-7671-9_6

waste. If cities become resource-efficient, they can combine greater productivity and innovation with lower costs and reduced environmental impacts, while providing sustainable lifestyles.

Cities play a critical role in a circular economy. Circular economies are based on the design principles of eliminating waste and pollution, circulating products and materials at their highest value, and enabling nature to regenerate (Ellen Macarthur Foundation 2017). Cities have a high concentration of resources, capital, data, and human resources spread over a contained geographical area. A report by the Ellen Macarthur Foundation (2017) outlines that a circular city has a built environment that mimics natural cycles; is supported by an energy system that is resilient, renewable, distributed, and localised such that costs are reduced; has a mobility system that is accessible, affordable, and effective; and has a bioeconomy that generates nutrients, reduces food waste by encouraging production locally through urban farming, and creates local value loops involving local manufacturing that incorporates digitalisation to support virtual engagements.

The circular economy has many definitions. The very nature of what circularity means at a city scale is contested (Paiho et al. 2020). The academic literature is rich with discussion of the pros and cons of various definitions, especially across the micro, meso, and macro scales (e.g., Kirchherr et al. 2017). This chapter explores the empirical approaches taken to transition to circularity at the city scale to understand the benefits of taking such a course of action (Williams 2021). Central are the principles of the circular economy across design and operational stages to ensure restorative and regenerative outcomes from environmental, social, and economic perspectives. Environmentally, the impacts are associated with material flows, waste, and waste management (Kozminska and Arch 2018; Savini 2019). Socially, a circular city promotes prosperity, digital opportunities, and technology that are optimised, with citizens engaged in the transition process (Davidescu et al. 2020; Marchesi and Tweed 2021). Economically, a city underpinned by a circular economy supports jobs and local economies (Sukhdev et al. 2018; Kannikar et al. 2021).

Understanding and optimising the metabolism of cities plays a crucial role in the transition towards a circular economy. Most cities could be understood to be linear-metabolism cities: resources flow in one direction through the system without much concern about their origin, or about the destination of wastes (Girardet 2014). The transformation of resources into waste in cities has a negative impact on the planet's life support systems. The vision for an urban circular economy would be to find practical ways to mimic nature's circular metabolic systems by relying predominantly on renewable resources and biological processes. Governments have a leading role to play in such a transition by defining the main challenge and establishing a regulatory framework, engaging in dialogue with citizens, and partnering with businesses to address these challenges (Webster 2017; Mazzucato 2021).

A review of the literature on this subject shows a greater concentration of circular economy strategies in Europe, whereas China shows a greater focus on environmental research (Petit-Boix and Leipold 2018; Verga and Khan 2021). There are some examples of a city-level understanding of the circular economy in Asia, Canada, and Europe, with European cities providing the most examples. Place-based studies have focused on Singapore (Carrière et al. 2020), Canada (Environment and Climate Change Canada 2021), Turku City in Finland (Turku n.d.), London (Turcu and Gillie 2020), Amsterdam and Rotterdam (Holland Circular Hotspot 2019; Russell et al. 2020), and Paris, London, and Amsterdam (Jones and Comfort 2018).

This chapter focuses on the role of *cities* in the transition to a circular economy. It undertakes ecosystem mapping to understand the current circular economy landscape in the State of Victoria with the intention of identifying opportunities and gaps to support a circular economy transition in that State. The chapter takes an exploratory, mixed-methods approach consisting of interviews and surveys to elicit a better understanding of the contemporary landscape. The scope of the material presented is restricted to those business organisations involved in the survey.

6.2   The Case for Circular Economy Transformation in Australia

For Australia's rapidly growing urban population, demand for resources for its built environment and travel are the main pressures on the environment (Cresswell et al. 2021). Unsustainable usage levels for resources such as water and energy were important considerations noted in the recent State of the Environment report (Cresswell et al. 2021). Australia's eight major cities are growing at rates faster than most developed cities internationally (Hill and Quintana 2021). While Australia's population concentration has risen from 2.9 people/km^2 in 2011 to 3.3 people/km^2 in 2020, these numbers are much lower than most other countries (e.g., 36 people/km^2 in the USA and 281 people/km^2 in the UK). Low-density settlement typically creates high vehicle-kilometres travelled, especially in countries with poor urban public transport networks and high car dependency, such as Australia, where transport accounts for nearly 20% of emissions (Iyer-Raniga and Gajanayake 2021). The concentration varies across Australia's cities, with Greater Sydney and Greater Melbourne having the highest population density.

It is anticipated that municipal waste will also increase with increases in population. Unrecyclable solid waste that is generated in cities is either disposed to landfill, causing land, air, and water pollution, or incinerated, causing air pollution. Australia has the second-highest per-person rate of waste generation, at 2.13 tonnes, lower than the USA (2.34 tonnes) but close to double that of Singapore (1.26 tonnes) (Pickin et al. 2022; Cresswell et al. 2021). Waste disposal per person is high in Australia, at 704 kg per person; again, second to the USA (771 kg per person). In 2018–2019, waste was generated from four main sectors: manufacturing (17%), construction (17%), household (16%), and electricity, gas, and water services (14.5%) (Cresswell et al. 2021).

Victoria's circular economy policy hinges on waste, as does the national focus. The waste targets are focused on waste avoidance and waste reduction. Figure 6.1 presents the waste flows in Australia based on source and final destination. The

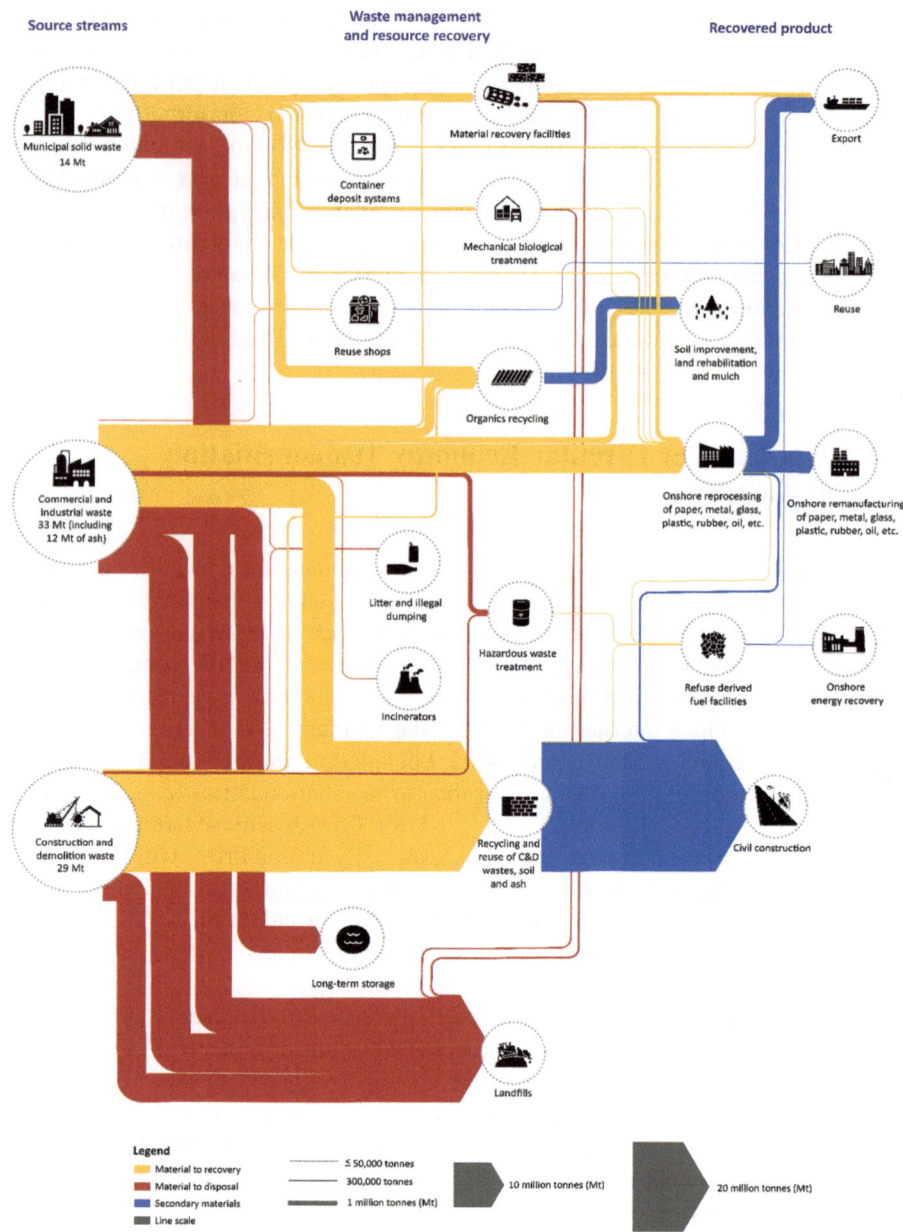

Fig. 6.1 Flows of core waste in Australia 2020–2021. (Source: Pickin et al. 2022)

Victorian government has a more ambitious target for waste generation, at 15% per capita by 2030 (DELWP 2020), compared to the national target of 10% per capita by 2030 (DCCEEW 2022a, b). This target focuses on municipal solid waste

(Iyer-Raniga et al. 2022). The federal government has regulated the export of e-waste, wastepaper, plastic, glass, and tyres since March 2020 (DCCEEW 2022a, b). For waste diversion, the federal government has a target for 80% average resource recovery rate from all waste streams following the waste hierarchy by 2030, and half the amount of organic waste sent to landfill by 2030 (DCCEEW 2022a, b). The national waste targets have been set by the federal government based on the national waste reports from individual states. Some waste items, such as electric vehicles, which are included in reports by other OECD countries, are not considered by Australia. Significantly increasing the use of recycled content by government and industry through the use of recyclate, phasing out problematic and unnecessary plastics by 2025, and making comprehensive, economy-wide data publicly available to support better consumer, investment, and policy decisions are all areas of interest to the federal government.

The Victorian government aims to divert 80% of waste from landfill by 2030, with an interim target of 72% by 2025 (DELWP 2020). Although per-capita waste generation in Australia declined by 3.3% between 2007 and 2020, construction and demolition waste increased by 32% per capita (Pickin et al. 2022). Additionally, the Victorian government seeks to halve the volume of organic material going to landfill between 2020 and 2030, with an interim target of a 20% reduction by 2025 (DELWP 2020). Furthermore, there is a focus in Victoria on ensuring that every household has access to food and garden organic waste recycling services or local composting options by 2030. Victoria's Big Build (Government of Victoria 2022) is interested in using recycled materials in its building projects and infrastructure. Recycled First Policy (EcologqiQ 2020) identifies opportunities for the use of recyclates in government projects in Victoria, in the absence of any proposed state targets.

Policy contexts both at national and state levels, however, do not recognise the nuances of the various R-strategies associated with a circular economy (Cramer 2022). The narrative used locally involves reduce waste being landfilled, reuse, and recycle; rather than the 10 Rs commonly found in EU policy: refuse (best option), rethink, reduce, reuse, repair, refurbish, remanufacture, repurpose, recycle, and recover. This narrative also includes regeneration, particularly in the Victorian policy context, such as programmes for timber regeneration (DEECA 2022) and asset regeneration (Parks Victoria 2023).

6.3 Enablers and Challenges for Cities Transitioning to a Circular Economy

Desktop research was undertaken in the form of an extensive international and national literature review to understand the circular economy context in cities by identifying the circular economy actors, policies, and regulatory environment, and the enablers of and barriers to circular economy transition. The Preferred Reporting

Items for Systematic Reviews and Meta-Analyses (PRISMA) method was used with several sources such as Web of Science, Scopus, and Google Scholar.

A range of issues arose from the literature review, with the main considerations being clearly defining the transition to circularity, setting targets, determining indicators, identifying enablers and barriers, and engaging with stakeholders (Paiho et al. 2020). Cities are centres of both production and consumption (Williams 2019a, b). 'If citizens do not "buy into" consuming circular products and services (e.g. recycled goods, renewable energy) or adopt circular practices (e.g. repairing or upcycling goods, composting organic waste), then a circular society is undeliverable' (Williams 2019a, b, p. 2751). Thus, the lifestyles and social practices of people living in cities, including their personal mobility, also need to transform.

Some authors suggest that the circular economy is still underdeveloped in terms of practical solutions and needs more attention (Williams 2019a, b; de Morais et al. 2021). Mies and Gold (2021) discuss the importance of social aspects for circular economies such as labour practices, human rights, and community well-being. They argue for a more balanced integration of the social sustainability dimension to elicit the best outcomes. They present a clear conceptual integration of the social dimension of circular economies, where collaboration is the main facilitator identified. Education, citizen participation, and legislative support are key points associated with collaboration for a circular economy transformation, according to the authors.

The role of digitalisation is critical to a circular economy (Woetzel et al. 2018; Pavlopoulou 2021). Innovative governance models and community structures can drive a maker movement towards circularity. Local maker champions may be used to drive the movement towards shared circular visions by using innovation diffusion and innovation ambassadors. Likewise, Marchesi and Tweed (2021) introduce the importance of social innovation for a circular economy, whereas Davidescu et al. (2020) investigate the role of citizens and their behaviours and actions in addressing climate change as part of the circular economy transition.

Much has also been written about strategic planning initiatives and a systematic approach to urban development (Bolger and Doyon 2019; Gravagnuolo et al. 2021). Research by Oral et al. (2020) and Katsou et al. (2020) shows how regenerative and restorative ecosystems may be used in circular cities through nature-based solutions (NbSs) combined with regulations and governance. In their project for addressing urban circularity challenges with NbSs, Langergraber et al. (2021) analysed various urban sectors that were responsible for reducing the use of resources and the production of waste. The authors considered built environment, urban water management, resource recovery, and urban farming as multiple sectors related to both NbSs and urban circularity challenges.

Paiho et al. (2020) categorise four main challenges for a city to transition to circularity from the literature:

- Policy: Administrative fragmentation, lack of proper policy and regulation, lack of long-term strategies, and need for subsidies and taxes that encourage resource reuse.

- Technical: Technological lock-in, need for additional innovation and technology, linear design of products, and limited waste treatment due to insufficient separation of technical nutrients (such as metals and plastics) from biological nutrients (such as food and wood), resulting in the lower quality of these nutrients and hindering their return to the value chain.
- Business: Insufficient market demand for secondary materials, insufficient funding for circular economy initiatives, high investment costs, vested interests of business actors, and product prices not taking environmental costs into account.
- Knowledge challenges: Existing linear modes of thinking, lack of consumer awareness and demand, limited availability of data, ambiguity of the concept of the circular economy, absence of performance metrics, and a narrow view of circular economies.

Paiho et al. (2020) also identified enablers of a circular economy in the same categories:

- Policy: Developing a long-term holistic vision, recognising barriers to circularity and addressing them, involving non-municipal stakeholders and encouraging cooperation between them, promoting coordination across government departments, and networking with other cities to share knowledge and lobbying for change.
- Technical: Applying circular principles in urban planning and supporting ICT solutions.
- Business: Using circular criteria in public procurement, facilitating locations and funding for innovation, identifying external sources of funding, facilitating a data economy including a wide range of initiatives, activities, and projects where business models developed rely on databases to generate products and services, and enabling fact-based decision-making in transitioning to a circular economy.
- Knowledge: Analysing the local conditions as a basis for developing a strategy, monitoring and evaluating circular projects continuously, and educating stakeholders about circular economy.

6.4 Pathways Towards a Circular Economy

Prendeville et al. (2018) discussed how cities were adopting circular economy as a strategy to embrace sustainability transitions. They found that political leadership, visioning, agility, experimentation, developing place-based responses, and engaging with diverse stakeholders were critical to supporting circular economy transitions. Their research focused on mapping the circular transitions of six cities, and their conclusions supported the policy and strategic actions outlined in Table 6.1.

It is clear from the literature that policy and regulatory schemes are an important part of the discussions for transitioning to a circular economy. Policies need to be supported with action-oriented strategies to enable outcomes. Technology and digitalisation can support circular outcomes, as the CSIRO Aspire programme is

Table 6.1 Synthesis of policy and strategies for circular economy derived from the literature

Policy	Strategies
Business support schemes	Fund entrepreneurs and start-ups
Set up collaborative platforms	Facilitate city-level collaborations with key stakeholders
Encourage cooperation between various levels of government and business and community	Foster visibility of initiatives through networking and publicity
Clear policy/vision for circular economy	Build public engagement through communicating a vision of adaptable urban futures
Mainstream successful examples where possible	Identify and bolster existing initiatives where possible
Support procurement and infrastructure	Seek commitments from major stakeholders
Support innovation and experimentation	Use urban living labs to facilitate experimentation where possible
Enable a policy of feedback loops so knowledge and skills can be continuously improved	Build knowledge through linking education, knowledge development, innovation, and collaboration
Develop circular procurement policies	Develop regulations, standards, and procurement guidelines for circular tendering
Capture feedback loops	Use data to understand, manage, and support resource flows

Source: Authors

demonstrating (https://aspiresme.com/), but financial outlay and public engagement are also important enablers.

6.5   Research Methods and Findings

The research presented in this chapter used a mixed-methods approach that combined interviews with key stakeholders and a survey specifically designed for small and medium enterprises (SMEs).

Interviews were undertaken with major actors, including state and local governments, industry bodies, and selected firms (for-profit and not-for-profit) working in a circular economy in Victoria. There were 23 interviews: 11 with participants from state and local governments, 2 with participants from industry associations, and 10 with participants from business. The participants from business were interviewed based on their circular economy practices that had been published in the media. The interviews were open-ended and focused on getting an understanding of the circular economy landscape within organisations, internal drivers, challenges, external enablers, barriers, and changes needed to transition to a circular economy. A high-level framework focusing on political, economic, social, and technological (PEST) underpinnings formed the basis of the interviews. The thematic areas arising from the interviews were considered from social, technological, economic, environmental, and political (STEEP) factors.

The surveys targeted businesses to gain an understanding of current business practices associated with a circular economy, internal and external drivers for circular economy practices, challenges to transitioning to a circular economy from an industry or business perspective, strategies to drive a circular economy transition in the business sector, and an understanding of the level of awareness about the concept of the circular economy within businesses. Surveys were deployed online via Qualtrics for a period of 5 months during 2021/2022. This timeline was adopted due to the COVID lockdown period within Victoria. Different channels were used to distribute the survey, such as peer-to-peer networks, newsletters through various channels, website posts, and snowball sampling. A total of 186 responses were collected; 33 responses were excluded, as some of the respondents were not from the target population. This meant that 82% were usable for data analysis.

The key findings arising from the research are presented below.

6.5.1 What Is a Circular Economy?

Awareness of the concept of the circular economy amongst those interviewed was high. A common understanding was, however, absent. Rather than asking participants to define the circular economy, the interview questions focused on participants' explanations of what they meant by the term. These explanations were analysed from a systems perspective, including the R-strategies (Cramer 2022), processes, and aims of a circular economy. Responses centred on waste and recycling. Some participants were able to align their understanding with broader academic definitions of the circular economy; however, most contextualised their understanding based on their organisation's mission and vision. Some interviewees were able to understand their organisation's position with respect to supply chains and their contribution to a circular economy; others noted the importance of collaborations and the benefit of partnerships.

In contrast to much of the academic literature, participants did not tend to view the circular economy as a broader economic system; rather, the narrative in the responses was more narrowly focused on the circular economy as part of an industrial system supporting a developed waste management system. The nuances of the R-strategies were not addressed; instead, the understanding of circular economy was predominantly represented by the lower-order strategies such as recycling and waste-to-energy (recovery). Regeneration of natural systems was not mentioned in the discussions, despite it being acknowledged as a critical aspect in the state policy. The processes associated with the circular economy were focused on production and consumption practices, not on a systemic understanding of the entire process from design to extraction, production, consumption, and disposal. Most participants equated the circular economy with achieving sustainable outcomes.

6.5.2 Drivers, Challenges, Enablers, and Barriers

For the purposes of the research, drivers and challenges were distinguished as being historical factors that had led organisations to adopt a circular economy approach. Enablers and barriers were considered to be more forward-looking, in that they sought to determine what steps can be taken in the present and future to support circular economy transitions. Indeed, enablers and barriers were two sides of the same coin, yet they were considered separate for the purpose of eliciting variety in the interview responses. STEEP was used to analyse key circular economy drivers, challenges, enablers, and barriers (Fig. 6.2).

Circular economy drivers identified centred on waste crises, market considerations, and financial benefits. Waste crises referred to waste strategy, waste management, and the ramifications of China's decision to stop taking Australia's waste from 2018. Market considerations included business mindsets, leadership, and market growth. Financial considerations focused on business and financial sustainability and commercial viability. Circular economy challenges identified were lack of awareness of the concept of the circular economy, financial considerations, and isolated and fragmented organisational structures that led to lack of an overall collaborative approach to achieving a circular economy.

Circular economy enablers identified included regulatory environments, collaboration, education, extended producer responsibility, and financial support. Circular economy barriers included lack of specific guidelines, negative perception of circular economies, and financial challenges, followed by waste infrastructure and market demand. It is noted that financial factors were present in almost every

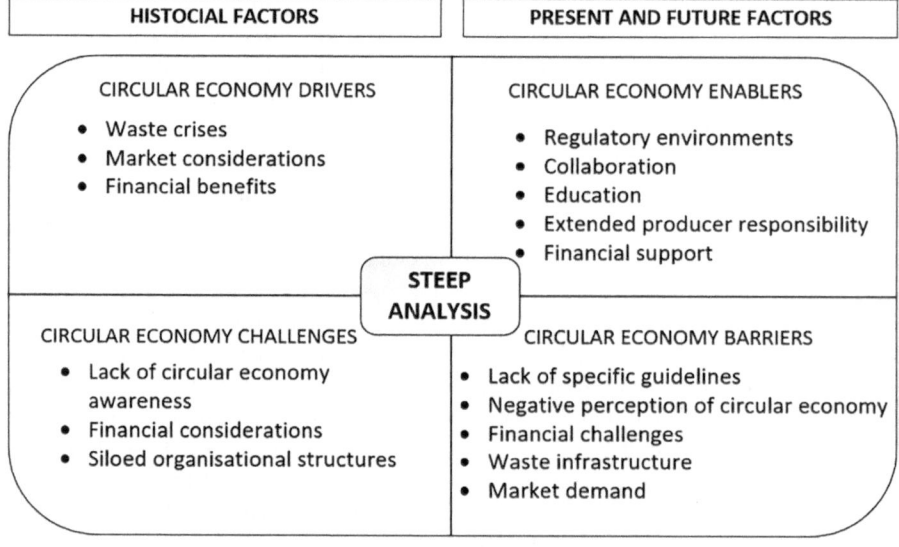

Fig. 6.2 Circular economy drivers, challenges, enablers, and barriers

consideration related to drivers, challenges, barriers, and enablers, which highlights the critical role of financial considerations across these factors and their influence on the decision-making process as well as organisational outcomes.

6.5.3   Survey Responses

Two-thirds of survey respondents were senior managers (67%). More than 80% had at least 1 year's experience in their business area, while over 45% had over 3 years' experience. People from a wide range of businesses from different sectors and business types responded to the survey, although the majority of businesses were SMEs. Businesses ranged in size from 1–5 employees (52%), 5–20 (24%), 20–200 (16%), and over 200 (8%). The principal sectors involved included manufacturing (21%), professional, scientific, and technical services (20.25%), retail (17%), accommodation and food services (9.52%), electricity, gas, water, and waste services (4.76%), wholesale trade (4.76), agriculture, forestry and fishing (1.19), and other services (13.09%).

Private sector owners reported that their main objective was to start their own business, whereas employees indicated that they joined the business because of an alignment between their own values and those of the organisations they worked in. In response to questions about major organisational objectives as the driver for the businesses, responses that ranked highly were 'sustainability impact', 'financial return', 'customer satisfaction', and 'product quality'. The lowest-ranked objectives were 'corporate social responsibility', 'shareholder returns', and 'exploit market opportunity'. The contrast between the low ranking of shareholder returns and the high ranking of financial return was identified.

When asked about the circular economy or sustainability as part of the considerations for business decision-making, 31% of participants aligned their responses to these issues, whereas only 7% indicated that they did not consider any environmental impacts or were unsure. Forty-nine percent of participants mentioned a specific environmental or circular economy-related strategy, such as recycling waste and reducing energy or water usage, resulting from business operations. Just 20% of organisations indicated that recycling was in the manufacturing or waste processing sectors, suggesting that waste recycling may be interpreted as a separation of waste rather than technically focusing on recycling operations or other forms of waste processing, thus pointing to the lower end of the waste recovery strategies.

Various drivers were identified as part of a multiple-choice question for implementing circular economy or environmental sustainability initiatives in a business. Seventeen drivers were provided in the survey. Participants had the option of checking more than one driver and ranking these according to their importance. The top three factors selected were 'right thing to do', 'climate/social conscience', and 'entrepreneurial/business opportunities' (Table 6.2).

Participants were also questioned about circular economy and environment-related initiatives in the short term (1–2 years). Top amongst the responses were

Table 6.2 Circular economy drivers

Drivers	Rank sum score	Rank
The right thing to do	4.3	1
Climate/social conscience	3.0	2
Entrepreneurial/business opportunities	2.1	3
Business strategy/organisational policies	2.0	4
Leadership/strategic commitment	2.0	5
Client or customer demand	1.7	6
Part of business objectives	1.5	7
Corporate social responsibility (CSR)/environmental social governance (ESG)	1.4	8
Part of the product/service model of the business	1.4	9
Technological opportunities/innovation	0.9	10
Government policies/regulations	0.8	11
Financial return	0.8	12
Keep up with industry trends	0.4	13
Supply chain imperatives/reverse logistics	0.4	14
Marketing	0.3	15
Employees' request	0.1	16

reducing energy or water use and improving awareness of environmental sustainability and the circular economy (61% for each). These were followed by developing sustainability or circular economy policy or strategy, at just over half (52%).

The most significant response to the question of barriers faced in implementing circular economy or environmental sustainability practices was 'upfront costs and financial considerations', followed by 'lack of viable business models' and 'infrastructure and networks'. Others identified were a lack of customer demand, limited coordination along the supply chain, and external or customer resistance to change. Enablers identified were financial incentives, as reported by 80% of participants, followed by business collaboration to close the loop (60% of participants) and relevant regulations (55% of participants). Enforcement of regulations came last at 29%.

When asked 'How far along the journey of circular economy or environmental sustainability is your organisation?', 40% claimed to be progressing well, 12% claimed to be industry leaders, 11% stated they had not commenced, and 37% claimed to be just starting their journey. Environmental issues relevant for individual businesses identified were waste management, followed by energy use and climate change. The least identified was biodiversity and ecology.

An optional part of the survey was about aspects of the circular economy related to application or practice in their businesses. Adoption of circular economy practices was not high, as demonstrated by over 50% of the organisations, despite participants mentioning that they were knowledgeable about the concept of the circular economy. The data clearly demonstrates 'not walking the talk' in the initial stage of circular economy transition, referring to a situation where there is a gap between knowledge and action or between understanding the circular economy and

Table 6.3 Maturity of circular economy in an organisation and changes for transitioning to circularity

How advanced is your organisation's circular economy strategy?	Responses (%)	What changes has your organisation made to move towards circular economy? (Top five responses)	Responses (%)
Very advanced—Circular economy as the core of organisational business	31	Educating staff about circular economy	58
Advanced—More than 50% of employees are aware of circular economy	16	Engaging in partnerships or networks that promote circular economy	55
Improving—More than 25% of employees are aware of circular economy	19	Using information about circular economy in communication with existing and/or future customers	45
Starting to integrate CE principles to organisational business	16	Updating corporate strategy to consider circular economy	30
Circular economy is still new and has not applied in my organisation yet	18	Being committed to net-zero emissions	30

implementing it in practice. When asked to select words that described a circular economy business strategy within their organisation, participants nominated 'sustainability' and 'environmental sustainability'. This was followed by 'looping materials' and 'recycling materials', showing the alignment with circular economy narratives from the interviews. Participants were also asked to add their own words under this question, and replied with the terms 'systemic', 'dynamic', and 'socioeconomic', demonstrating that they considered the circular economy to be more holistic in approach.

Participants were asked to comment on the maturity of circular economy strategies and awareness in their organisation. Thirty-one percent of participants indicated that 'circular economy was very advanced' (core to the business), and 16% stated that 'circular economy was advanced' (i.e. more than 50% of employees were aware). When participants were asked about changes that were part of a circular economy transition from an organisational perspective, the top three selected were 'educating staff about circular economy' (58%), 'engaging in partnerships or networks that promote circular economy' (55%), and 'using information about circular economy in communication with existing and/or future customers' (45%) (Table 6.3).

Participants indicated that the *Recycling and Waste Reduction Act (2020)* was the most important policy to affect their organisation (47%). The Commonwealth Act on *National Waste* (DCCEEW 2022a, b) policy scored almost equally with *Recycling Victoria: A new economy,* (DELWP 2020) (46% and 44%, respectively).

6.6 Discussion

The findings show that in Australia the interest in circular economies commenced from the country's waste crises, from which the drivers for developing most governmental circular economy policies stemmed. In line with this, the policy goals mainly focus on reducing the amount of waste going to landfill and reducing the reliance on foreign markets to process waste. The waste export bans for selected waste categories are a driver for increased local processing of waste. The need for increased local processing of waste and the use of recyclable materials within the Australian economy needs to be used as a lever to encourage a migration from conventional waste management to more circular approaches. However, continual and increased use of recyclable materials in local manufacturing will only be a sustainable solution if manufacturing capacity within the state is expanded and financial incentives support this driver.

While the participants' responses suggested that, in general, Australian businesspeople have a high level of understanding of circular economy concepts, the translation of these concepts into action is still limited. The common circular economy principles cited across the policies focused on designing out waste, circulating products and materials at their highest value, and regenerating nature. Despite this narrative, the action and targets associated with implementing these ideas are found wanting. In comparison to the R-strategies, the focus was on the lower-order circular practices such as recycling and waste-to-energy, rather than focusing on the extension of a product's life with the principles of 'rethink, reuse, repair, refurbish, remanufacture, and repurpose'. Closer attention needs to be paid to achieving a more nuanced understanding of the variations in the R-strategies. The implementation of a circular economy can be expanded by potential practices such as rethinking product design, promoting refurbishment and remanufacturing, and exploring opportunities for applying various R-strategies together, such as repurposing, repairing, and reusing. With a comprehensive understanding of R-strategies, it becomes possible to develop a more sustainable system and facilitate the transition towards circularity. This approach encourages products' longevity and value retention, reduces waste generation, and promotes a more sustainable and resource-efficient economy.

It is quite clear from the surveys and interviews in this study that the focus of various actors in the State is on recycling, commencing with the newly formed government department Recycling Victoria in accordance with the Recycling Act. A recycling economy is the sustainable management of waste generated through a traditional linear industrial system, while a circular economy sensu stricto has an emphasis on design principles that decouple environmental and traditional economic considerations (e.g. the use of renewable energy sources such as solar and hydrogen as alternatives to fossil fuels). Another related issue is measurements or impact metrics associated with the linear versus the circular economy.

According to the survey results, 'corporate social responsibility (CSR)', 'shareholder returns', and 'exploit market opportunity' were ranked as the lowest objectives of environmental sustainability or a circular economy. The low rank given to shareholder returns as opposed to the high rank given to financial return may be due to the low number of publicly listed companies responding to the survey. All the publicly listed companies gave a rank of 2 or 3 to shareholder returns, emphasising that financial returns were an important consideration. The top three factors selected for implementing a circular economy in the business were 'right thing to do', 'climate/social conscience', and 'entrepreneurial/business-opportunities'. Therefore, moral and ethical considerations appeared top of mind for the participants, rather than typical business drivers such as financial return, industry trends, and marketing.

Measures for (linear) economic performance include GDP and jobs growth, which are linked to production and consumption practices. A circular economy closes loops in production and consumption practices to ensure that materials stay longer in the system, less or no virgin material is used, and there are major attempts to avoid landfill and incineration. It was clear from some of the interviews that the voluntary sector is critical for supporting the reuse and recycling of items sold in in the clothing and bric-a-brac reuse sector. Related to this is the role of carbon and CO_2 emission targets. In a circular economy, it is critical to include not just Scope 1 emissions (within boundary and caused directly by the company), but also Scope 2 (indirect emissions, through the energy purchased and used) and Scope 3 (indirect emissions upstream and downstream, caused by the company's activities that it is indirectly responsible for, such as embodied energy/carbon in the products it purchases), as well as those related to the wider production and consumption through extended life spans and smarter forms of manufacture.

As indicated by Paiho et al. (2020), when targeting circularity as an outcome, cities need to consider the definition of circular cities, transition processes, target-setting, and circularity indicators. Rather than focusing only on circular economy policy, it makes sense to incorporate related government departments that are also aligned, such as climate change, water, and sewerage, and to support implementation actions. Fragmented policies lead to silo mentalities, making the broader circular economy goals much harder to achieve. This applies at all levels of government where integrated approaches are central to addressing major challenges.

6.7 Conclusion

Australia is a federation of states. States set their own policies and targets, which may or may not be aligned with the federal government's directives. This has been the case particularly in the waste management sector. The aim of the research presented in this chapter has been to understand the current circular economy landscape in Victoria and identify the opportunities that can be nurtured and the gaps that can be addressed over the short term to support a transition to a circular economy.

Taking an exploratory approach, the study uncovered challenges that are not dissimilar to those reported in the international literature. The importance of collaboration was a recurrent theme from the interviews.

The interviews and surveys show that from a business perspective, financial incentives are needed to pivot businesses to support the move to a circular economy. As reported by other authors, knowledge gaps are evident, which include ambiguity in how the concept of circular economy is defined (e.g. equating it to recycling), and lack of systemic thinking overall. A common understanding of what a circular economy is was absent. So, increasing knowledge and awareness about circular economies is critical to unlocking the value they offer.

The survey clearly showed that drivers of a circular economy or sustainability are important for businesses in this sector, underpinned by a clear sustainability- or circular economy-related strategy. Reducing energy and water use was important for businesses, but the upfront costs required to put this in place were a major barrier. Collaborations with other businesses and having relevant regulations would support businesses to move in the direction of a circular economy. The adoption of circular economy practices in businesses was not widespread, but those business strategies that were adopted to achieve a circular economy demonstrated an understanding of circular economy practices.

The policy focus in this area is ambiguous: it is called the circular economy, but the focus has been on waste, and if this is not acknowledged, there is a danger of 'lock-in' to a focus only on waste management. In terms of technical considerations, the current Victorian government's Innovation Fund acts as a catalyst to support changes in the waste ecosystem. However, these initiatives still need to be mainstreamed. Furthermore, there is a need for more engagement across various government departments, to enable a more holistic, systematic, and systemic transition for a circular economy. Moving from fragmentation and a silo mentality to genuine engagement such as linking planning and building regulations, for instance (among others), is critical for the long-term transition. Waste legislation can be supportive of stockpiling disassembled building and construction materials so they can be reused and recycled. Both solid and liquid waste need to be in scope. Technical, financial, and social underpinnings to formulate holistic circular economy solutions are required. Clear circular economy metrics are needed to allow policy, business, and community to be aligned in their vision and goal for the circular economy transition. This is truly a mission-scale challenge for the twenty-first century.

Acknowledgements The authors acknowledge that the research was funded through the Victorian Government's Circular Economy Innovation Fund as part of the Circular Economy Business Innovation Centre and Sustainability Victoria. The authors also acknowledge the project team: Victorian Circular Activator, Planet Ark, Swinburne University of Technology, RMIT Activator, and the City of Melbourne. Input by the members of the VCA Steering Committee, Circular Economy Advisory Committee, Victorian government, and the survey respondents are gratefully acknowledged.

References

Architecture 2030 (2021) Architecture 2030. https://architecture2030.org

Australian Government (2017) Australian 2030—prosperity through innovation. Australia

Bai X, Dawson RJ, Ürge-Vorsatz D, Delgado GC, Barau AS, Dhakal S, Dodman D, Leonardsen L, Masson-Delmotte V, Roberts DC (2018) Six research priorities for cities and climate change. Nature Publishing Group

Batten D (2000) Discovering artificial economics. Westview Press, Boulder, CO

Bellinson R (2022) Mobilising local action to address 21st century challenges: considerations for mission-oriented innovation in cities. Policy Brief series (IIPP PB 19). UCL Institute for Innovation and Public Purpose, Policy Brief series (IIPP PB 19)

Bolger K, Doyon A (2019) Circular cities: exploring local government strategies to facilitate a circular economy. Eur Plan Stud 27(11):2184–2205

Bos JJ, Brown RR (2012) Governance experimentation and factors of success in socio-technical transitions in the urban water sector. Technol Forecast Soc Chang 79:1340–1353

Boschetti F, Gaffier C, Moglia M, Walke I, Price J (2017) Citizens' perception of the resilience of Australian cities. Sustain Sci 12(3):345–364

Boschma R (2005) Proximity and innovation: a critical assessment. Reg Stud 39(1):61–74

Boschma RA, Frenken K (2006) Why is economic geography not an evolutionary science? Towards an evolutionary economic geography. J Econ Geogr 6(3):273–302

Bottero M, Bragaglia F, Caruso N, Datola G, Dell'Anna F (2020) Experimenting community impact evaluation (CIE) for assessing urban regeneration programmes: the case study of the area 22@ Barcelona. Cities 99:102464

Bouckaert S, Pales AF, McGlade C, Remme U, Wanner B, Varro L, D'Ambrosio D, Spencer T (2021) Net zero by 2050: a roadmap for the global energy sector. In: International Energy Agency, Paris. https://www.iea.org/reports/net-zero-by-2050

Brundiers K, Eakin HC (2018) Leveraging post-disaster windows of opportunities for change towards sustainability: a framework. Sustainability (Switzerland) 10(5)

Buchtmann M, Wise R, O'Connell D, Crosweller M, Edwards J (2022) Reforming Australia's approach to hazards and disaster risk: national leadership, systems thinking, and inclusive conversations about vulnerability. Disaster Prev Manag 32:49

Bugge M, Coenen L, Marques P, Morgan K (2017) Governing system innovation: assisted living experiments in the UK and Norway. Eur Plan Stud 25(12):2138–2156

Bulkeley H, Castán Broto V (2013) Government by experiment? Global cities and the governing of climate change. Trans Inst Br Geogr 38(3):361–375

Buurman J, Babovic V (2016) Adaptation pathways and real options analysis: an approach to deep uncertainty in climate change adaptation policies. Polic Soc 35(2):137–150

Byrne J, Taylor M, Ambrose M, Berry S, Sproul AB (2019) Guide to low carbon residential buildings—new build. Low Carbon Living CRC. https://apo.org.au/node/239826

Campbell I, Sachar S, Meisel J, Nanavatty R (2021) Beating the heat: a sustainable cooling handbook for cities. United Nations Environment Programme. https://www.unep.org/resources/report/beating-heat-sustainable-cooling-handbook-cities

Carbon Neutral Cities Alliance (2022) Mobilizing transformative climate action in cities. https://carbonneutralcities.org/

Carrière S, Weigend Rodríguez R, Pey P, Pomponi F, Ramakrishna S (2020) Circular cities: the case of Singapore. Built Environ Project Asset Manag 10(4):491–507

Colloff MJ, Gorddard R, Abel N, Locatelli B, Wyborn C, Butler JRA, Lavorel S, van Kerkhoff L, Meharg S, Múnera-Roldán C, Bruley E, Fedele G, Wise RM, Dunlop M (2021) Adapting transformation and transforming adaptation to climate change using a pathways approach. Environ Sci Policy 124:163–174

CoM (2018) Urban realm action plan: Melbourne innovation districts City North 2018–2023

CoM (2020) MID City North opportunities plan 2020

CoM (2021a) Budget 2021–22

CoM (2021b) Council plan 2021–2025

CoM (2021c) Economic Development Strategy 2031

CoM (2022a) Climate and biodiversity emergency. CoM

CoM (2022b) Budget 2022–23

CoM (2022c) Inclusive Melbourne Strategy 2022–32

Cork S, Alexandra C, Alvarez-Romero JG, Bennett EM, Berbés-Blázquez M, Bohensky E, Bok B, Costanza R, Hashimoto S, Hill R, Inayatullah S, Kok K, Kuiper JJ, Moglia M, Pereira L, Peterson G, Weeks R, Wyborn C (2023) Exploring alternative futures in the Anthropocene. Annu Rev Environ Resour 48:25

Cramer J (2022) Building a circular future: ten takeways for global changemakers. Amsterdam Economic Board and Holland Circular Hotspot. https://circulareconomy.europa.eu/platform/sites/default/files/building-a-circular-future-jacqueline-cramer-amsterdam-economic-board.pdf. Accessed 20 Mar 2022

CRCLCL (2019) Cooperative Research Centre for Low Carbon Living. https://www.lowcarbon-livingcrc.com.au/research

Cresswell I, Janke T, Johnston E (2021) Australia state of the environment 2021: overview, independent report to the Australian Government Minister for the Environment. Commonwealth of Australia, Canberra

Crona B, Folke C, Galaz V (2021) The Anthropocene reality of financial risk. One Earth 4(5):618–628

Crosweller M, Tschakert P (2021) Disaster management leadership and policy making: a critical examination of communitarian and individualistic understandings of resilience and vulnerability. Clim Pol 21(2):203–221

Cuthill M (2010) Strengthening the 'social' in sustainable development: developing a conceptual framework for social sustainability in a rapid urban growth region in Australia. Sustain Dev 18(6):362–373

David Tàbara J, Frantzeskaki N, Hölscher K, Pedde S, Kok K, Lamperti F, Christensen JH, Jäger J, Berry P (2018) Positive tipping points in a rapidly warming world. Curr Opin Environ Sustain 31:120–129

Davidescu AA, Apostu S-A, Paul A (2020) Exploring citizens' actions in mitigating climate change and moving toward urban circular economy. A multilevel approach. Energies 13(18):4752

Davidson K, Håkansson I, Coenen L, Nguyen TMP (2023) Municipal experimentation in times of crises: (re-)defining Melbourne's innovation district. Cities 132:104042

DCCEEW (2022a) National Waste Policy Action Plan 2019. https://www.dcceew.gov.au/environment/protection/waste/publications/national-waste-policy-action-plan. Accessed 13 Sept 2022

DCCEEW (2022b) Waste exports. https://www.dcceew.gov.au/environment/protection/waste/exports. Accessed 26 Apr 2023

de Jong M, Joss S, Schraven D, Zhan C, Weijnen M (2015) Sustainable–smart–resilient–low carbon–eco–knowledge cities; making sense of a multitude of concepts promoting sustainable urbanization. J Clean Prod 109:25–38

de Morais LHL, Pinto DC, Cruz-Jesus F (2021) Circular economy engagement: altruism, status, and cultural orientation as drivers for sustainable consumption. Sustain Prod Consum 27:523–533

DEECA (2022) Additional environmental protections. https://www.deeca.vic.gov.au/futureforests/what-were-doing/additional-environmental-protections. Accessed 26 Apr 2023

Deloitte Access Economics (2021) Special report: update to the economic costs of natural disasters in Australia

DELWP (2020) Recycling Victoria: a new economy. https://www.vic.gov.au/sites/default/files/2020-02/Recycling%20Victoria%20A%20new%20economy.pdf. Accessed 18 Nov 2021

Dick H, Rimmer P (1999) Privatising climate: first world cities in third world settings. In: East West perspectives on 21st century urban development. Ashgate, Aldershot

Ding L, He B, Craft W, Petersen H, Osmond P, Santamouris M, Prasad D, Koc CB, Derksema C, Midlam N (2019a) Cooling Sydney strategy: planning for Sydney 2050. Cooperative Research Centre for Low Carbon Living

Ding L, Petersen H, Craft W (2019b) Microclimate and urban heat Island mitigation decision-support tool (project short report). CRC for Low Carbon Living. https://apo.org.au/node/246691

EcologqiQ (2020) Recycled First Policy March 2020. Purposely Greener Infrastructure Victorian State Government. https://bigbuild.vic.gov.au/about/ecologiq/recycled-first-policy. Accessed 15 Feb 2022

Ellen Macarthur Foundation (2017) Cities in a circular economy: an initial exploration. https://emf.thirdlight.com/link/6geje0hxj9n1-2aoa77/@/preview/1?o. Accessed 13 Sept 2022

Environment and Climate Change Canada (2021) Discussion paper & event summary: circular North America: accelerating the transition to a thriving and resilient low-carbon economy. January 2021, Cat. No.: En4*413/2020E-PDF, ISBN: 978-0-660-36205-2

European Union (2022a) Excess mortality hits +16%, highest 2022 value so far. Excess mortality in July 2022: https://ec.europa.eu/eurostat/web/products-eurostat-news/-/ddn-20220916-20220911

European Union (2022b) NetZeroCities Pilot Cities Programme guidebook. NZC Consortium coordinated by EIT Climate-KIC. https://netzerocities.eu/wp-content/uploads/2022/2006/Pilot-Cities-Guidebook.pdf

Fastenrath S, Coenen L (2020) Future-proof cities through governance experiments? Insights from the Resilient Melbourne Strategy (RMS). Reg Stud

Fastenrath S, Coenen L, Davidson K (2019) Urban resilience in action: the Resilient Melbourne Strategy as transformative urban innovation policy? Sustain For 11(693):1–10

Feduzi A, Runde J, Schwarz G (2022) Unknowns, black swans, and bounded rationality in public organizations. Public Admin Rev 82(5):958–963

Frantzeskaki N, McPhearson T, Collier MJ, Kendal D, Bulkeley H, Dumitru A, Walsh C, Noble K, Van Wyk E, Ordóñez C, Oke C, Pintér L (2019) Nature-based solutions for urban climate change adaptation: linking science, policy, and practice communities for evidence-based decision-making. Bioscience 69(6):455–466

Freychet N, Hegerl GC, Lord NS, Lo YTE, Mitchell D, Collins M (2022) Robust increase in population exposure to heat stress with increasing global warming. Environ Res Lett 17(6):064049

Friedman TL (2017) Thank you for being late: an optimist's guide to thriving in the age of accelerations (version 2.0, with a new afterword). Picador/Farrar Straus and Giroux

Fuenfschilling L, Frantzeskaki N, Coenen L (2019) Urban experimentation & sustainability transitions. Eur Plan Stud 27(2):219–228

Furtado BA, Fuentes MA, Tessone CJ (2019) Policy modeling and applications: state-of-the-art and perspectives. Complexity 2019

Girardet H (2014) Creating regenerative cities. Routledge

GlobalABC (2020) GlobalABC regional roadmap for buildings and construction in Africa: towards a zero-emission, efficient and resilient buildings and construction sector. IEA, Paris. https://globalabc.org/roadmaps-buildings-and-construction

Gorddard R, Colloff MJ, Wise RM, Ware D, Dunlop M (2016) Values, rules and knowledge: adaptation as change in the decision context. Environ Sci Pol 57:60–69

Government of Victoria (2022) Victoria's Big Build. Roal and rail projects to transform how you travel. https://bigbuild.vic.gov.au/. Accessed 15 Feb 2022

Gravagnuolo A, Girard LF, Kourtit K, Nijkamp P (2021) Adaptive re-use of urban cultural resources: contours of circular city planning. City Cult Soc 26:100416

Groffman PM, Cadenasso ML, Cavender-Bares J, Childers DL, Grimm NB, Grove JM, Hobbie SE, Hutyra LR, Darrel Jenerette G, McPhearson T, Pataki DE, Pickett STA, Pouyat RV, Rosi-Marshall E, Ruddell BL (2017) Moving towards a new urban systems science. Ecosystems 20(1):38–43

Haddad CR, Nakic V, Bergek A, Hellsmark H (2022) Transformative innovation policy: a systematic review. Environ Innov Soc Trans 43:14–40

He B-J, Wang J, Liu H, Ulpiani G (2021a) Localized synergies between heat waves and urban heat islands: implications on human thermal comfort and urban heat management. Environ Res 193:110584

He B-J, Zhao D, Xiong K, Qi J, Ulpiani G, Pignatta G, Prasad D, Jones P (2021b) A framework for addressing urban heat challenges and associated adaptive behavior by the public and the issue of willingness to pay for heat resilient infrastructure in Chongqing, China. Sustain Cities Soc 75:103361

Heilmann S (2008) From local experiments to national policy: the origins of China's distinctive policy process. China J 59:1–30

Hijmans RJ, Cameron SE, Parra JL, Jones PG, Jarvis A (2005) Very high resolution interpolated climate surfaces for global land areas. Int J Climatol 25(15):1965–1978

Hill S, Cumpston Z & Quintana Vigiola, G (2021) Australia state of the environment 2021: urban independent reprt to the Australian Government Minister for the Environment, Commonwealth of Australia, Canberra, https://doi.org/10.26194/G1G4-4J51.

Holland Circular Hotspot (2019). https://hollandcircularhotspot.nl/wp-content/uploads/2019/04/HCH-Brochure-20190410-web_DEF.pdf. Accessed 3 Dec 2019

Howlett M, Rayner J (2013) Patching vs packaging in policy formulation: assessing policy portfolio design. Polit Gov 1(2):170–182

Huovila P, Ala-Juusela M, Melchert L, Pouffary S, Cheng C-C, Ürge-Vorsatz D, Koeppel S, Svenningsen N, Graham P (2009) Buildings and climate change: summary for decision-makers. United Nations Environment Programme: Sustainable Buildings & Climate Initiative. https://www.uncclearn.org/wp-content/uploads/library/unep207.pdf

IEA (2021a) Empowering cities for a net zero future: unlocking resilient, smart, sustainable urban energy systems. International Energy Agency. https://iea.blob.core.windows.net/assets/4d5c939d-939c937-490b-bb953-932c930d923f932cf933d/G920EmpoweringCitiesforaNetZeroFuture.pdf

IEA (2021b) Net zero by 2050: a roadmap for the global energy sector. International Energy Agency. https://www.iea.org/reports/net-zero-by-2050, pp 3–24

Infrastructure Australia (2021) A pathway to infrastructure resilience. Advisory paper 1: opportunities for systemic change. Infrastructure Australia, Canberra

Infrastructure NSW, Infrastructure Australia (2022) City Deals. https://www.infrastructure.gov.au/territories-regions-cities/cities/city-deals. Accessed 16 Aug 2022

IPCC (2018) Summary for policymakers. In: Masson-Delmotte V, Zhai P, Pörtner H-O, Roberts D, Skea J, Shukla PR, Pirani A, Moufouma-Okia W, Péan C, Pidcock R, Connors S, Matthews JBR, Chen Y, Zhou X, Gomis MI, Lonnoy E, Maycock T, Tignor M, Waterfield T (eds) Global warming of 1.5°C. An IPCC special report on the impacts of global warming of 1.5°C above pre-industrial levels and related global greenhouse gas emission pathways, in the context of strengthening the global response to the threat of climate change, sustainable development, and efforts to eradicate poverty. Cambridge University Press, Cambridge, pp 3–24

IPCC (2023) Synthesis report of the IPCC sixth assessment report (AR6): summary for policymakers. Intergovernmental Panel on Climate Change, 2023. https://report.ipcc.ch/ar6syr/pdf/IPCC_AR6_SYR_SPM.pdf

Iyer-Raniga U, Gajanayake A (2021) Infrastructure financing for climate change adaptation: Chapter 8-Australia. Available at SSRN 4189988

Iyer-Raniga U, Gajanayake A, Ho O and Trewick S (2022) Exploring Victoria's emerging CE ecosystem. from https://www.circularactivator.com/circular-economy-ecosystems. Accessed 13th September 2022

Jain M, Rohracher H (2022) Assessing transformative change of infrastructures in urban area redevelopments. Cities 124:103573

Janssen MJ, Torrens J, Wesseling JH, Wanzenböck I (2021) The promises and premises of mission-oriented innovation policy—a reflection and ways forward. Sci Public Policy 48(3):438–444

Jones P, Comfort D (2018) Winning hearts and minds: a commentary on circular cities. J Public Aff 18(4):e1726

Jones RN, Patwardhan A, Cohen SJ, Dessai S, Lammel A, Lempert RJ, Mirza MMQ, von Storch H (2014) Foundations for decision making. Cambridge University Press, Cambridge
Jorgenson S, Stephens JC (2022) Action research for energy system transformation. Educ Action Res 30:655
Kanda W, Kuisma M, Kivimaa P, Hjelm O (2020) Conceptualising the systemic activities of intermediaries in T sustainability transitions. Environ Innov Soc Trans 36:449–465
Kannikar K-N, Lampong K, Prateep P, Chainarong K-N (2021) A digital circular economy for smart cities. Psychol Educ 58(1):1432–1439
Kasemir B, Van Asselt MBA, Dürrenberger G, Jaeger CC (1999) Integrated assessment of sustainable development: multiple perspectives in interaction. Int J Environ Pollut 11(4):407–425
Katsou E, Nika C-E, Buehler D, Marić B, Megyesi B, Mino E, Babí Almenar J, Bas B, Bećirović D, Bokal S (2020) Transformation tools enabling the implementation of nature-based solutions for creating a resourceful circular city. Blue-Green Syst 2(1):188–213
Keath NA, Brown RR (2009) Extreme events: being prepared for the pitfalls with progressing sustainable urban water management. Water Sci Technol 59(7):1271–1280
Keele S, Coenen L (2019) The role of public policy in critical infrastructure resilience. University of Melbourne and The Resilience Shift, London
Kemter M, Fischer M, Luna LV, Schönfeldt E, Vogel J, Banerjee A, Korup O, Thonicke K (2021) Cascading hazards in the aftermath of Australia's 2019/2020 Black Summer wildfires. Earth's Future 9(3)
Kern F, Howlett M (2009) Implementing transition management as policy reforms: a case study of the Dutch energy sector. Policy Sci 42(4):391–408
Kirchherr J, Reike D, Hekkert M (2017) Conceptualizing the circular economy: an analysis of 114 definitions. Resour Conserv Recycl 127:221–232
Kivimaaa P, Boon W, Hyysalo S, Klerkx L (2019) Towards a typology of intermediaries in sustainability transitions: a T systematic review and a research agenda. Res Policy 48:1062–1075
Kjaersgaard SP, Evans CB, Harris MS (2019) Guide to low carbon landscapes. CRC for Low Carbon Living, Canberra
Kozminska U, Arch PE (2018) Circular cities: determinants of closed circulation of building materials. In: Circular economy distruptions, past, present and future. University of Exeter
Kuhlmann S, Rip A (2018) Next-generation innovation policy and grand challenges. Sci Public Policy 45:448–454
Kundurpi A, Westman L, Luederitz C, Burch S, Mercado A (2021) Navigating between adaptation and transformation: how intermediaries support businesses in sustainability transitions. J Clean Prod 283:125366
Laakso S, Berg A, Annala M (2017) Dynamics of experimental governance: a meta-study of functions and uses of climate governance experiments. J Clean Prod 169(8–16):8
Langergraber G, Castellar JA, Andersen TR, Andreucci M-B, Baganz GF, Buttiglieri G, Canet-Martí A, Carvalho PN, Finger DC, Griessler Bulc T (2021) Towards a cross-sectoral view of nature-based solutions for enabling circular cities. Water 13(17):2352
Larrue P (2022) Do mission-oriented policies for net zero deliver on their many promises?
LETI (2021) LETI climate emergency retrofit guide. London, UK/Nairobi
Marchesi M, Tweed C (2021) Social innovation for a circular economy in social housing. Sustain Cities Soc 71:102925
Massey D, Quintas D, Wield D (2003) High-tech fantasies: science parks in society, science and space. Routledge
Matschoss K, Heiskanen E (2017) Making it experimental in several ways: the work of intermediaries in raising the ambition level in local climate initiatives. J Clean Prod 169(15):85–93
Matschoss K, Repo P (2018) Governance experiments in climate action: empirical findings from the 28 European Union countries. Environ Polit 27(4):598–620
Mayan M, Pauchulo AL, Gillespie D, Misita D, Mejia T (2019) The promise of collective impact partnerships. Commun Dev J 55(3):515–532

Mazzucato M (2018a) Mission-oriented innovation policies: challenges and opportunities. Ind Corp Chang 27(5):803–815

Mazzucato M (2018b) Mission-oriented research & innovation in the European Union. A problem-solving approach to fuel innovation-led growth. European Commission Directorate-General for Research and Innovation, Brussels

Mazzucato M (2019) Governing missions: governing missions in the European Union. European Commission

Mazzucato M (2021) Mission economy: a moonshot guide to changing capitalism. Penguin, London

Meerow S, Newell JP (2016) Urban resilience for whom, what, when, where, and why? Urban Geogr 40:1–21

Meerow S, Newell JP, Stults M (2016) Defining urban resilience: a review. Landsc Urban Plan 147:38–49

Meharg S (2020) Catalysing change agents through research for development. PhD, The Australian National University

Meharg S (2022) Critical change agent characteristics and competencies for ensuring systemic climate change interventions. Sustain Sci 18:1445

Midgley G, Rajagopalan R (2021) Critical systems thinking, systemic intervention, and beyond. In: Metcalf GS, Kijima K, Deguchi H (eds) Handbook of systems sciences. Springer, Singapore, pp 107–157

Miedzinski M, Mazzucato M, Ekins P (2019) A framework for mission-oriented innovation policy roadmapping for the SDGs: the case of plastic-free oceans. In: Working Paper Series (IIPP WP 2019–03). UCL Institute for Innovation and Public Purpose, pp 1–61

Mies A, Gold S (2021) Mapping the social dimension of the circular economy. J Clean Prod 321:128960

Moglia M, Cork SJ, Boschetti F, Cook S, Bohensky E, Muster T, Page D (2018a) Urban transformation stories for the 21st century: insights from strategic conversations. Glob Environ Change 50:222–237

Moglia M, Cork SJ, Boschetti F, Cook S, Bohensky E, Muster T, Page D (2018b) Urban transformation stories for the 21st century: insights from strategic conversations. Glob Environ Chang 50:222–237

Moglia M, Cork S, Cook S, Muster T, Bohensky E (2019) The future of Sydney: scenarios to guide collaboration by the Sydney Common Planning Assumptions Group. CSIRO, Sydney

Moglia M, Frantzeskaki N, Newton P, Pineda-Pinto M, Witheridge J, Cook S, Glackin S (2021) Accelerating a green recovery of cities: lessons from a scoping review and a proposal for mission-oriented recovery towards post-pandemic urban resilience. Dev Built Environ 7:100052

Morisson A (2020) Framework for defining innovation districts: case study from 22@ Barcelona. In: Bougdah H, Versaci A, Sotoca A et al (eds) Urban and transit planning. Springer, Cham

Morisson A, Bevilacqua C (2019) Balancing gentrification in the knowledge economy: the case of Chattanooga's innovation district. Urban Res Pract 12(4):472–492

Muiderman K, Zurek M, Vervoort J, Gupta A, Hasnain S, Driessen P (2022) The anticipatory governance of sustainability transformations: hybrid approaches and dominant perspectives. Glob Environ Chang 73:102452

Naderpajouh N, Matinheikki J, Hills M (2019) The role of legislation in critical infrastructure resilience. RMIT University and Resilience Shift, UK

National Infrastructure Commission (2020) Anticipate, react, recover. Resilient infrastructure systems. Report prepared by the UK National Infrastructure Commission, London

New York State (2020) Carbon neutral buildings. https://www.nyserda.ny.gov/All-Programs/Carbon-Neutral-Buildings/Carbon-Neutral-Buildings-State-Fair#:~:text=What%20is%20a%20carbon%20neutral,meet%20the%20State's%20climate%20goals

Newton P (2019) The performance of urban precincts: towards integrated assessment. In: Newton P, Prasad D, Sproul A, White S (eds) Decarbonising the built environment: charting the transition. Springer, Singapore, pp 357–384

Newton PW, Doherty P (2014) The challenges to urban sustainability and resilience. In: Resilient sustainable cities: a future, pp 7–18

Newton P, Frantzeskaki N (2021) Creating a National Urban Research and Development platform for advancing urban experimentation. Sustain For 13(2)

Newton PW, Rogers BC (2020) Transforming built environments: towards carbon neutral and blue-green cities. Sustainability 12:4745

Newton P, Prasad D, Sproul A, White S (2019) Decarbonising the built Environment: charting the transition. Springer, Singapore

Newton P, Glackin S, Garner S, Witheridge J (2020) Beyond small lot subdivision: pathways for municipality-initiated resident supported precinct scale residential infill regeneration in grey-field suburbs. Urban Policy Res 38:338

Nielsen JS, Farrelly M (2019) Conceptualising the built environment to inform sustainable urban transitions. Environ Innov Soc Trans 33:231–248

O'Connell D, Wise RM, Doerr V, Grigg N, Williams R, Meharg S, Dunlop M, Meyers J, Edwards J, Osuchowski M, Crosweller M (2018) Approach, methods and results for co-producing a systems understanding of disaster. Technical report supporting the development of the australian vulnerability profile. CSIRO, Canberra

O'Connell D, Maru Y, Grigg N, Walker B, Abel N, Wise R, Cowie A, Butler J, Stone-Jovicich S, Stafford-Smith M, Ruhweza A, Belay M, Duron G, Pearson L, Meharg S (2019) Resilience, adaptation pathways and transformation approach. A guide for designing, implementing and assessing interventions for sustainable futures (version 2). CSIRO, Canberra

OECD (2022a) Lessons for tackling complex and systemic societal challenges. GGSD Forum

OECD (2022b) Cities and environment. https://www.oecd.org/cfe/cities/cities-environment.htm. Accessed 13 Sept 2022

Oral HV, Carvalho P, Gajewska M, Ursino N, Masi F, Hullebusch ED, Kazak JK, Exposito A, Cipolletta G, Andersen TR (2020) A review of nature-based solutions for urban water management in European circular cities: a critical assessment based on case studies and literature. Blue-Green Syst 2(1):112–136

Osmond P, Sharifi E (2017) Guide to urban cooling strategies. Low Carbon Living CRC

Paiho S, Mäki E, Wessberg N, Paavola M, Tuominen P, Antikainen M, Heikkilä J, Rozado CA, Jung N (2020) Towards circular cities—conceptualizing core aspects. Sustain Cities Soc 59:102143

Palin M (2017) Urban Island heat effect: rising temperatures in Aussie cities could create death traps. Australia News. https://www.news.com.au/technology/environment/climate-change/urban-island-heat-effect-rising-temperatures-in-aussie-cities-could-create-death-traps/news-story/0b035c4707ea4708f4781e4732ee4700df4704fa4546bf

Pancost RD (2016) Cities lead on climate change. Nat Geosci 9(4):264

Parks Victoria (2023) Asset regeneration program. https://www.parks.vic.gov.au/projects/statewide-projects/asset-regeneration-program. Accessed 26 Apr 2023

Patterson J, Schulz K, Vervoort J, van der Hel S, Widerberg O, Adler C, Hurlbert M, Anderton K, Sethi M, Barau A (2017) Exploring the governance and politics of transformations towards sustainability. Environ Innov Soc Trans 24:1–16

Pavlopoulou Y (2021) Methodologies of stakeholders' engagement in circular collaborative ecosystems. Deliverable D, Leuven

Pereira P, Baró F (2022) Greening the city: thriving for biodiversity and sustainability. Sci Total Environ 817:153032

Petit-Boix A, Leipold S (2018) Circular economy in cities: reviewing how environmental research aligns with local practices. J Clean Prod 195:1270–1281

Pfotenhauer S, Jasanoff S (2017) Panacea or diagnosis? Imaginaries of innovation and the 'MIT model' in three political cultures. Soc Stud Sci 47:1–28

Pickin J, Wardle C, O'Farrell K, Stovell L, Nyunt P, Guazzo S, Lin Y, Caggiati-Shortell G, Chakma P, Edwards C, Lindley B (Cirq Solutions), Latimer G (Ascend Waste and Environment), Downes J (BehaviourWorks), Axiö I (RMCG) 2022. National Waste Report 2022. The department of climate change, energy, the environment and water, Blue Environment Pty Ltd. Accessed from: https://www.dcceew.gov.au/environment/protection/waste/national-waste-reports/2022

Prasad D, Ding L, Yenneti K, Fan H, Craft W, Sanchez AX, Li X, Arnold P, Bouhmad K, Earley R, Wang Y, Wu J, Li F, You N, Fox J, Aragon JD (2017) Guidelines for sustainable cities and communities." Cooperative Research Centre (CRC) for Low Carbon Living, United Nations Environment Programme, Jia Cui Environment Promotive Center 2(9):805–814

Prasad D, Kuru A, Oldfield P, Ding L, Dave M, Noller C, He B (2021) Race to net zero carbon: a climate emergency guide for new and existing buildings in Australia. Low Carbon Institute. https://www.unsw.edu.au/content/dam/pdfs/unsw-adobe-websites/arts-design-architecture/built-environment/net-zero-guide/2023-2001-2013-Net-Zero-guide-online-version.pdf

Prasad D, Kuru A, Oldfield P, Ding L, Dave M, Noller C, He B (2023) The climate emergency and the built environment. In: Delivering on the climate emergency: towards a net zero carbon built environment. Springer, pp 1–27

Prendeville S, Cherim E, Bocken N (2018) Circular cities: mapping six cities in transition. Environ Innov Soc Transit 26:171–194

Quigley EV (2019) Five deep decarbonization strategies. Clean Energy Transition Institute. https://www.cleanenergytransition.org/post/five-deep-decarbonization-strategies

Raymond CM, Frantzeskaki N, Kabisch N, Berry P, Breil M, Nita MR, Geneletti D, Calfapietra C (2017) A framework for assessing and implementing the co-benefits of nature-based solutions in urban areas. Environ Sci Policy 77:15–24

RIBA (2021) RIBA 2030 challenge. Version 2. London

Ritchie H, Roser M (2022) CO$_2$ and greenhouse gas emissions. Our World in Data. https://ourworldindata.org/co2-and-other-greenhouse-gas-emissions

Rogge KS, R. K. (2016) Policy mixes for sustainability transitions: an extended concept and framework for analysis. Res Policy 45(8):1620–1635

Røpke I (2012) The unsustainable directionality of innovation—the example of the broadband transition. Res Policy 41:1631–1642

Rosenzweig C, Solecki W (2018) Action pathways for transforming cities. Nat Clim Chang 8(9):756–759

Rosenzweig C, Solecki W, Romero-Lankao P, Mehrotra S, Dhakal S, Ibrahim SA (2018) Pathways to urban transformation. In: Rosenzweig C, Romero-Lankao P, Mehrotra S et al (eds) Climate change and cities: second assessment report of the Urban Climate Change Research Network. Cambridge University Press, Cambridge, pp 3–26

Roth M (2012) Urban heat islands. In: Handbook of environmental fluid dynamics, vol 2. CRC Press, pp 162–181

Russell M, Gianoli A, Grafakos S (2020) Getting the ball rolling: an exploration of the drivers and barriers towards the implementation of bottom-up circular economy initiatives in Amsterdam and Rotterdam. J Environ Plan Manag 63(11):1903–1926

Salamanca F, Georgescu M, Mahalov A, Moustaoui M, Wang M (2014) Anthropogenic heating of the urban environment due to air conditioning. J Geophys Res Atmos 119(10):5949–5965

Santamouris M, Kolokotsa D (2016) Urban climate mitigation techniques. Routledge, London

Santamouris M, Haddad S, Fiorito F, Osmond P, Ding L, Prasad D, Zhai X, Wang R (2017) Urban heat island and overheating characteristics in Sydney, Australia. An analysis of multiyear measurements. Sustainability 9. https://doi.org/10.3390/su9050712

Savini F (2019) The economy that runs on waste: accumulation in the circular city. J Environ Policy Plan 21(6):675–691

Schipper ELF (2020) Maladaptation: when adaptation to climate change goes very wrong. One Earth 3(4):409–414

Schot J, Steinmueller WE (2018) Three frames for innovation policy: R&D, systems of innovation and transformative change. Res Policy 44:1554–1567

Schweizer P (2019) Governance of systemic risks for disaster prevention and mitigation. Contributing Paper to the UNDRR Global Assessment Report

Schweizer P, Renn O (2019) Governance of systemic risks for disaster prevention and mitigation. Contributing paper to the Global Assessment Report on Disaster Risk Reduction. Institute for Advanced Sustainability Studies, Potsdam

Scoones I (2016) The politics of sustainability and development. Annu Rev Environ Resour 41(1):293–319

Sengers F, Berkhout F, Wieczorek AJ, Raven R (2016) Experimenting in the city—unpacking notions of experimentation for sustainability. In: Evan J, Karvonen A, Raven R (eds) The experimental city. Routledge, New York, pp 15–31

Sengers F, Wieczorek AJ, Raven R (2019) Experimenting for sustainability transitions: a systematic literature review. Technol Forecast Soc Chang 145:153–164

Smith A, Stirling A (2010) The politics of social-ecological resilience and sustainable socio-technical transitions. Ecol Soc 15(1)

Smith MS, Horrocks L, Harvey A, Hamilton C (2011) Rethinking adaptation for a 4°C world. Philos Trans R Soc A Math Phys Eng Sci 369(1934):196–216

Snowden DJ, Boone ME (2007) A leader's framework for decision making. Harv Bus Rev 2007:69–76

Sovacool BK, Turnheim B, Martiskainen M, Brown D, Kivimaa P (2020) Guides or gatekeepers? Incumbent-oriented transition intermediaries in a T low-carbon era. Energy Res Soc Sci 66:101490

Steffen W, Broadgate W, Deutsch L, Gaffney O, Ludwig C (2015) The trajectory of the Anthropocene: the great acceleration. Anthropocene Rev 2(1):81–98

Stirling A (2014) Transforming power: social science and the politics of energy choices. Energy Res Soc Sci 1:83–95

Stirling AC, Scoones I (2009) From risk assessment to knowledge mapping: science, precaution, and participation in disease ecology. Ecol Soc 14(2)

Sukhdev A, Vol J, Brandt K, Yeoman R (2018) Cities in the circular economy: the role of digital technology. Ellen MacArthur Foundation, Cowes

Talmar M, Walrave B, Raven R, Romme AGL (2022) Dynamism in policy-affiliated transition intermediaries. Renew Sust Energ Rev 159:112210

Tetlock PE (2005) Expert political judgment United States. Princeton University Press, Princeton

The Resilience Shift (2022) The global hub for resilience best practice. Lloyd's Register Foundation

Thomson G, Newton P, Newman P, Byrne J (2019) Guide to low carbon precincts

Turcu C, Gillie H (2020) Governing the circular economy in the city: local planning practice in London. Plan Pract Res 35(1):62–85

Turku C (n.d.) Circular Turku—a roadmap toward resource wisdom. https://circulars.iclei.org/resource/circular-turku-a-roadmap-toward-resource-wisdom/?gclid=CjwKCAjwo7iiBhAE EiwAsIxQEfL58EawRjXsej4lbLZB369Jgk197W5Mkuip20n0x6suqP4Z9yVpGBoCYrwQ-AvD_BwE. Accessed 1 May 2023

UN Environment (2022) Goal 11: sustainable cities and communities. https://www.unep.org/es/node/2037. Accessed 13 Sept 2022

UN SDG (2015) The 17 goals. Department of Economic and Social Affairs: Sustainable Development, United Nations. https://sdgs.un.org/goals

UNEP (2020) Facts about the climate emergency. United Nations Environment Programme. https://www.unep.org/facts-about-climate-emergency

UNEP (2021) 2021 global status report for buildings and construction: towards a zero-emission, efficient and resilient buildings and construction sector. Nairobi

UNEP (2023) Sustainable cities. Home regions Asia and the Pacific Regional initiatives Supporting resource efficiency. https://www.unep.org/regions/asia-and-pacific/regional-initiatives/supporting-resource-efficiency/sustainable-cities

United Nations (2015) Transforming our world: the 2030 agenda for sustainable development. United Nations. pp 1–41

United Nations (2022a) Cities and climate change. Our World in Data. https://www.unep.org/explore-topics/resource-efficiency/what-we-do/cities/cities-and-climate-change

United Nations (2022b) Race to zero campaign. United Nations Framework Convention on Climate Change. https://unfccc.int/climate-action/race-to-zero-campaign

Ürge-Vorsatz D, Rosenzweig C, Dawson RJ, Rodriguez RS, Bai X, Barau AS, Seto KC, Dhakal S (2018) Locking in positive climate responses in cities. Nat Clim Chang 8(3):174

US DOE (2015) A common definition for zero energy buildings. U.S. Department of Energy. https://www.energy.gov/sites/prod/files/2015/2009/f2026/A%2020Common%2020Definition%2020 20for%2020Zero%2020Energy%2020Buildings.pdf

Uyarra E, Ribeiro B, Dale-Clough L (2019) Exploring the normative turn in regional innovation policy: responsibility and the quest for public value. Eur Plan Stud 27(12):2359–2375

van Asselt MBA, van't Klooster SA, Veenman SA (2014) Coping with policy in foresight. J Futur Stud 19(1):53–76

Veelen B, v. (2019) Caught in the middle? Creating and contesting intermediary spaces in low-carbon transitions. EPC: Polit Space 38(1):116–133

Verga GC, Khan AZ (2021) Crafting insights on urban circularity: two case studies of inclusive socio-ecological practices in two types of public open-air space. IBA Crossing Boundaries

Victoria State Government (2021) Innovation Victoria. Victoria

Voß JP, Newig J, Kastens B, Monstadt J, Nölting B (2007) Steering for sustainable development: a typology of problems and strategies with respect to ambivalence, uncertainty and distributed power. J Environ Policy Plan 9(3–4):193–212

Walker B (2020) Finding resilience: change and uncertainty in nature and society. CSIRO Publishing, Canberra

Wang J, Meng Q, Zou Y, Qi Q, Tan K, Santamouris M, He B-J (2022) Performance synergism of pervious pavement on stormwater management and urban heat Island mitigation: a review of its benefits, key parameters, and co-benefits approach. Water Res 221:118755

Wanzenböck I, Wesseling JH, Frenken K, Hekkert MP, Weber KM (2020) A framework for mission-oriented innovation policy: alternative pathways through the problem–solution space. Sci Public Policy 47(4):474–489

Webb R, Bai X, Smith MS, Costanza R, Griggs D, Moglia M, Neuman M, Newman P, Newton P, Norman B, Ryan C, Schandl H, Steffen W, Tapper N, Thomson G (2018) Sustainable urban systems: co-design and framing for transformation. Ambio 47(1):57–77

Weber KM, Rohracher H (2012) Legitimizing research, technology and innovation policies for transformative change: combining insights from innovation systems and multi-level perspective in a comprehensive 'failures' framework. Res Policy 41:1037–1047

Webster K (2017) The circular economy: a wealth of flows. Ellen MacArthur Foundation Publishing, Cowes

WGBC (2020) Whole life carbon vision. World Green Building Council. https://www.worldgbc.org/advancing-net-zero/whole-life-carbon-vision

Williams J (2019a) Circular cities. Urban Stud 56(13):2746–2762

Williams J (2019b) Circular cities: challenges to implementing looping actions. Sustainability 11(2):423

Williams J (2021) Circular cities: what are the benefits of circular development? Sustainability 13(10):5725

Wise RM (2018) Key capabilities for long-term development strategies in the face of unprecedented and uncertain large-scale global change. World Resources Institute (WRI). https://www.wri.org/climate/expert-perspective/key-capabilities-transformational-long-term-development-strategies. Accessed 27 Jun 2019

Wise RM, Fazey I, Smith MS, Park SE, Eakin H, Van Garderen EA, Campbell B (2014a) Reconceptualising adaptation to climate change as part of pathways of change and response. Glob Environ Chang 28:325–336

Wise RM, Fazey I, Stafford Smith M, Park SE, Eakin HC, Archer Van Garderen ERM, Campbell B (2014b) Reconceptualising adaptation to climate change as part of pathways of change and response. Glob Environ Chang 28:325–336

Wise RM, Marinopoulos J, O'Connell D, Mesic N, Tieman G, Gorddard R, Chan J, Flett D, Lee A, Meharg S, Helfgott A (2022) Enabling resilience investment guidance version 1. Report prepared by CSIRO, Value Advisory Partners, University of Adelaide. CSIRO, Canberra

Woetzel J, Remes J, Boland B, Lv K, Sinha S, Strube G, Means J, Law J, Cadena AS, von der Tann V (2018) Smart cities: digital solutions for a more livable future. In: McKinsey Global Institute, New York, pp 1–152

Wolfram M, Frantzeskaki N, Maschmeyer S (2016) Cities, systems and sustainability: status and perspectives of research on urban transformations. Curr Opin Environ Sustain 22:18–25

WWF (2020) Australia's 2019–2020 bushfires: the wildlife toll: interim report. New WWF report: 3 billion animals impacted by Australia's bushfire crisis. https://www.wwf.org.au/news/news/2020/2023-billion-animals-impacted-by-australia-bushfire-crisis

Zukin S (2020) New York tech dossier: "innovation districts" in New York: contentious geographies of growth. Metropolitics

Usha Iyer-Raniga has combined experience working in government, private practice, and as a researcher and academic. She is Professor in sustainable built environment at the School of Property, Construction and Project Management, RMIT University. She is affiliated with the United Nations Environment Programme in France, co-leading the United Nations One Planet Network's (OPN) 10 Year Framework's (10YFP) Sustainable Buildings and Construction (SBC) Programme, arising from the Rio+20 Conference. This programme is responsible for Sustainable Development Goals (SDG) 12 on responsible consumption and production, targeting SDGs 12.1 and 8.4 associated with production and consumption, and the critical role of circular economies.

Oanh Thi-Kieu Ho is a Lecturer at the Faculty of Architecture, Building, and Planning at the University of Melbourne. She previously served as a research fellow in Circular Economy and the Built Environment at RMIT University. Prior to that, she worked as an estimator and contract administrator on various commercial and residential projects. Olivia was awarded a PhD in sustainability and built environment, focusing on developing a decision-making framework to assist in the selection of green features and technologies for office projects. Her research interests lie in circular economy, sustainability, construction technologies, and project procurement.

Akvan Gajanayake is a Lecturer at the School of Property, Construction and Project Management at RMIT University. He holds a PhD in sustainable engineering and a Master's in economics. He co-leads the policy and governance workstream of the Asia-Pacific Circular Economy Researchers Network. His research interests include systems thinking for a circular economy and the economics of CE. Before joining academia, he worked in the corporate and not-for-profit sectors for over a decade, where he oversaw waste management and resource recovery for an apparel manufacturer and the setting-up of community-led integrated waste management networks.

Part III
Green Cities and Renaturing Cities

Chapter 7
Remaking Cities: Applying New Urban-Transition Concepts and Processes to Regenerate Greyfield Suburbia

Peter W. Newton and Stephen Glackin

Abstract The mission of remaking cities to become more sustainable, productive, liveable, resilient, and inclusive is a twenty-first-century grand challenge. This chapter reports on the application of urban transition frameworks and processes in the development and implementation of a new planning model for regenerating and re-urbanising Australia's low-density, car-dependent greyfield suburbs: the established, ageing, but well-located middle-ring suburbs built in the post-war era on larger lots. Most housing in these areas has now reached the end of its service life and is prime for redevelopment. Since greyfields comprise most residentially zoned land in cities, this positions them as the critical entry points for regenerative, medium-density, compact city redevelopment. But the wrong planning models are being used. Most infill redevelopment in greyfields is fragmented, piecemeal, small-lot subdivision, delivering a low yield of new housing, significant loss of greenspace, and no added services, infrastructure, or residential amenity.

This chapter introduces greyfield precinct regeneration (GPR), the product of a set of innovative, transition-oriented planning concepts, models, tools, and processes capable of regenerating established, ageing precincts in occupied greyfields: a mission-scale challenge. It provides a blueprint for mainstreaming GPR, illustrated with a case study from a middle-suburban municipality in Melbourne that charts the urban transition from concept to implementation.

Keywords Regenerative urban redevelopment · Transition management · Municipalism · District greenlining · Greyfield precinct regeneration (GPR) · Place-activated GPR · Transit-activated GPR · Urban infill · Missing middle · Medium-density housing

P. W. Newton (✉) · S. Glackin
Centre for Urban Transitions, Swinburne University of Technology, Hawthorn, VIC, Australia
e-mail: pnewton@swin.edu.au

© The Author(s) 2025 147
N. Frantzeskaki et al. (eds.), *Future Cities Making*, Theory and Practice of
Urban Sustainability Transitions, https://doi.org/10.1007/978-981-97-7671-9_7

7.1   Introduction

A confluence of factors makes living in large, fast-growing cities a challenging proposition in the twenty-first century. They are both exogenous (external) and endogenous (local origin), and while each is important (Table 7.1, "Situation"; Newton and Glackin 2014), it is their compounding effects that will increasingly confront city governments now and into the future (Westman et al. 2022). The "great acceleration" in exogenous pressures since the middle of the last century, led by unsustainable growth in greenhouse gas emissions and the climate change it is forcing (Steffen et al. 2015), is accompanied by a set of associated threats to human settlements that include urban heating, megafires, intensified storms and rainfall, flooding, and sea level rise (Norman et al. 2021). To these can be added an accelerating growth in population and consumption of resources by cities' industries, residents, and infrastructures, reflected in expanding ecological and urban footprints (IRP 2018), financial shocks, health pandemics, forced migration, and rapid technological change associated with digitalisation and automation.

Endogenous pressures are clearly seen through life-cycle processes associated with the ageing of urban infrastructures and resident populations. Socio-demographic composition and patterns of segregation within cities are changing as waves of immigration continue to alter urban social fabrics (Newton et al. 2022a, b). While cities are the primary engines of modern economies, their negative externalities are evident in the form of social and spatial inequalities among resident populations and in the loss of urban environmental quality. Urban challenges are many and persistent (Table 7.1). In summary, cities are proving increasingly difficult to plan for, locally and globally. They are complex, dynamic places.

In this context, a recent United Nations review of international planning practices and urban governance processes in relation to sustainable spatial development (UNDESA 2021, p. 1) concluded:Many, if not most, governmental planning systems are not up to the task of addressing the economic, social, and environmental trends and issues that have emerged over time, such as in sustainable transport, net zero-carbon green and blue economies, urban farming and nature-based solutions to climate change, biodiversity loss and the general well-being of urban dwellers, just to name a few. The monitoring of the state of cities and urban areas informs us that population and economic growth or decline are not too well anticipated nor mitigated by sustainable territorial planning policies, plans and designs. Hence territorial planning and policymaking processes will need to (better and more quickly) adapt to those changes to ensure that no one, and no place, is left behind.

Clearly, overhauling urban planning is a critical mission for city futures.

Australia's series of five-yearly State of Environment Reports between 1996 and 2022 (https://www.dcceew.gov.au/science-research/soe) also highlight the growing challenges to sustainable urban development, as does the latest report on Australia's progress in achieving the UNSDGs (MSDI 2020) where all indicators for SDG11 (Sustainable Cities and Communities) are "off track" (especially material use, carbon emissions, and access to affordable housing). Academic studies of the urban

Table 7.1 Transition framework and key concepts for delivering regenerative precinct-scale medium density suburban infill in greyfield suburbs

Situation[a] Societal challenges and global urban "landscape" problems[b]	Societal-scale challenges (exogenous)	Persistent problems in large, fast-growing, low-density cities (endogenous)	Problems with urban infill policies and practices
	• UNSDGs • Greenhouse gases, climate change • Sustainable cities • Health pandemics • Hegemonic conflicts in countries and regions • Mass refugee migration	• Affordable housing • Consumption, waste generation, greenhouse gases • Urban sprawl, lack of integrated land use and transport (public/active), car dependency • Ageing centralised infrastructures • Congestion • Suburbanisation of social disadvantage • Growing urban socioecological deficits	• Failure of current greyfield planning strategies, lack of integrated vertical and horizontal planning • Lack of land supply in established suburbs • Dominance of small-lot subdivision – Lack of dwelling diversity – Loss of greenspace, tree canopy – Increased hard surfaces, runoff, heat • Limited public transport/place activation and increased density • COVID demonstrated need for localisation of urban services
Complication[a] Regimes, barriers[b]	Aligning three-tier government policy with international urban goals	State/metropolitan planning policy, land use zoning	Lack of transformative capacity in local government
	• UNSDGs • IPCC/COP26 • WHO • UNHCR • OECD • UNDESA • UNEP	• Blanket zoning removes potential for innovative suburban regeneration • Current planning schemes discourage sustainable, regenerative, and medium- and high-density builds • Developer lobby groups resist planning for regenerative urban redevelopment • "Lack of precedent" for innovative GPR projects	• Vertical fiscal and governance imbalance inhibits municipalisation of innovative planning tools • Focus primarily on zoning and development approvals, lack of tools for assessing precincts • Developer-led for profit, not community benefit • Community resistance (NIMBY) • No demonstration products or processes for NIMBY-to-YIMBY transition for projects beneficial to community

(continued)

Table 7.1 (continued)

Solution / Niche innovations[b]	Greyfield precinct regeneration (GPR): New planning model for redeveloping the "missing middle"	Transition management process—Key phases:

Transformative concepts for urban regeneration that can be implemented locally and globally
 	
• *Greyfields* as a distinct urban development arena (cf. greenfields and brownfields)
• *Greyfield precinct regeneration* as a new planning model for urban redevelopment in ageing middle suburbs
• *Scaling up and mainstreaming* following planning precedent and an established blueprint
	• "Missing middle" comprises:
	  – Established, ageing, well-located low-density *middle suburbs* with high redevelopment potential
	  – *Precinct-scale* redevelopment projects
	  – *Medium-density* housing
	• GPR comprises three linked models:
	  – *District greenlining:* Government and cross-sectoral planning and retrofitted infrastructure
	  – *Place-activated GPR:* Redesigned precinct with increased housing and green infrastructure
	  – *Transit-activated GPR:* New, quality transit technology on arterial corridors linking place-activated GPR precincts possessing increased capacity for more active and low-carbon travel
		1. *Strategic thinking, problem structuring:*
		  – Identify arena and scale (greyfields + precincts)
		  – Engage opinion leaders and front-runners
		  – Initiate shadow process for GPR
		2. *Develop transition agenda and coalition:*
		  – Assemble supporters and co-design team
		  – Scope new models and processes for GPR
		  – Develop narratives and images (what *could be*)
		3. *Co-design experiment—Pilot neighbourhoods:*
		  – Locate clusters with high redevelopment potential
		  – Devise alternative design scenarios for precinct renewal
		  – Precinct assessment: Additionality and feasibility
		  – Engage local and state governments on planning instruments
		  – Encourage citizen engagement
		  – Develop playbooks for all stakeholder groups
		4. *Implementation, monitoring, evaluation, and learning*
		  – Zones approved by minister and council
		  – Scaling up and mainstreaming to multiple jurisdictions

[a] S-C-S is a well-established framework often used in business planning presentations

[b] Landscape contexts, regimes, and niche innovations are key urban-sustainability transition concepts (after Geels 2002)

and ecological footprints of Australian cities confirm the magnitude of the challenge (Newton 2012; Sobels and Turner 2022). They highlight the fact that much post-World War II urban planning was, until relatively recently, undertaken in an era when politicians, populations, and city development practitioners alike foresaw little or no resource, economic, or environmental constraints on continued business-as-usual urban development in high-income societies (Rees and Roseland 1991; Newton 2011) and paid only lip service to environmental sustainability. Consequently, significant levels of resource consumption and inequality of access have been designed into Australian cities in the form of their housing stock (mostly detached, with large floor areas), land-use mix (significant separation of uses via zoning), transport systems (dominated by private car use), and low-density development (reflected in sprawling, car-dependent suburbs). The magnitude of the sustainability challenge of our cities, coupled with their rapid rate of growth in the twenty-first century, has resulted in an outstripping of governments' current capacity to plan appropriately for future urban development (House of Representatives 2018). The COVID-19 pandemic unleashed a further set of shocks to Australia's economic, health, employment, and urban systems. A significant shift to work from home has introduced a new dynamic in workplace-residence preferences for populations in major cities that has yet to be normalised for the longer term, but that adds stimulus for accelerated regenerative urban redevelopment of the established residential suburbs (Moglia et al. 2021; Glackin et al. 2022). For some time, the Council of Australian Governments has established high-level city performance goals for Australia's cities (Department of Infrastructure and Transport 2011)—competitive, productive, sustainable, liveable, inclusive, and resilient—and they feature as high-level performance objectives in all capital cities' strategic plans. Transformative changes are needed, however, to deliver on these objectives, and business-as-usual planning is proving inadequate.

This chapter outlines a new approach to urban redevelopment that attempts to integrate some of the more static, rigid, and well-embedded contemporary "top-down" urban planning concepts, instruments, and processes with the more dynamic, agile, and adaptive "bottom-up" participative approaches and methods characteristic of the rapidly emerging field of urban transition studies. A long-term programme of applied mission-oriented research into *Greening the Greyfields* (GtG) (Newton et al. 2022a, b) was initiated in 2011 to create a new model of planning for regenerative urban redevelopment in the greyfields: the third (along with brownfields and greenfields) and most challenging arena for urban development at the precinct scale in Australian cities (Newton 2010).

The sections that follow outline key phases of this research that have created a new model and pathway for greyfield precinct regeneration (GPR). GPR represents an attempt to redress a failure of contemporary compact-city and infill-planning strategies in the greyfields—defined as the established, ageing, well-located, car-dependent, low-density, middle-ring suburbs built in the post-war era on larger lots (Newton et al. 2020). Most housing in these areas has now reached the end of its service life and is prime for redevelopment. Since the greyfields comprise most residentially zoned land in cities, this positions them as the critical entry points for regenerative, medium-density redevelopment. But the wrong planning models are being used. Most infill redevelopment in the greyfields is fragmented, piecemeal, small-lot subdivision, delivering a low yield of new housing, significant loss of greenspace, and no added services, infrastructure, or residential amenity. GPR is a new model resulting from the application of a set of new transition-oriented planning concepts, models, tools, and processes capable of precinct-scale regeneration in the occupied greyfields. It delivers the following benefits over current business-as-usual residential redevelopment:

- Linked "bottom-up" and "top-down" planning processes.
- Increased supply of medium-density housing with a mix of typologies and price points.
- Significant environmental enhancements associated with zero-carbon energy and water-sensitive design.
- Improved non-car-based mobility and neighbourhood walkability.
- Enhanced climate resilience associated with greater surface permeability and reduced stormwater runoff and greater tree canopy coverage and greenspace.

The key transition concepts employed in the GtG research programme are summarised in Table 7.1. They are unpacked under the three well-established and readily understood headings employed in business planning presentations, generally attributed to McKinsey: "situation", "complication", and "solution". These headings closely align with the more academic multilevel transition concepts of "landscape", "regime", and "niche innovation" that were introduced by Geels (2002, 2011, 2012) and subsequently amplified into a growing field of urban transition studies, of which this chapter is part. The research reported in this chapter is among the first to have employed transition management (TM) concepts from the outset in framing a significant transformative urban planning challenge that is central to city sustainability in the twenty-first century (see Loorbach 2007 for the formative framework). As noted by Wolfram (2018), transition management represents an innovative conceptual framework that has emerged from academia but has yet to be applied in practice to cities beyond a small set of pilot-scale Living Lab studies (Marvin et al. 2018). More recently, however, the European Commission's (2022) Urban Transitions Mission is targeting 300 cities worldwide to demonstrate "integrated pathways towards holistic, people-centred urban transitions built around clean energy and innovative net-zero carbon solutions". This represents a step change in the application of transition management in practice. To date, many of the roles in transition studies are not commonly undertaken in municipal or state planning authorities, as they often challenge existing practices (e.g. requiring a focus on transformative change involving system innovation, adoption of co-creation processes providing pathways to visionary futures, and use of urban experimentation in multiple domains such as urban design, performance assessment, and stakeholder engagement) (Frantzeskaki et al. 2018).

The "landscape" in transition theory (Geels 2002) identifies the wider set of political and socio-economic trends exerting significant pressure on current urban systems. As illustrated in Fig. 7.1, the landscape defines the prevailing issues and

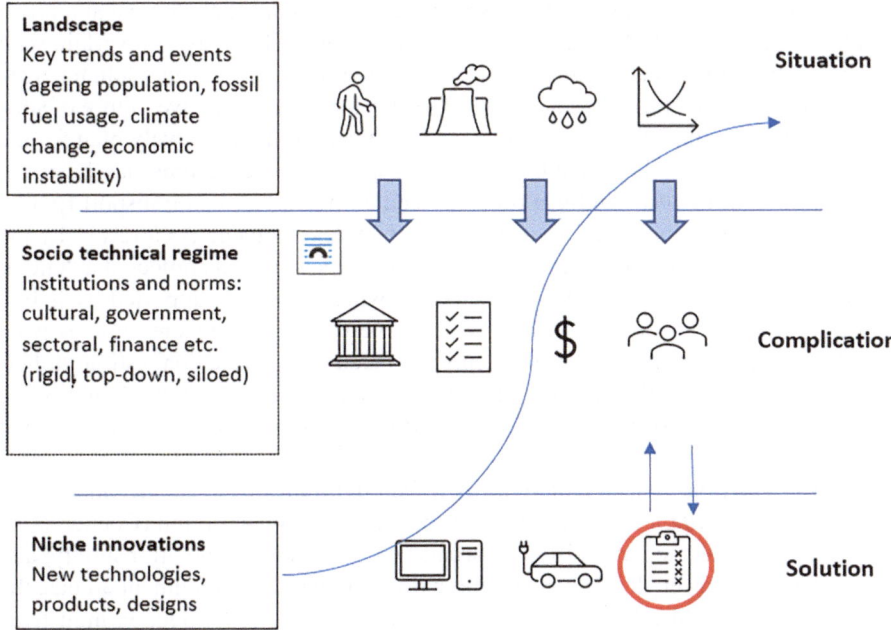

Fig. 7.1 Frameworks for representing parallel "transition" and "business change" concepts and processes

pressures that are triggering the need for change and paradigm shift. The "socio-technical regime" represents the cultural, governance, and business institutions that are supporting the current state of planning and are mostly compatible with business-as-usual practices. "Niche innovation" represents transformative technologies and processes of change capable of initiating a significant sustainability transition. Their adoption typically confronts resistance from business, government, and community regimes and commonly fails (evidence of the complete logistic S-curve is relatively rare when charting the diffusion of innovation in sustainable urban development). For the purposes of this chapter, we have labelled these three arenas "Situation", "Complication", and "Solution", signalling a parallel with transition concepts.

7.2 Situation

The most persistent urban problems have been well identified in the literature and are outlined briefly in the introduction to this chapter and in more detail in Newton et al. (2022a, b). The cluster of urban development challenges specific to this chapter encompass rapid population growth in Australia's largest capital cities that has outstripped governments' capacity to successfully plan, resulting in a lack of affordable housing, lack of land supply accessible to jobs and urban services, continued low density, car-dependent greenfield sprawl, an unsustainable ecological and urban footprint, and a decentralisation of social disadvantage. These are all overlain with an acceleration of climate change impacts that is now testing urban resilience (see Newton et al. 2018; Norman et al. 2021). There has been a successive failure of state and federal governments since the mid-twentieth century to effectively plan for the growth of Australia's largest cities; foremost among these failures has been the absence of an *integrated* land-use (housing and employment) and transport (public as well as private) blueprint for sustainable urban development. As a result, by the beginning of the twenty-first century, Australia's largest capitals reflected a "tale of two cities": the higher-amenity inner suburbs, which are well supported by public transport and access to new-economy jobs, versus car-dependent, low-density, middle-to-outer suburbia, where residents are faced with traffic congestion in their daily commute to work, schools, and shops (Latz 2021). The latter regions are becoming the focus for urban retrofitting and regeneration as they age, as outlined in the sections that follow.

Urban-transitions research also needs to be cognisant of the *geographical context* and *spatial scale* for any significant urban planning intervention (Coenen et al. 2012). Three fundamentally different urban fabrics—walking city, transit city, and auto city (Thomson et al. 2017)—and three different urban development arenas, the peri-urban greenfields, the brownfields (generally large, well-located, ex-industrial, manufacturing, or commercial sites in inner-city locations, prime for redevelopment), and the greyfields (the ageing middle-ring suburbs, where redevelopment property value lies primarily in the land rather than the buildings) (Newton 2010), should be recognised. Both greenfield and brownfield arenas have well-established sets of planning models to guide new development at the precinct scale—the scale at which the building blocks of cities have traditionally been designed (Newton 2018a, b; Thomson et al. 2019). The most challenging arena is the greyfields, given that all property here is occupied and there is significant local resistance to redevelopment, especially at high densities. All metropolitan planning strategies have a limited set of designated zones where intensified greyfield (re)development is encouraged; these generally include a hierarchy of activity centres and the land along major road corridors. The remaining greyfield residential areas are restrictively zoned to inhibit densification, essentially locking up approximately half the land in cities such as Melbourne that would otherwise have high potential for regenerative redevelopment. Consequently, knock-down-rebuild and small-lot subdivisions, a scale of redevelopment permitted in "neighbourhood" and "general" residential zones, are spreading like a virus, with their now-well-recognised negative impacts (Newton and Glackin 2014; Newton and Glackin 2020; Newton et al. 2022a, b). To transition to a more sustainable urban future, GPR is presented as a niche innovation for regenerative urban redevelopment.

In short, GPR is a new planning concept, model, and process capable of delivering more sustainable urban redevelopment in the greyfields. But as a niche innovation, the new *product* requires additional collaborative and applied R&D to overcome the multiple regime barriers that are currently in operation in order to for it to be mainstreamed.

7.3 Complication

No planning precedents existed to support the introduction of GPR projects in the bulk of residentially zoned middle suburbs in Australian cities when the *Greening the Greyfields* programme began in 2011. New and innovative urban development concepts and proposals that do not "fit" with existing planning schemes are typically confronted by legislation and controls that dictate what development activity can be undertaken where. They are the instruments of established *regimes* led by governments and are typically supported by the property industry and their industry associations, a majority of whose members are averse to change—especially transformational change (in innovation-diffusion nomenclature they are classed as "followers" or "laggards" rather than "leaders"). Residents also use local zoning and building codes to oppose change in their local neighbourhoods, a "not in my back yard" (NIMBY) response to proposals for increased housing density and population (Newton et al. 2017).

The initial challenge for the GtG research programme was to establish a transition management process and team to undertake the tasks associated with fully scoping the challenges and translating the GPR concept and vision into an implementable project. This drew significantly on the work of Loorbach (2007), who established the following phases and categorisation of tasks:

- *Strategic*: structuring the problem, identifying innovation arenas, envisioning future outcomes, engaging front-runners, and conducting strategic discussions
- *Tactical*: envisaging policy changes, conducting experiments, designing new model(s) for intervention, assessing future scenarios, devising new narratives for change, managing risk, and establishing a transition agenda and coalitions
- *Operational*: mobilising a co-design/co-production team, executing (pilot) projects, monitoring, and evaluating
- *Communicating*: publicising, scaling up, and mainstreaming

The initial team comprised a small core of Australia's leading researchers in urban planning, design, and construction capable of winning competitive grants to fund the early phases of the research (see Acknowledgements). They instituted a "shadow process" often employed in transition studies that involved a "network of people who are working both inside and outside of the dominant system to explore alternatives that replace the dominant system when there is a window of opportunity" (Wutich et al. 2020, p. 28; also, Newton 2018a, b). The first grant, from the Australian Housing and Urban Research Institute (AHURI), funded a 12-month series of investigative panels that would engage widely with key stakeholders from government, the built environment and property industry sector, and community organisations in a series of workshops to more deeply explore the new planning, design, and construction concepts associated with GPR. The resulting report (Newton et al. 2011) contains significant feedback on such topics as defining greyfields (a new concept at that stage), determining what constitutes a "precinct", the current inhibitors to regeneration of the middle suburbs at precinct scale, a

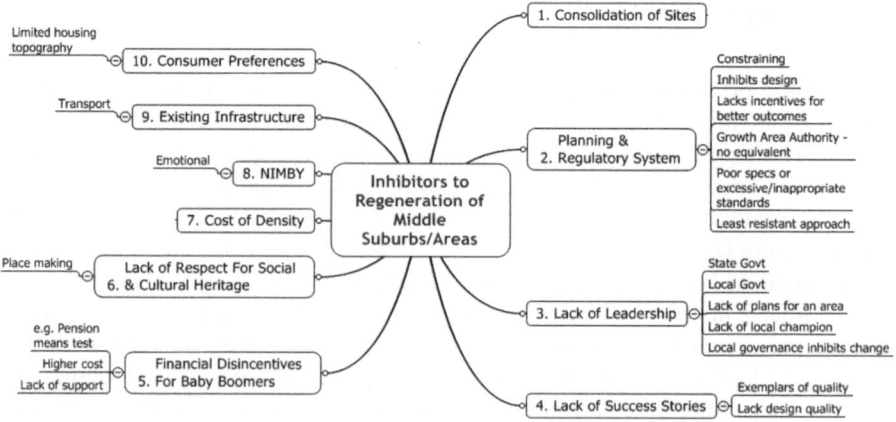

Fig. 7.2 Inhibitors to regeneration in the middle suburbs. (Source: Newton et al. 2011, p. 111)

household needs analysis (who is wanting to live in medium-density housing in the middle suburbs), how to ensure change happens, viable solution pathways, investor and resident concerns, creating a narrative, key delivery agencies, current planning barriers, and financial challenges. The key challenges identified from the first stakeholder engagement are shown in Fig. 7.2 and column 2 of Table 7.1.

While regimes can often be inhibitors of change, Loorbach (2010) correctly points out that they can also represent the enabling environment for facilitating and legitimating a transition. Within the context of a regime, strategies need to be developed with key stakeholders and "front-runners" in respect of the transition arena under examination and the nature of transformation envisaged (in this case, the transformation of greyfields through GPR). Four *innovation arenas* were identified from this first transition management phase that would need further applied research to validate and reduce the inherent risk in the GPR planning model and to provide clear pathways for its promotion and implementation by the state government's metropolitan planning authority, local government planning officers and elected councillors, local residents, and property developers. The innovation arenas central to GPR implementation posed the following key questions: *Where* to focus? *What* to design into the precincts identified for regeneration? *How* to achieve this urban transition? The answers and outcomes provided a solution to the challenge of tackling greyfield regeneration at the precinct scale. A fourth innovation arena (*Who?*) was the focus of an extensive survey of residents in greyfield suburbs of Melbourne and Sydney (see Newton et al. 2017). The survey confirmed that residents in Australia's largest cities are ready for this transition, with equivalent proportions of households preferring to live in medium-density accommodation in established areas well served by public transport versus those preferring detached housing in car-dependent outer suburbs with service deficits (Newton et al. 2017, 2022a, b). The principal challenge is how to satisfy the demand for more sustainable, medium-density housing in the occupied, greyfield, middle suburbs.

7.4   Solution

The objective of the *Greening the Greyfields* applied research programme has been to create new concepts, processes, and instruments capable of initiating a suburban-to-urban sustainability transition in locations where infrastructure and housing are ageing, the land has high redevelopment potential, and there is a significant capacity to accommodate regenerative precinct-scale projects—not business-as-usual, fragmented, suboptimal, single-lot, knock-down-rebuild, as is currently the case (Fig. 7.3). By strategically and purposefully remaking the city "one neighbourhood at a time" (Thomson et al. 2019), GPR can become the preferred new product and process for local governments, property developers, and residents in established suburbs to transform the places where most urban populations live and increasingly work. With an explicit medium-density focus, greyfield regeneration also increases housing supply. This is the vision that has driven this transition study from the outset (Newton 2010).

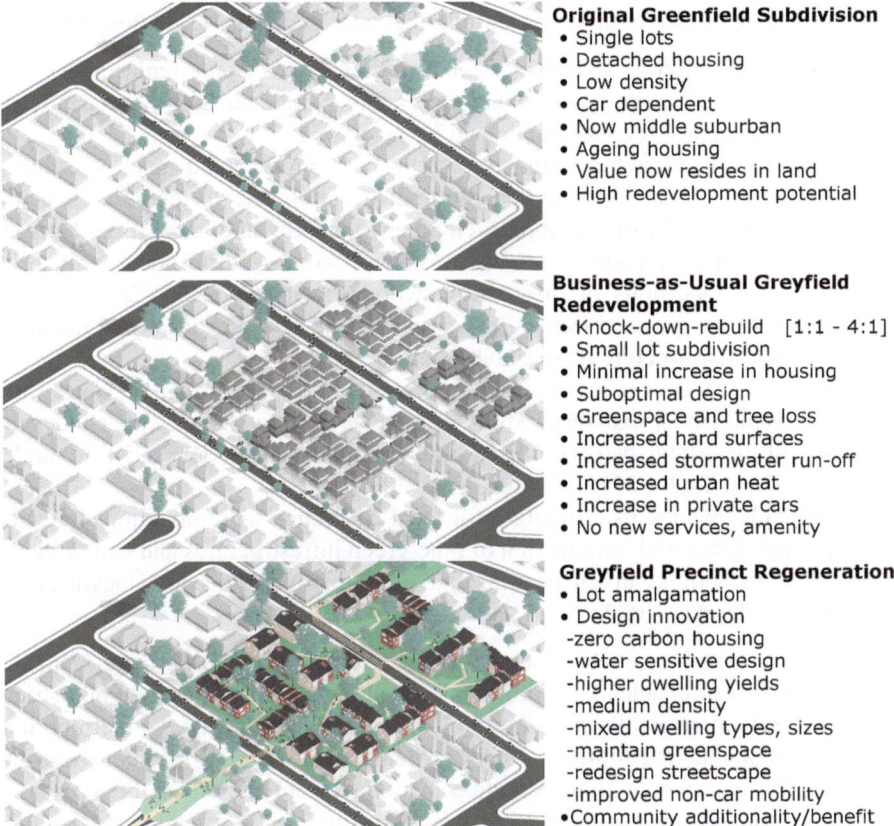

Original Greenfield Subdivision
- Single lots
- Detached housing
- Low density
- Car dependent
- Now middle suburban
- Ageing housing
- Value now resides in land
- High redevelopment potential

Business-as-Usual Greyfield Redevelopment
- Knock-down-rebuild   [1:1 - 4:1]
- Small lot subdivision
- Minimal increase in housing
- Suboptimal design
- Greenspace and tree loss
- Increased hard surfaces
- Increased stormwater run-off
- Increased urban heat
- Increase in private cars
- No new services, amenity

Greyfield Precinct Regeneration
- Lot amalgamation
- Design innovation
 -zero carbon housing
 -water sensitive design
 -higher dwelling yields
 -medium density
 -mixed dwelling types, sizes
 -maintain greenspace
 -redesign streetscape
 -improved non-car mobility
- Community additionality/benefit

Fig. 7.3 Suburban-to-urban transition with greyfield precinct regeneration

7.4.1 Where to Intervene?

Successfully producing long-term metropolitan policies, strategies, and plans capable of directing future urban development and redevelopment in an integrated fashion remains a challenge in terms of both horizontal planning (across provision of housing, transport, energy, water, waste, and social services) and vertical planning (across three tiers of government and local communities). Identification of where and how to intervene and at what scale is especially challenging in greyfields.

7.4.1.1 District Greenlining

Opportunities for GPR involving local housing and infrastructure redesign and regeneration are ideally signalled by the concept of *district greenlining* in metropolitan and municipal strategic plans (Newton et al. 2022a, b). District greenlining represents a first step in outlining the intention to regenerate a particular locality or series of localities, a process requiring vertically and horizontally integrated planning. It is an approach to urban retrofitting that addresses several challenges:

- Infrastructure ageing and obsolescence requiring plans for replacement that reflect the service-life performance of system components and the nature of asset-maintenance regimes (Iselin and Lemer 1993).
- Increased demand on local infrastructures—both physical and social—resulting from changes in land use and population distribution such as increased housing density associated with urban redevelopment generally and, most recently, from COVID-induced impacts of increased rates of working from home in suburbs with high proportions of information workers (Glackin et al. 2022). These changes have exposed a lack of infrastructure and service capacity in favoured locations (Infrastructure Australia 2020).
- Emergence of distributed infrastructures associated with energy, water, and waste management that successfully operate at precinct scales, either independently of the established centralised infrastructure systems or as a hybrid model (Newton and Taylor 2019; Newton 2019, 2022a, b).
- Climate change and its challenges for local adaptation and mitigation strategies requiring more rapid introduction of blue-green infrastructures and nature-based services to address urban heating and flooding (Newton and Glackin 2020; Frantzeskaki et al. 2022).

District greenlining enables neighbourhood regeneration to be targeted in the middle suburbs. It enables GPR projects to be attracted to and nest within districts that have been strategically identified in larger-scale and longer-term metropolitan and municipal planning strategies for urban densification and infrastructure retrofitting. Ideally, district greenlining should be undertaken collaboratively between state and municipal planning authorities and major utilities as a necessary first step in identifying future strategies and timetables for major infrastructure retrofitting across the metropolitan area. District greenlining will require better cross-sector

Fig. 7.4 Greyfield precinct regeneration with its core components: district greenlining, with nested place-activated regeneration precincts and transit-activated corridor regeneration precincts—subject to rezoning

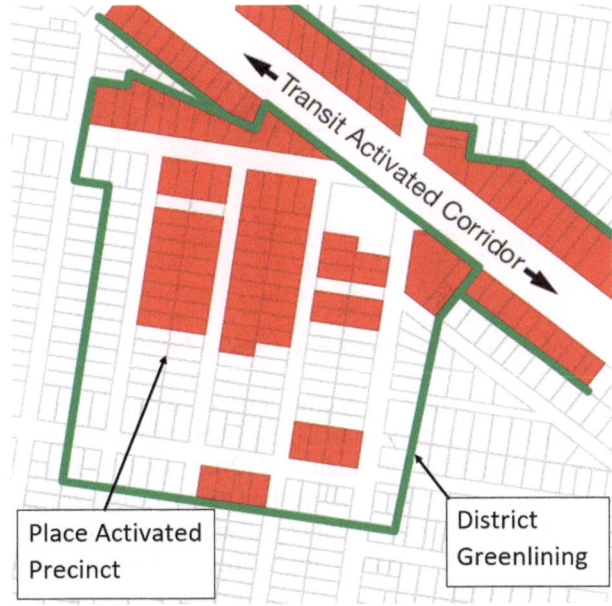

Place Activated Precinct

District Greenlining

collaboration and new modes of urban governance. In response to this challenge, the CRC for Water Sensitive Cities has developed principles and strategies capable of integrating urban and water planning in any given spatial context (see Malekpour et al. 2020; Chesterfield et al. 2021). In the absence of state-municipal-utility-level collaboration, however, future strategic planning by local governments needs to incorporate a district greenlining process to identify localities where change is required to increase housing density within their jurisdiction and where place-activated and transit-activated GPR (PA-GPR and TA-GPR, respectively) projects are to be encouraged (Fig. 7.4).

7.4.1.2 Transit-Activated Corridors and Greyfield Regeneration

Transit-activated corridors represent a second strategic planning plank in the regeneration of post-World War II, car-dependent suburbia. Major transport corridors have been advanced as a focus for medium-rise, higher-density development. The requirements for this to work, as set out by Adams (2009), include prescriptive zoning controls over key aspects of corridor development, including upfront "as of right" development to levels of between four and eight storeys. Much of this model is being implemented along the inner tram corridors of Melbourne and is now moving into middle suburbs along major arterial roads.

However, large parts of inner, middle, and outer suburbs remain without quality transit options. Main roads (often created by the removal of original tram lines following the end of World War II) are usually heavily congested with traffic and have had reduced urban value as a result.

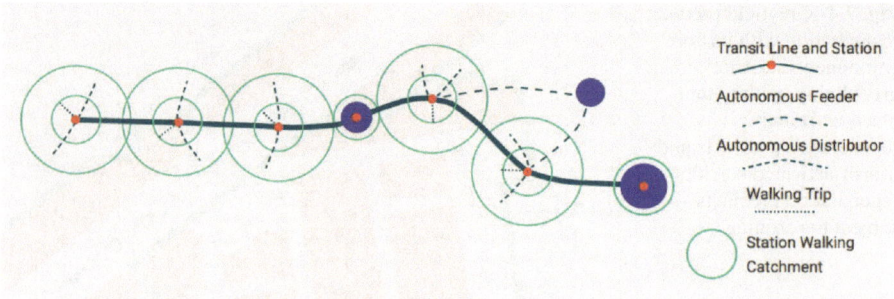

Fig. 7.5 Transit-activated corridor. (Source: Thomson et al. 2019)

The need to both regenerate the mobility and redevelop the land along such roads is the next big agenda in transport and urban policy (Newton et al. 2022a, b). A solution advanced by (Newman et al. 2021; Caldera et al. 2022) for regenerating main roads in the car-based established suburbs uses transit activated corridors (TACs) *and* transit-activated greyfield precinct regeneration (T-A GPR around the resulting station precincts. It integrates quality mid-tier transit technology such as trackless trams (Newman et al. 2019)) with quality precinct-scale land development on, in, and around transit stops, including last-mile integration (Fig. 7.5). TACs are thus a corridor created from current car-oriented activity centres by linking them like pearls on a string. The key to unlocking T-A GPR is to emphasise the benefits that flow to local communities: they get more than just infill housing—they get a transit service and other urban services within the transit-activated precinct (such as net-zero distributed electricity generation). This is "additionality", a critical factor recognised as missing to date in recent greyfield infill and a key to creating more extensive greyfield precinct regeneration.

7.4.1.3 Locating Precincts with High Potential for Regenerative Residential Redevelopment

The next step for creating more extensive GPR involves identification of prospective precincts: concentrations of residential property where economic value lies predominantly in the land rather than the ageing built asset and where there are prospects of delivering a dividend to individual property owners *and* local communities from lot assembly and neighbourhood activation. In metropolitan Melbourne, there are over 600,000 dwellings where 70% or more of the value of the property is vested in the land. A software tool (ENVISION; Newton and Glackin 2013) was specifically developed to identify the location of properties with high redevelopment potential, employing multi-criteria analyses that incorporate the following:

- The employment of market-based analytics related to financial feasibility and housing yield. These constitute the principal instruments for developers in preliminary analyses.
- Local planning priorities, such as proximity to schools, public transport, and parks, depending on assessment criteria selected (Fig. 7.6). The tool provided the

Fig. 7.6 ENVISION process for assessing the redevelopment potential of localities: (**a**) Market analysis of property uplift potential. (**b**) Multi-criteria local government planning assessment of where to focus increased densification and regeneration (via GPR development overlays)

evidence base to assist with the identification of the planning amendments (C134maro and 136maro available at https://www.planning.vic.gov.au/schemes-and-amendments) for the two GPR pilot precincts in the City of Maroondah (a middle-suburban municipality 25 km from Melbourne's CBD). This was a key phase in the statutory underpinning for GPR.

- The determination of propensity to collaborate as part of lot amalgamation among neighbouring property owners required for the full implementation of GPR. This requires the GPR broker (either an arm of local government under municipalism strategies or the private sector working closely with local government) to acquire an additional layer of information regarding resident's future mobility/location intentions.

7.4.2 What to Design into Place-Activated Greyfield Regeneration Precincts

There is significant community resistance to change in the established greyfield suburbs of Australian cities, especially those associated with densification, typically eliciting a NIMBY response. Residential zoning systems currently operated by state and municipal planning authorities provide significant barriers to regenerative redevelopment initiatives, as outlined in detail in Newton et al. (2022a, b)) and summarised below. Instead, planning authorities support a knock-down-rebuild approach that is piecemeal in scope, scale, and performance across much of suburbia. To be successful, the urban designs developed for PA-GPR projects need to be demonstrably superior to those currently being employed in knock-down-rebuild, providing clear evidence of multiple community and property-owner benefits—encouraging a "yes in my back yard" (YIMBY) response. The innovation arenas involved in PA-GPR include the following:

- *Mid-rise, medium-density dwelling typologies.* This represents the original definition of the "missing middle" (Parolek 2020), providing a more acceptable housing transition for populations living in suburbs currently dominated by ageing detached dwellings. Apartment developments above four storeys deliver a level of densification that is not in harmony with a majority of residents in these areas and typically attracts a NIMBY response. New medium-density designs, however, demonstrate what *could be* for the future, in that they offer greater liveability and sustainability: zero-carbon, water-sensitive (rainwater harvesting and wastewater recycling), healthy, quiet, smart, secure, and accessible—long-promised performance objectives (Newton 2002), now realisable (Newton et al. 2022a, b).
- *Precinct-scale residential redevelopment.* This is relatively rare in greyfields, compared to greenfields (dominated by low-density "master-planned" precincts) and brownfields (where high-rise apartment precincts concentrate) (see Newton et al. 2020). *Precinct-scale* medium-density housing projects, however, are a

necessary requirement for greyfield *regeneration* and constitute what we consider to be the second dimension of "missing middle" development.

- *Reconfigured precinct streetscapes.* This enables redistribution to community use of space previously devoted to cars, via a meshing of previously "private" and "public" spaces (footpaths, verges, and sections of road), thus reactivating the neighbourhood for local residents (Murray et al. 2015) and compensating for some loss of open space and tree canopy due to densification (Newton et al. 2020). This increases both environmental resilience (to increased heat and rainfall) and local amenity.
- Introduction of *distributed green infrastructures* capable of operating at the precinct scale (see Smith et al. 2021). These include renewable energy systems; integrated water systems (potable plus rainwater harvesting, wastewater recycling, and stormwater capture and recycling); zero waste to landfill as a result of ensuring waste separation for glass, food, and garden organics (FOGO), mixed recyclables, and household rubbish; and shared low/zero-carbon mobility services. This delivers the regenerative capacity of GPR and its connection to a city's circular economy.
- *Visualisation* of "what could be" versus "what is", as a critical part of community engagement (Dixon and Tewdwr-Jones 2021; Fig. 7.3; see also the illustrations at https://greyfields.com.au/; Glackin and Newton 2016).
- *The application of a suite of digital precinct-assessment tools* capable of examining proposed designs against benchmarks that can establish the extent to which they improve upon the built environment that is being replaced. Such assessments provide an evidence base for local councils and communities to evaluate the level of dividend (benefit) that can flow from proposed precinct-scale regenerative redevelopment and will be instrumental in determining the level of support and incentive that local (and state) governments can contribute. A suite of eco-efficiency performance-assessment and visualisation tools for use at a precinct sketch-planning stage have been developed by the CRC for Low Carbon Living (Newton et al. 2013; Newton and Taylor 2019; Xing et al. 2019) and the CRC for Water Sensitive Cities (Moravej et al. 2022) and have been applied to new GPR pilot projects in Melbourne, demonstrating significant levels of community additionality compared to business-as-usual at the building and neighbourhood scale (Newton et al. 2020). These results were central to the City of Maroondah proceeding with rezoning these precincts (as discussed in the next section). These tools represent a new generation of object-based, precinct information modelling-enabled tools that will be critical in driving an urban-sustainability transition more generally (Newton 2019).

7.4.3 How to Deliver Greyfield Regeneration at the Precinct Scale

Realising GPR as a new urban-infill planning model (a niche innovation) involved a TM process that employed a *new governance framework*. The stakeholders

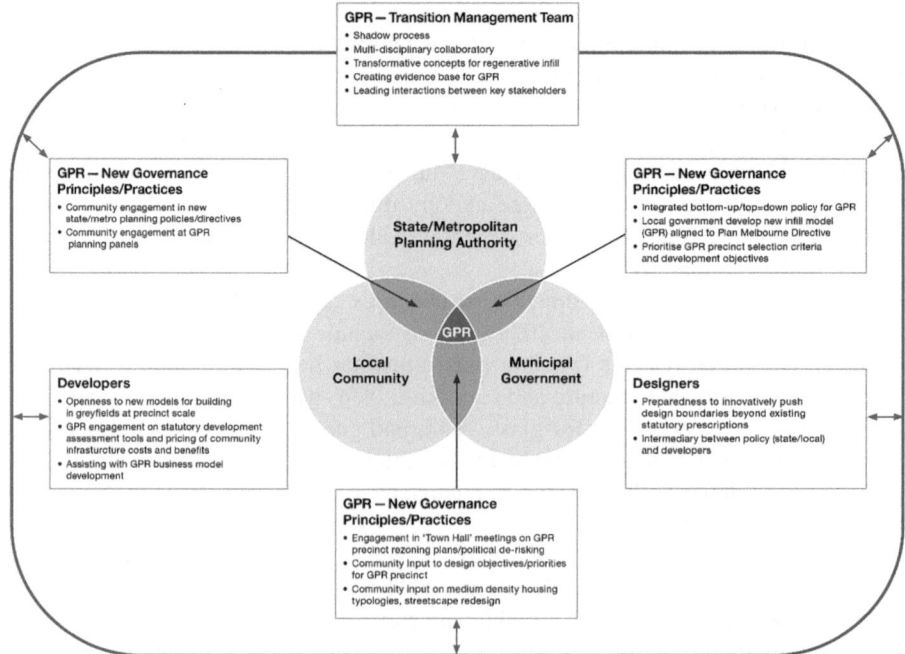

Fig. 7.7 New governance model involving key actors, principles, and practices in greyfield precinct regeneration, as developed for Greening the Greyfields

represented were not new, but several of the principles and practices that emerged between these actors, in the course of TM engagements, were (outlined in sketch form in Fig. 7.7 and in the sections below). A TM Team was formed following the publication of the Greyfields concept paper (Newton 2010). It created a shadow network for GPR: an "informal network of people who are working both inside and outside of the dominant system, who facilitate information flows, create nodes of expertise, identify knowledge gaps, engage in social learning, and explore alternatives that could replace the dominant system when there is a window of opportunity" (Wutich et al. 2020, p. 28). The team comprised leading urban researchers, thought leaders, and "front-runners" in the field who had a shared vision of "what needed to change" in Australia's largest cities to sustainably plan for and accommodate high rates of population growth and associated housing development. Early tasks involved scoping a programme of applied research (see Newton et al. 2011) capable of attracting funding and providing an evidence base for GPR that demonstrated its fitness for purpose.

Extensive publication in peer-reviewed academic journals and books (most of these are listed in Newton et al. 2022a, b) as well as presentations at conferences and workshops saw the concepts of greyfields and GPR appearing in federal (Australia State of the Environment Report 2016; House of Representatives 2018) and state government publications (MAC 2015; Department of Environment, Land-Use, Water and Planning 2017), recognising them as new planning concepts. The remaining challenge was for GPR to materialise as a viable, implementable product: facilitating precinct-scale regenerative redevelopment projects in the greyfields.

7.4.4 Engaging with Metropolitan Strategic Planning

Regular contact was maintained from the outset with senior planners in Victoria's Ministry of Planning who had been associated with most of the recent "long-term" strategic plans for Melbourne (Newton 2018a, b). These involved both informal and formal meetings and research presentations, the most significant of which was to a Ministerial Advisory Committee established to advise on planning policies and strategies for potential inclusion in the next metropolitan plan. Consequently, "greyfield renewal" was identified as an Option for Discussion in the next Plan Melbourne (MAC 2015, Recommendation 21, p. 57). The ensuing *Plan Melbourne 2017–2050* (2017, p. 72) contained a new Policy Directive (2.2.4) that was focused on greyfield redevelopment:

> Greyfield sites are residential areas where building stock is near the end of its useful life and land values make redevelopment attractive. Melbourne has many residential areas that qualify as greyfield sites, particularly in established middle and outer suburbs.
>
> These areas often have low-density, detached housing on suburban-sized allotments that have good access to public transport and services.
>
> Up until now, the redevelopment of these areas has been generally uncoordinated and unplanned. That must change. Greyfield areas provide an ideal opportunity for land consolidation and need to be supported by a coordinated approach to planning that delivers a greater mix and diversity of housing and provides more choice for people already living in the area as well as for new residents.
>
> Methods of identifying and planning for greyfield areas need to be developed. A more structured approach to greyfield areas will help local governments and communities achieve more sustainable outcomes.

This represented a clear message to local governments (and their resident communities) regarding areas where change needed to happen and the importance of municipalities undertaking strategic planning to become more proactive change agents in regenerative suburban densification.

7.4.5 Engaging with Local Government Strategic and Statutory Planning

In parallel with its engagement with Victoria's metropolitan planning agencies, the TM Team explored potential GtG partnerships with local governments in Melbourne's middle suburbs to ensure a "joined up" approach to GPR and a willingness on the part of councils to institute precinct-scale rezonings and develop pilot projects. Without the full collaboration of a municipal government, where the planning officers, CEOs, and councillors are all aligned, there is little prospect of introducing GPR in their jurisdiction. Currently, for many metropolitan local governments in Australia, councillors are elected by residents on platforms of "opposing overdevelopment" and "protecting neighbourhood character". In 2013 the TM Team began a partnership with the City of Maroondah as a collaborator in a research consortium extending to 2022 (see Acknowledgements for the list of financial supporters of this $3 million research programme). This represented an attempt to drive what has been called "New Municipalism" in local government planning (Sareen and Waagsaether 2022), where councils take ownership of issues and interventions, and, as demonstrated in this chapter, can facilitate co-created lot assembly among neighbouring property owners, providing the basis for GPR projects (potentially also co-designed and co-developed along development pathways appropriate to the neighbourhood, its resident population, and its spatial context).

The City of Maroondah is a middle-ring municipality of Melbourne with a population of 120,000 and a significant stock of greyfield housing (24,402 out of 52,998 properties, 46%, where the total value is 70% or more in the land compared to the building, making this housing stock ripe for redevelopment). To take advantage of this (i.e. to transform individual lot-by-lot redevelopment into GPR) requires significant changes to the planning regime as well as to current development business processes. However, there is a lack of transformative capacity (Wolfram et al. 2019) in Australia's local governments, reflecting the significant vertical fiscal imbalance between the three tiers of government involved in city development (Productivity Commission 2012). The TM Team was able to fill this gap by simultaneously engaging with state and council planning authorities on a range of issues and tasks (such as advanced urban analytics, participatory stakeholder workshops, and urban design charettes). A high-level overview of the timeline of key tasks is presented in Fig. 7.8.

These processes guided the creation of new statutory outcomes through a Development Plan Overlay (DPO), combined with a Developer Contribution Plan (DCP) necessary to capture a proportion of development/project value uplift to be used to deliver broader-based benefits linked to the immediate local community. A precinct plan that incorporated dwelling design and the precinct additionality elements was spatialised as a visual plan for the overlay. The schedule to the overlay (the text defining the rules and obligations) enshrined the preferred types of development. The design guides and other relevant information were placed into the scheme as incorporated documents and reference documents. The package of all documents was drafted within the (state provided) planning-provisions template and presented to the state planning authority to be considered as an amendment to the Municipal Planning Scheme.

Together these documents cover the explicit outcomes developers must deliver to comply with the desired planning outcomes. This is largely a coded version of the design guide. Should developers comply with the code, and if lot amalgamation has occurred, one additional storey is provided (on lots over 2000 m², enabling the development of four-storey buildings to 14 m), and third-party objections are removed. The removal of third-party rights will be granted once residents have been engaged and have had the opportunity to object during the advertising process of the new overlay. Objections after its passing are therefore considered invalid. These

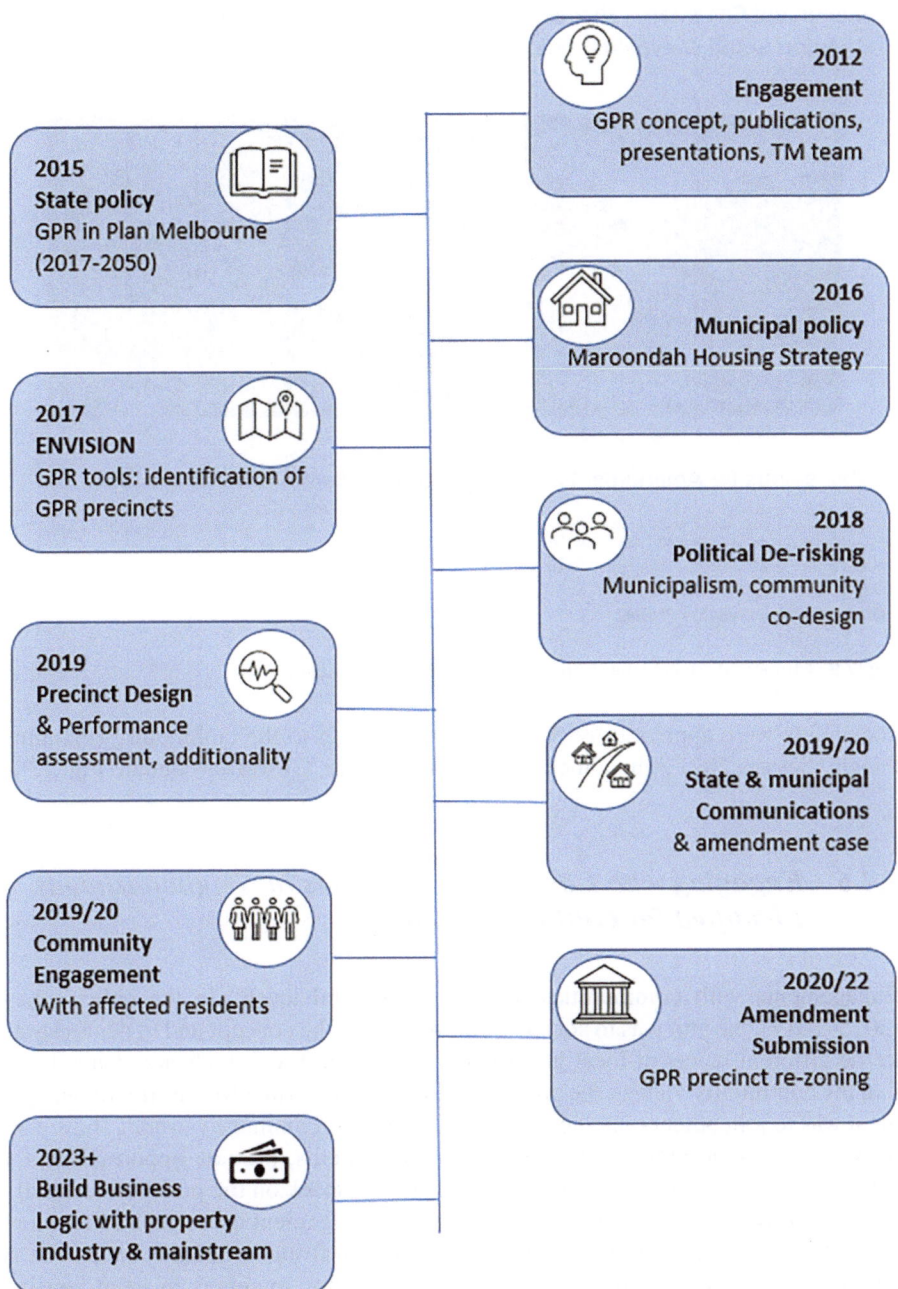

Fig. 7.8 Key stages in the Greening the Greyfields programme and implementation of greyfield precinct regeneration

Amendment C134maro - Ringwood Greyfield Precinct and Amendment C136maro - Croydon South Greyfield Precinct

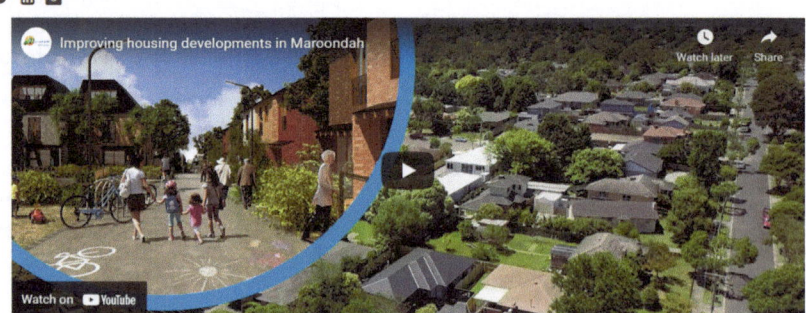

Submissions for Amendments C134maro and C136maro have now closed.

Council has been working with the Centre for Urban Transitions, Swinburne University since 2015 to apply the greyfield renewal concept in two selected precincts within Maroondah (in Ringwood and Croydon South).

During this time, several community engagement workshops have been conducted, building design guides prepared and feasibility analysis undertaken, to help determine new planning provisions.

Greening the Greyfields Project

Fig. 7.9 Municipal advertising of the GPR planning amendments

documents were approved by council for advertising to the community (Planning Panels Victoria 2021 Amendments C134-Maroondah, C136-Maroondah; Fig. 7.9).

7.4.6   Engaging with Community Residents in Neighbourhoods Identified for GPR and Rezoning

Engagements with communities occurred at several levels. In the policy years (2015–2017), the aim was to socialise community to the concept and to demonstrate to the different layers of local government (senior staff, CEO, elected councillors) that the community viewed the project positively. The format for this tier of engagement was to join other council programmes at various community events, show residents the strategic aims of the project and its benefits, provide opportunities for Q&A, and ask community members to provide feedback on the project to council.

In the move to project activation, the form of engagement necessitated a far more concerted approach, requiring significant co-design from community members. A community advisory group was established, including members of local housing and sustainability community groups, business lobbyists, developers, real estate agents, and a few concerned individuals who responded to a municipally published expression of interest. The TM Team worked with this group for a year, meeting once per month to tackle a range of issues, including jointly agreeing on pilot precincts and context-specific additionality, validating municipal decisions on the project, discussing built-form and open-space objectives, and determining how best to deliver the project to local landowners with respect to narrative, communications, and outreach. The outcomes from this phase were a clearly aligned set of community and municipal objectives, pilot areas, and delivery mechanisms. This phase also created community representatives and champions for the project, many of whom later provided support at open-house events and planning-panel hearings.

The last phase of engagement occurred at open-house events held prior to the submission of the planning amendment; the aim was to provide all information to landowners in the affected precincts and to address issues at the personal level rather than the legal level. This was the most complex set of events, as, being a voluntary municipal activity outside of legislated engagement, as well as the last opportunity to ease the passage of the amendments, it required significant consideration by council communications and legal departments. Furthermore, it required delivery tailored to answer often quite explicit questions by a potentially irate constituency. The outcome from this phase was generally positive, with many of the community concerns answered. However, a small number of landowners voiced considerable concern about the ongoing "overdevelopment" of their locale, regardless of future business-as-usual developments delivering inferior performance without GPR. However, these events ensured that all residents were aware of the project and of the opportunity to participate in a voluntary amalgamation of their land parcels with neighbours to realise better outcomes.

Given that there were several "holdouts" among residents in the two designated precincts following the advertising of the planning scheme amendments, the case was submitted by Maroondah Council to Planning Panels Victoria (PPV), a body that independently assesses planning proposals and major projects by considering submissions, conducting hearings, and preparing a report for the State Government's Planning Minister. Specifically, this case concerned an application by the City of Maroondah to rezone two precincts in their municipality to permit more sustainable, medium-density redevelopment. The resulting PPV Report (2021, p. i) concluded:The Panel supports the Amendments overall as they provide for net community benefit and sustainable development. The precincts are suitable locations for medium density renewal and the urban form sought responds to context, encourages housing diversity and manages amenity. Part of the Panel's consideration, though not underpinning its conclusion, is the pilot nature of the Amendments, lack of precedents to reference, and risks association with the vision not eventuating. The vision relies on landowners collaborating to assemble lots and developers gaining sufficient economic return. The Panel considers there is demonstrable support for the concept to suggest it is implementable and that the risks associated with the vision failing to be realised are acceptable.

In November 2022, the Minister for Planning in the Victorian Government agreed with the Planning Panel's recommendations and approved amendments to rezone both greyfield precincts (C134 and C136) in the City of Maroondah (Victorian Government Gazette 2022). A precedent has now been established in municipal planning whereby local government planners can adopt a more proactive leadership role in co-creating the basis for a more sustainable future for the nation's established middle suburbs, which are the prime targets for redevelopment at medium density.

7.5   Conclusion

In statutory planning, lack of precedent represented the principal complication for implementation of a niche urban innovation, specifically greyfield precinct regeneration. This has now been addressed with the creation of a government-endorsed blueprint and set of tangible processes and tools capable of guiding this mission-scale urban transition process.

The *Greening the Greyfields* programme of collaborative applied research established a positive vision of potential changes to the urban form and fabric of middle-suburban low-density greyfield suburbs, which possess high redevelopment potential. These changes can encourage a transformation of attitudes from NIMBY to YIMBY, suburban re-urbanisation, and better prospects for the evolution of more resilient 20-min neighbourhoods.

The application and extension of urban-transition concepts and transition management processes played significant roles in helping frame solutions to the mission of planning more compact cities via more regenerative urban infill projects that target well-located greyfield suburbs that are prime for redevelopment. The window of opportunity for a sustainable urban transition is narrowing, however, so long as small-lot subdivision and knock-down-rebuild remains the sole model for greyfield redevelopment. The pace of subsequent scaling up of GPR within Melbourne and other cities will dictate the extent to which this new model of urban infill will be able to make a positive impact on the situation that currently faces large, fast-growing, low-density cities: their future sustainability, liveability, and resilience.

We have been successful in developing pathways for reducing the risk of GPR in the eyes of the principal actors in the state of Victoria: state and metropolitan planners, municipal authorities, and local communities (Fig. 7.10). This enables the remaining key players in the property-development industry—urban developers, designers, and financiers—to be in a better position to engage and focus on developing and delivering actual GPR products on the ground (beyond the pilot projects described in this chapter). Tailored to the opportunities and urban contexts offered in different urban jurisdictions, the next phase of *Greening the Greyfields* will be to mainstream GPR.

Greyfield regeneration constitutes a considerably larger, more complex, and more pervasive challenge for Australian cities in the twenty-first century than brownfield regeneration did four decades previously. The federal government realised then that it needed to intervene and take a leading role in developing a national programme and framework for brownfield renewal. *Building Better Cities* was the programme to emerge, operating from 1991 to 1996. It included all three tiers of government, across all states and territories together with partnerships and projects involving the building, construction, and property-development industry. A new multi-stakeholder model was created and implemented, and brownfield revitalisation is now a reality in all major cities across Australia (see Newton and Thomson 2017).

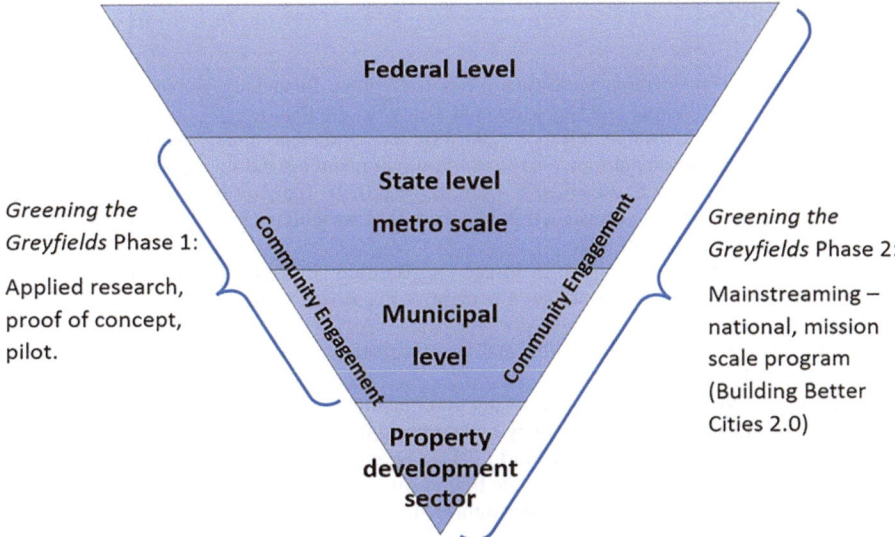

Fig. 7.10 Shift from niche to mission-scale planning of greyfield regeneration

The next step for *Greening the Greyfields* will involve mainstreaming GPR with federal government and additional state government involvement, as well as engagement with a larger group of progressive municipalities prepared to embark on longer-term strategic planning involving precinct-scale *regeneration*—not business-as-usual, piecemeal, single-lot *redevelopment* (Fig. 7.10). This will provide opportunities for the emergence of a new category of property developer specialising in GPR within a public-private partnership arrangement that can reduce the risk associated with the much needed larger, precinct-scale projects in the greyfields. *Building Better Cities 2.0* would constitute a mission-scale response to delivering much needed housing and urban regeneration in the ageing middle suburbs of Australia's fast-growing cities.

Acknowledgements We acknowledge the research funding that Greening the Greyfields has received since its inception in 2011 from the Australian Housing and Urban Research Institute, CRC for Spatial Information, Australian Urban Research Infrastructure Network, CRC for Low Carbon Living, Australian Government Smart Cities and Suburbs Program, Victoria's Ministry of Planning, and the City of Maroondah. We have also greatly benefitted from the involvement of editors and reviewers associated with leading journals and monographs who have published our work (a comprehensive list is in the 2022 Palgrave Macmillan open-access book: Greening the Greyfields: New Models for Regenerating the Middle Suburbs of Low-Density Cities; https://link.springer.com/book/10.1007/978-981-16-6238-6). Along the way, we have benefited from the contributions of several of Australia's leading academics (Professor Peter Newman, Professor Shane Murray, Professor Ron Wakefield), and a visit to Melbourne in 2012 from Professor Derk Loorbach, funded by the Project. Finally, we gratefully acknowledge the Planning Institute of Australia (Victoria) for awarding Greening the Greyfields its 2021 President's Award for Research Excellence.

References

Adams R (2009) Transforming Australian cities for a more financially viable and sustainable future. City of Melbourne and Department of Transport, Melbourne

Australia State of the Environment Report (2016) Built environment. https://soe.environment.gov.au/theme/built-environment/topic/2016/current-urban-planning-and-management

Caldera S, Desha C, Reid S, Newman P, Mouritz M (2022) Applying a place making sustainable centres framework to transit activated corridors in Australian cities. J Sustain Dev Energy Water Environ Syst 10(2):1–23

Chesterfield C, Tawfik S, Malekpour S, Murphy C, Bertram N (2021) Guiding integrated urban and water planning: project overview. Cooperative Research Centre for Water Sensitive Cities, Melbourne

Coenen L, Benneworth P, Truffer B (2012) Toward a spatial perspective on sustainability transitions. Res Policy 41(6):968–979

Department of Environment, Land-Use, Water and Planning (2017) Plan Melbourne 2017–2050. https://www.planning.vic.gov.au/policy-and-strategy/planning-for-melbourne/plan-melbourne

Department of Infrastructure and Transport (2011) Our cities, our future: a national urban policy for a productive, sustainable and liveable future. Australian Government, Canberra

Dixon T, Tewdwr-Jones M (2021) Urban futures. Planning for city foresight and city visions. Bristol University Press, Bristol

European Commission (2022) Driving the global green transition: call now open to select the first city cohort under the Urban Transitions Mission of Mission Innovation. https://research-and-innovation.ec.europa.eu/news/all-research-and-innovation-news/driving-global-green-transition-call-now-open-select-first-city-cohort-under-urban-transitions-2022-09-29_

Frantzeskaki N, Holscher K, Bach M, Avelino F (eds) (2018) Co-creating sustainable urban futures. Springer, Cham

Frantzeskaki F, Oke C, Barnett G, Bekessy S, Bush J, Fitzsimons M, Ignatieva M, Kendal D, Kingsley J, Mumaw L, Ossola A (2022) A transformative mission for prioritising nature in Australian cities. Ambio 51:1433. https://doi.org/10.1007/s13280-022-01725-z

Geels FW (2002) Technological transitions as evolutionary reconfiguration processes: a multi-level perspective and a case-study. Res Policy 31(8–9):1257–1274

Geels F (2011) The multi-level perspective on sustainability transitions: responses to seven criticisms. Environ Innov Soc Trans 1(1):24–40

Geels F (2012) A socio-technical analysis of low-carbon transitions: introducing the multi-level perspective into transport studies. J Transp Geogr 24:471–482

Glackin S, Newton P (2016) Engaging the greyfields: community engagement and co-design in residential redevelopment of public housing. In: Legacy C, Leshinsky R (eds) Instruments of planning: tensions and challenges for delivering more equitable and sustainable cities. Routledge, London

Glackin S, Moglia M, Newton P (2022) Working from home as a catalyst for urban regeneration. Sustainability 14(19):12584. https://doi.org/10.3390/su141912584

House of Representatives (2018) Building up and moving out. Australian Government, Canberra

Infrastructure Australia (2020) Infrastructure beyond COVID-19. A national study on the impacts of the pandemic on Australia. An interim report for the 2021 Australian Infrastructure Plan. Australian Government, Sydney

IRP (2018) The Weight of Cities. Resource Requirements of Future Urbanisation. A Report by the International Resource Panel, UNEP, Nairobi

Iselin D, Lemer A (1993) The fourth dimension in building: strategies for minimising obsolescence. National Academy Press, Washington, DC

Latz P (2021) A tale of two cities: Micromobility report. https://micromobilityreport.com.au/infrastructure/planning-design-education/a-tale-of-two-cities/

Loorbach D (2007) Transition management. International Books, Utrecht

Loorbach D (2010) Transition management for sustainable development: a prescriptive, complexity-based governance framework. Governance 23(1):161–183

MAC (2015) Plan Melbourne 2015 review: report by the Ministerial Advisory Committee. https://www.planmelbourne.vic.gov.au/__data/assets/pdf_file/0006/377133/MAC_2015_Final_Report.pdf

Malekpour S, Tawfik S, Chesterfield C (2020) Designing cross-sectoral collaborations for integrated urban and water planning. Cooperative Research Centre for Water Sensitive Cities, Melbourne

Marvin S, Bulkeley H, Mai L, McCormick K, Palgan YV (eds) (2018) Urban living labs: experimenting with city futures. Routledge, London

Moglia M, Frantzeskaki N, Newton P, Pineda Pinto M, Witheridge J, Cook S, Glackin S (2021) Accelerating a green recovery of cities: lessons from a scoping review and a proposal for mission-oriented recovery towards post-pandemic urban resilience. Dev Built Environ 7:100052. https://doi.org/10.1016/j.dibe.2021.100052

Moravej M, Renouf MA, Kenway S, Urich C (2022) What roles do architectural design and on-site water servicing technologies play in the water performance of residential infill? Water Res 213(118109):1–14. https://doi.org/10.1016/j.watres.2022.118109

MSDI (2020) Transforming Australia SDG progress report. Monash Sustainable Development Institute, Melbourne. https://www.sdgtransformingaustralia.com/

Murray S, Bertram N, Khor L-A, Rowe D, Meyer B, Newton P, Glackin S, Alves T, McGauran R (2015) Processes for developing affordable and sustainable medium density housing models for greyfield precincts. AHURI Final Report No. 236. Australian Housing and Urban Research Institute, Melbourne

Newman P, Hargroves K, Davies-Slate S, Conley D, Verschuer M, Mouritz M, Yangka D (2019) The trackless tram: is it the transit and city shaping catalyst we have been waiting for? J Transport Technol 9:31–55. https://doi.org/10.4236/jtts.2019.91003

Newman P, Hargroves K, Desha C, Izadpanahi P (2021) Introducing the 21st century boulevard: a post-COVID response to urban regeneration of main road corridors. Curr Urban Stud 9:831–854. https://doi.org/10.4236/cus.2021.9404

Newton P (2002) Urban Australia 2001: review and prospect. Aust Plan 39(1):37–45

Newton PW (2010) Beyond greenfields and greyfields: the challenge of regenerating Australia's greyfield suburbs. Built Environ 36(1):81–104

Newton PW (2011) Consumption and environmental sustainability. In: Newton PW (ed) Urban consumption. CSIRO Publishing, Melbourne

Newton PW (2012) Liveable and sustainable? Socio-technical challenges for 21st century cities. J Urban Technol 19(1):81–102

Newton P (2018a) Transitioning urban greyfields. In: Moore T, de Hahn F, Horne R, Gleeson B (eds) Urban sustainability transitions: Australian cases—international perspectives. Routledge, London

Newton P (2018b) Framing new retrofit models for regenerating Australia's fast growing cities. In: Eames M, Dixon T, Hunt M, Lannon S (eds) Retrofitting cities for tomorrow's world. Wiley-Blackwell, Oxford

Newton P (2019) The performance of urban precincts: towards integrated assessment. In: Newton P, Prasad D, Sproul A, White S (eds) Decarbonising the built environment: charting the transition. Palgrave Macmillan, Singapore

Newton P, Glackin S (2013) Using geo-spatial technologies as stakeholder engagement tools in urban planning and development. Built Environment 39(4):473–501

Newton P and Glackin S (2014) The challenges to urban sustainability and resilience. In: Pearson L, Newton P, Roberts P (eds) Resilient Sustainable Cities. Routledge, London

Newton P, Glackin S, Witheridge J & Garner L (2020) Transforming built environments: towards carbon neutral and blue-green cities. Sustain For 12:4745. https://doi.org/10.3390/su12114745

Newton P, Taylor M (eds) (2019) Precinct design assessment: a guide to smart sustainable low carbon urban development. CRC for Low Carbon Living, Sydney

Newton P, Thomson G (2017) Urban regeneration in Australian cities. In: Roberts P, Sykes H, Granger R (eds) Urban regeneration: a handbook (rev. ed.). Sage, London

Newton P, Murray S, Wakefield R, Murphy C, Khor L, Morgan T (2011) Towards a new development model for housing regeneration in greyfield residential precincts. AHURI Final Report No. 171. Australian Housing and Urban Research Institute Limited, Melbourne/. https://www.ahuri.edu.au/research/final-reports/171

Newton P, Marchant D, Mitchell J, Plume J, Seo S, Roggema R (2013) Performance assessment of urban precinct design: a scoping study. CRC for Low Carbon Living, Sydney

Newton P, Meyer D, Glackin S (2017) Becoming urban: exploring the capacity for a suburban-to-urban transition in Australia's low-density cities. Sustain For 9(10):1718. https://doi.org/10.3390/su9101718

Newton P, Bertram N, Handmer J, Tapper N, Thornton R, Whetton P (2018) Australian cities and the governance of climate change. In: Tomlinson R, Spiller M (eds) Australia's metropolitan imperative. An agenda for governance reform. CSIRO Publishing, Melbourne

Newton P, Glackin S, Witheridge J, Garner L (2020) Beyond small lot subdivision: pathways for municipality-initiated and resident-supported precinct-scale medium-density residential infill regeneration in greyfield suburbs. Urban Policy Res 38(4):338–356. https://doi.org/10.1080/08111146.2020.1815186

Newton P, Glackin S, Meyer D (2022a) The role of immigration in changing the social fabric of Australian cities. In: Levin I, Nygaard CA, Gifford SM, Newton PW (eds) Migration and urban transitions in Australia. Palgrave Macmillan, Singapore, pp 91–123

Newton PW, Newman PW, Glackin S, Thomson G (2022b) Greening the greyfields: new models for regenerating the middle suburbs of low-density cities. Palgrave Macmillan, Singapore

Norman B, Newman P, Steffen W (2021) Apocalypse now: Australian bushfires and the future of urban settlements. NPJ Urban Sustain 1:2. https://doi.org/10.1038/s42949-020-00013-7

Parolek D (2020) Missing middle housing. Island Press, Washington, DC

Planning Panels Victoria (2021) Maroondah planning scheme amendments C134maro and C136maro: greening the greyfields. Panel Report, State Government Victoria, Melbourne, Aug 9

Productivity Commission (2012) Performance benchmarking of Australian business regulation: the role of local government. Australian Government, Canberra

Rees W, Roseland M (1991) Sustainable communities: planning the 21st century. Plan Canada 31:15–26

Sareen S, Waagsaether K (2022) New municipalism and the governance of urban transitions to sustainability. Urban Stud 1(19):2271

Smith M, Newton P, Pears A, Denis-Ryan A, Ahuja E (2021) Buildings and precincts. In: Baldwin K, Howden SM, Lindesay J, Hussey K, Smith M (eds) Transitioning to a prosperous, resilient carbon-free economy: a guide for decision makers. Cambridge University Press, Cambridge

Sobels J, Turner G (2022) There are no sustainable cities in Australia. In: Levin L, Nygaard C, Gifford S, Newton P (eds) Migration and urban transitions in Australia. Palgrave Macmillan, Singapore

Steffen W, Richardson K, Rockström J, Cornell SE, Fetzer I, Bennett EM, Biggs R, Carpenter SR, De Vries W, De Wit CA, Folke C (2015) Planetary boundaries: guiding human development on a changing planet. Science 347(6223):1259855. https://doi.org/10.1126/science.1259855

Thomson G, Newton P, Newman (2017) Urban regeneration and urban fabrics in Australian cities. J Urban Regen Renew 10 (2):169–190

Thomson G, Newton P, Newman P (2019) Sustainable precincts: transforming Australian cities one neighbourhood at a time. In: Newton P, Prasad D, Sproul A, White S (eds) Decarbonising the built environment: charting the transition. Palgrave Macmillan, Singapore

UNDESA (2021) Long term territorial planning and spatial development. In: UN Department of Economic and Social Affairs Committee of Experts on Public Administration Guidance note, New York

Victorian Government Gazette (2022) Notice of approval of A=amendment No. S 647 and No. S 657. http://www.gazette.vic.gov.au/gazette/Gazettes2022/GG2022S647.pdf; http://www.gazette.vic.gov.au/gazette/Gazettes2022/GG2022S657.pdf

Westman L, Patterson J, Macrorie R, Orr CJ, Ashcraft CM, Castán Broto V, Dolan D, Gupta M, van der Heijden J, Hickmann T, Hobbins R (2022) Compound urban crises. Ambio 51(6):1402–1415. https://doi.org/10.1007/s13280-021-01697-6

Wolfram M (2018) Urban planning and transition management: rationalities, instruments and dialectics. In: Frantzeskaki N, Holscher K, Bach M, Avelino F (eds) Co-creating sustainable urban futures. Springer, Cham, pp 103–125

Wolfram M, Borgström S, Farrelly M (2019) Urban transformative capacity: from concept to practice. Ambio 48:437–448

Wutich A, DeMyers C, Bausch JC, White DD, Sullivan A (2020) Stakeholders and social influence in a shadow network: implications for transitions toward urban water sustainability in the Colorado River basin. Ecol Soc 25(1):28. https://doi.org/10.5751/ES-11451-250128

Xing K, Wiedmann T, Newton P, Huang B, Pullen S (2019) Development of low carbon urban forms—concepts, tools and scenario analysis. In: Newton P et al (eds) Decarbonising the built environment: charting the transition. Palgrave Macmillan, Singapore

Peter W. Newton is an Emeritus Professor in the Centre for Urban Transitions at Swinburne University of Technology in Melbourne, Australia. From 2007 to 2021, he held the position of Research Professor, following 15 years as Chief Research Scientist with the Commonwealth Scientific and Industrial Research Organisation (CSIRO). He has had a distinguished career in built-environment research and was elected a Fellow of the Academy of Social Sciences in Australia in 2014. His principal fields of research have focused on the technology of planning, sustainability science, and urban transitions. He has published over 25 books, including *Greening the Greyfields* and *Migration and Urban Transitions in Australia* in 2022.

Stephen Glackin is a Senior Research Fellow in urban planning and urban geography at the Centre for Urban Transitions, Swinburne University of Technology in Melbourne, Australia. His fields of expertise include strategic planning, planning law, community engagement, and geospatial analysis. He focuses on applied research with government and industry partners, covering topics such as urban regeneration, asset management, housing, infrastructure, community engagement, and decision-support systems for more-compact urban forms.

Chapter 8
Pathways for Restoring and Connecting with Nature in Australian Cities

Niki Frantzeskaki, Judy Bush, Dave Kendal, Clare Adams, Loretta Bellato, Alessandro Ossola, and Cathy Oke

Abstract With increasing focus on the importance of integrating nature spaces and nature-based solutions into our cities, what are the key priorities and pathways for action in Australian cities? Australia, a highly urbanised settler colonial country, has a rich biodiversity and cultural heritage, the result of thousands of years of custodianship and care for Country by the First Peoples. Their deep cultural and ecological knowledge and ongoing care and connection with Country, including urban Country, underpins approaches to restoring and connecting with nature in Australian cities. With continuing urbanisation and urban change, as well as increasing impacts of climate change, including heatwaves, wildfires, flooding, and extreme weather, we identify four pathways for just transitions with and for urban nature in cities. The pathways focus on ways of thinking, organising, acting, and knowing for prioritising urban nature. We highlight evidence-based planning for nature (ways of knowing); inclusive governance for just transitions (ways of organising); conserving,

N. Frantzeskaki (✉)
Faculty of Geosciences, Human Geography and Spatial Planning, Utrecht University,
Utrecht, The Netherlands
e-mail: n.frantzeskaki@uu.nl

J. Bush
Faculty of Architecture, Building and Planning, University of Melbourne,
Parkville, VIC, Australia

D. Kendal
Future in Nature Pty Ltd, Melbourne, VIC, Australia

C. Adams · L. Bellato
Centre for Urban Transitions, Swinburne University of Technology, Hawthorn, VIC, Australia

A. Ossola
Department of Plant Sciences, University of California Davis, Davis, CA, USA

School of Agriculture, Food and Ecosystem Sciences, University of Melbourne,
Parkville, VIC, Australia

C. Oke
Melbourne Centre for Cities, University of Melbourne, Parkville, VIC, Australia

© The Author(s) 2025   177
N. Frantzeskaki et al. (eds.), *Future Cities Making*, Theory and Practice of
Urban Sustainability Transitions, https://doi.org/10.1007/978-981-97-7671-9_8

restoring, and maintaining nature (ways of acting); and emphasising First People's and local communities' knowledges and practices (ways of knowing). Our pathways, or stepping stones, point to interlinked and interrelated priorities for ensuring nature is actively and effectively integrated into Australian city planning and practice.

Keywords Nature-based solutions · Urban · Metropolitan · Planning · Policy · Indigenous knowledges

8.1 Introduction

Australia is a nation of cities and towns.[1] With fast-paced urbanisation, Australian cities continue expanding into areas of high biodiversity, challenging planners to account for and recognise the multiple impacts of this, alongside other competing values for these growth areas. Australian cities are located on unceded First Peoples' lands; through their custodianship for millennia, unique and rich ecosystems have been nurtured. Australian cities today are home to rich suites of biodiversity, with many endemic species, including threatened species, of flora and fauna (Oke et al. 2021). For some threatened species, urban environments are their only habitats, and so urban biodiversity habitat conservation and restoration is essential for species' survival and resilience (Oke et al. 2021). Urban nature and biodiversity also provide essential functions, services, and contributions to urban sustainability, liveability, and resilience (Frantzeskaki and McPhearson 2021). Yet urban development is often prioritised over nature conservation and restoration, with economic pressures and other imperatives competing with the need to protect natural landscapes. Underpinned by assumptions of the separateness of nature and culture (or nature and cities), urban development often proceeds according to a human-centric approach, with greening relegated to the margins of planning and design (Daniels et al. 2020). We need new approaches to urban planning and development (Australian Academy of Science 2021) that place nature up front and centred in processes of urbanisation, renewal, and densification.

Recent academic debates have highlighted the importance of rethinking and prioritising the *values of* nature over the *profit from* nature (IPBES 2022), including the need to account for First Peoples' cultural practices, knowledges, and values for the environment—for Country (Janke 2021). Governments, businesses, and communities are increasingly recognising "nature-positive" approaches that seek to go beyond a goal of minimising harm to instead restore and expand nature conservation (World Economic Forum 2020). Responding to these debates could help

---

[1] We use the term "cities" in the Australian context to indicate metropolitan areas rather than jurisdictional boundaries, as most large Australian cities are governed by multiple local governments as well as state governments.

Australian planners transform the ways that policies and programmes are formulated and support a greater prioritisation of nature across urban-policy portfolios and sectors.

These shifts of discourses at the global and national levels intersect with two context-driven and context-shaping developments. First, climate change is a macro-driver of change, creating a context in which current planning approaches are stressed and necessitating new ways to foster resilience as well as ensure a quality of life that is equitable for all. At this macro scale, Australia must accelerate work to mitigate climate change by switching to renewable energy and nature-based solutions and moving away from its reliance on its substantial reserves of fossil fuels (coal and gas) (Australian Academy of Science 2021). Australia must also comprehensively address climate risks and adaptation imperatives, with its increasing exposure to climate change impacts of droughts, bushfires, heatwaves, and sea-level rise that affect both people and infrastructure (IPCC 2023; Norman et al. 2021).

Urban development processes must plan for climate changed futures that take account of these impacts. With much of the Australian population living in coastal cities and towns, the liveability and character of Australian cities is in jeopardy if "climate-ready" interventions and policies are not promptly implemented (Ossola and Lin 2021). Climate readiness or resilience must also embed climate justice, which has entered the policy discourse more prominently after the IPCC's Sixth Assessment Report (IPCC 2023), pointing to the need for climate solutions that enable and facilitate "just transitions" in systems and processes that contribute to climate change mitigation without generating or deepening injustices.

Urban nature, in the form of nature-based solutions and ecosystem-based approaches, provide multiple benefits and functions critical for climate resilience. Trees and waterways cool urban areas; mangroves protect coastlines from storm surges; water-sensitive urban design approaches address water quality and urban flooding; and biodiverse plantings foster mental health and well-being for stressed city dwellers (Frantzeskaki et al. 2019; Keeler et al. 2019; Kabisch et al. 2017; Ignatieva 2018, Ignatieva et al. 2020; Garrard et al. 2018; Coutts et al. 2013). Yet it is also imperative that nature-based solutions are designed for climate changed urban environments of the future, with increased focus on tolerance thresholds to inform species selection and landscape design (Ossola and Lin 2021). Urban nature is both vulnerable to the impacts of climate change and a key element in better adapting Australian cities towards sustainability, liveability, and resilience; thus, transformations in ways of planning, governing, and relating with nature are required (Australian Academy of Science 2021).

The second context driver is the increasing awareness and recognition of First Peoples' rights, cultural practices, knowledges, and custodianship over land, water, and nature, all conceptualised as "Country" in Australia (Langton 2018). As a group of non-Indigenous scholars, we respectfully acknowledge that Australia's cities are built on unceded Aboriginal and Torres Strait Islander lands and that these up to 250 different Nations have deeply held relationships with the environment, both lands and waters, through connection to and caring for Country that have existed for tens of thousands of years and are ongoing. When it comes to restoring or bringing nature back into Australian cities, many local governments, educational institutions, and citizens are listening to, learning with, and co-designing strategies with Traditional Owners. This is not uniquely Australian: in many countries across the globe, rights for nature have been gaining traction through citizen assemblies and new state (constitutional) legislative actions. In New Zealand, nature has statutory recognition; in Ireland, a Peoples' Assembly has proposed to the government to recognise nature as a legal actor, and similar debates and citizens-to-policy proposals are underway in Canada, Iceland, South American nations, and the Arctic. As these unfold, the recognition of First People's sophisticated knowledge systems about nature, society, and culture (and, relatedly, resilient responses to climate challenges) (Kingsley et al. 2013) come hand in hand with the paradox of the systemic and pervasive injustices faced by First Peoples, an entry point to understand and deal with urban injustices when developing climate-resilient solutions with and about nature in Australian cities.

Against this ecological, societal, and political landscape, our chapter considers pathways for bridging scientific and societal/policy aspirations for just transitions with and for (urban) nature in Australian cities that are striving for climate resilience. The exploratory research question that guides us in this chapter is: *What are the required shifts in ways of thinking, organising, acting, and knowing for prioritising urban nature in Australian cities while enabling just transitions in the context of climate change?* Our conceptual grounds are drawn from sustainability transition theories, which point to the need for historically understanding, and thereafter informing through active shaping and co-creating, interventions for fundamental shifts in systems of action, power, and service to achieve more sustainable outcomes (Loorbach et al. 2020). We adopt the conceptualisation of transitions as fundamental shifts in "ways of thinking, organising and acting" (Frantzeskaki and de Haan 2009; Loorbach et al. 2020) and of knowing (Avelino et al. 2019) to guide our proposed pathways for just transitions for Australian cities. These pathways act as *stepping stones* or guides for action, rather than rigid or prescriptive directives (Bush et al. 2023; Tozer et al. 2022). The *pathways* concept reinforces that integrated or "assembled" approaches (Tozer et al. 2022) are needed to address complex challenges and that these approaches must be adapted to respond to place-based conditions rather than be framed as one-size-fits-all responses. Together, these pathways point to new directions that could underpin transformations in ways of planning, governing, and managing urban nature and biodiversity, based on a commitment to climate resilience and just transitions.

The chapter is organised as follows: we first present our vision for how nature in cities contributes to the dual challenges of *nature restoration* in the Decade of Ecosystem Restoration and *transitions* for urgent climate change action (UNEP 2021a). Following this, we elaborate on the four pathways for new approaches to *thinking, organising, acting,* and *knowing* urban nature. We conclude by highlighting the key roles that cities can play in contributing to #GenerationRestoration in this critical decade and global efforts to reframe our relationships with nature.

8.2   Transformative Pathways for Nature in Australian Cities

The UN Decade on Ecosystem Restoration highlights the urgency of action for the conservation and restoration of species, landscapes, and ecosystems as essential for climate change action, for food security, and for slowing species extinctions (UNEP 2021a). Calls to become #GenerationRestoration recognise that everyone, everywhere must play their part in addressing these challenges and imperatives, not least in cities and urban areas. Indeed, the UN's Environment Programme urges that, to address the interconnected global challenges of species extinctions, climate change, and pollution, we must "make peace with nature" and that everyone has a role to play (UNEP 2021b).

Cities can ensure that urban nature is not only retained but becomes an integral piece in the urban mosaic that can boost resilience and adaptability to old and new global and climate challenges (Daniels et al. 2020; Davidson and Gleeson 2018; Hobbie and Grimm 2020; Ossola and Lin 2021), support biodiversity conservation and ecosystem function, and provide a plethora of benefits for cities (Almenar et al. 2021; Dumitru et al. 2020; Padma et al. 2020; Kabisch et al. 2017). To achieve this, urban planning and design play key roles (Shade et al. 2020). These include building the evidence base, developing effective planning mechanisms as well as removing the physical and planning barriers that impede the integration of nature-based solutions, and prioritising the retention and restoration of nature in place, as well as creating new spaces for nature (Lin et al. 2021). While cities are diverse and complex, they offer plentiful opportunities to plan, retrofit, enhance, and design urban nature at different scales, from city masterplan to urban neighborhood, private yard, public park, or verge garden (Frantzeskaki et al. 2019; Ignatieva 2010; Kingsley et al. 2021a).

We propose four pathways as interconnected priorities (or stepping stones) to steer urban planning and governance for prioritising urban nature while enabling just urban futures:

- Establish evidence-based planning for nature in cities to incorporate both climate and biodiversity crises (shifts in ways of thinking about urban nature).
- Strengthen inclusive governance for nature in cities to ensure just processes of and outcomes for urban agendas and plans (shifts in ways of organising).
- Focus on creating and maintaining nature-based solutions, ecosystem-based approaches, and ongoing habitat management practices for restoring nature in Australian cities (shifts in ways of acting).
- Foreground local communities and First Peoples' knowledges and practices to innovate with nature (shifts in ways of knowing).

All the proposed pathways consider the contextual complexities of Australian cities, especially the local communities of Aboriginal and First Peoples. Australia is a settler-colonial state on the unceded lands of at least 250 Aboriginal and Torres Strait Islander Peoples language groups (Jones et al. 2018; Porter 2019). Recognised as the Traditional Custodians of this continent, they have been caring for and innovating with and as a part of nature for over 65,000 years (Janke 2021). Through thousands of years of careful and meticulous observation and experimentation and dissemination, Aboriginal and Torres Strait Islanders have developed their knowledges from the lands and waterways of the places to which they belong (Woodward et al. 2020). Despite the devastating effects of colonisation, Aboriginal and Torres Strait Islander Peoples continue to develop, revive, and evolve their knowledges. Furthermore, efforts are being made to "reset the relationship" between the First Peoples of Australia and settler-colonial governments to create the conditions for overcoming colonial harms (tebrakunna country and Lee 2019).

8.2.1 Pathway 1: Evidence-Based Planning for Nature in Cities (Ways of Thinking)

The first pathway to drive urban agendas and actions is to embed evidence-based metropolitan and urban planning for valuing, prioritising, and maintaining nature in cities. Evidence-based planning can manifest through research, database creation, and urban experimentation and through the embedding and integration of this information and knowledge into policy and planning mechanisms. Evidence is produced through research and experimentation, by trialling or piloting new solutions in the city, such as urban living labs (Bulkeley et al. 2016; Kronsell and Mukhtar-Landgren 2018; Willems et al. 2022) and pilots for new approaches (Wickenberg et al. 2022; Willems and Giezen 2022). This experimentation-informed evidence base can be led by government, university, environmental NGO, community, or the private sector. An example is the social enterprise The Climate Factory (https://climatefactory. com.au/), which creates and manages community-led micro-forests in Canberra communities. Evidence-based planning is underpinned by the active and responsive creation, use, and updating of research and data to inform decision-making and improve knowledge about the state of the environment. Evidence-based planning can also develop and use policy tools to inform development assessment and for visualisations, such as urban heat vulnerability mapping.

Evidence-based planning for nature requires a nature-positive shift in the institutional platforms, statutory planning rules such as zoning and overlays, and development assessment or design code requirements that enable and accelerate transition processes to transform the way cities are planned. Research can provide the rigorous foundation for these new policy approaches to create defensible positions that are more resistant to opposition from vested interests. Research is also required to test, evaluate, and interrogate existing policy mechanisms to ensure they are achieving their stated nature-oriented objectives. For example, planning tools and instruments, such as regulatory requirements and voluntary biodiversity offsetting mechanisms (Hanson and Olsson 2023), are increasingly relied on by governments and developers to "balance" or compensate the impact of development (Pascoe et al. 2019; Söderqvist et al. 2021). These planning instruments can come in the form of strategic plans (such as data and maps), process tools (such as guidelines and policies), and counting tools (such as the measurement and evaluation of outcomes) (Hanson and Olsson 2023). While governments and developers may view offsetting as useful planning mechanisms, there are substantial research questions, uncertainties, and knowledge gaps regarding their effectiveness and suitability in broader conservation and restoration programmes (Droste et al. 2022; Hanson and Olsson 2023; Kalliolevo et al. 2021; Lindenmayer et al. 2017).

The need for evidence-based planning for urban nature-based solutions is important for biodiversity conservation (Palmer et al. 2008; Fitzsimons et al. 2011), especially in terms of urban planning that considers ecological and biodiversity evidence, research, and monitoring, and that taps into interdisciplinary knowledge of planners and ecologists (Bohnet 2010; Williams et al. 2021). New urban planning tools and policies need to be integrated into existing planning practices and processes, for example, to build the evidence base for urban biodiversity and connect that knowledge to experimentation and innovation in the design of urban nature-based solutions. Recent examples of transdisciplinary collaborations for evidence-based planning include connecting data on tree canopy cover and health (Tabassum et al. 2021) and other environmental indicators and outcomes to future planning decisions such as the Green Factor Tool (Bush et al. 2021) and the Guidelines for Biodiversity Green Roofs (Schiller et al. 2023).

Box 8.1 Guidelines for Biodiversity Green Roofs: City of Melbourne

We need to emphasise the (scientific) knowledge that enables evidence-based planning for and with urban nature-based solutions to promote and inform sound decision-making and innovation for understanding and embracing nature in the city. The City of Melbourne's Guidelines for Biodiversity Green Roofs have been recently developed to fill a knowledge gap regarding the potential of green roofs to improve urban biodiversity habitats (Schiller et al. 2023). The guidelines outline deliberate design considerations and management plans for successful outcomes for biodiversity on green roof installations (Schiller et al. 2023). They include consideration of suitable habitat for the biodiversity "target" species (e.g. for insects, small mammals, and birds); connectivity between roof and ground habitats, pests, and ecosystem (dis) services; and ongoing maintenance and management requirements for biodiversity green roofs (Schiller et al. 2023).

This contributes the scientific evidence base both for biodiversity green roofs in Australia—that is, for place-based context such as drought-tolerant species—and for how policy can be created in this space. It further identifies continuing knowledge needs to direct research to inform and strengthen the evidence and processes of evidence-based planning. Evidence-based planning needs to be understood as an ongoing process of (re)defining what is important information on which to base planning decisions and procedures to plan for and with nature-based solutions, in other words, to support an awareness that planning is a social and political process in which nature-positive intentions are actively developed by multiple actors across multiple sectors that can inform transformations in how nature is considered and is prioritised in the city. Tools such as these guidelines are important for developing an evidence base to plan for and with urban nature-based solutions; most importantly, it is an accessible evidence base that can be useful to other metropolitan Melbourne Councils.

8.2.2 Pathway 2: Inclusive Governance for Just Transitions (Ways of Organising)

Our second pathway highlights the necessity for new ways of organising for nature-positive cities, through a focus on inclusive governance. Inclusive governance seeks to broaden participation and power in the processes, structures, and stakeholders involved in planning and managing urban nature (Fors et al. 2021). Inclusive governance considers whose voices are heard and develops ways to indeed include the voices, perspectives, and aspirations of those of different ages, abilities, and cultural and socio-economic backgrounds (Giachino et al. 2021; Moloney and Doyon 2021; Tozer et al. 2020). Likewise, inclusive governance must develop new approaches to listening for the voices of more-than-humans—of nature and biodiversity (Apfelbeck et al. 2020; Bush and Doyon 2023; Maller 2021; Pineda Pinto et al. 2021). Inclusive governance for the planning and implementation of nature-based solutions and climate resilience may contribute towards just transitions, so that nature-positive responses do not deepen existing inequalities (Tozer et al. 2020).

Inclusive governance involves collaboration between different sectors, levels of government, interest groups, businesses, and residents. Australian cities operate under policy and planning cycles of state and federal governance levels that are not always synchronised with the needs and pressure at the local level. Therefore, new approaches must prioritise ways to open communication channels, build trust, and demonstrate the value and benefit from effective transdisciplinary and cross-sectoral working relationships. In this it is critical to adopt a cross-sectoral approach that bridges interests and brings together different forms of knowledge (Malekpour et al. 2021) and that specifically prioritises Indigenous knowledges and knowledge systems (see Box 8.2).

The concept of mosaic governance can explore and explain such collaborative relationships between local government authorities and urban communities (Buijs et al. 2019). Mosaic governance describes relationship building and bridging among (active) citizens and other actors, such as the private sector and local governments, to produce opportunities for place-based collaboration and coordination for creating, implementing, and maintaining urban green infrastructure (Buijs et al. 2017; van der Jagt et al. 2021). These relationships are important for shifting towards governance models that are more open and inclusive to build trusting relationships between local citizens and local governments (Gentin et al. 2022).

These partnerships between urban communities and local governments can be seen in urban gardening and urban agriculture as examples of stewardship activities across public and private land tenures (Kingsley et al. 2021a); examples include local governments and residents embracing programmes such as Gardens for Wildlife (2021) and (Mumaw 2017). Urban communities require a supportive policy environment to be empowered to act in collaboration with local governments. Support can include removing regulations that prevent innovation as well as encouraging and providing explicit support for culturally and socio-economically diverse communities to participate in nature-positive practices, such as community gardens (Oke et al. 2021).

Box 8.2 Giving a Voice to the River

Nature's multifunctionality is widely recognised and celebrated, from biodiversity habitat to cooling cities, treating water and air, and providing space for social connections and mental and physical health and well-being. We need to shift our current monofunctional governance and management arrangements to developing new approaches that can both accommodate nature's multifunctionality and actually make the most of this multifunctionality (Bush 2020). Participatory approaches are one of these mechanisms. An example is the Victorian *Yarra River Protection (Wilip-gin Birrarung murron) Act 2017*, which aims to give a voice to nature through the establishment of the Birrarung Council statutory body, with representatives from Traditional Custodians, environmental and agricultural industry groups, and local community groups (O'Bryan 2019).

These approaches foreground Indigenous knowledges and Custodianship: the Birrarung Council is bicultural, requiring the coming together of diverse voices and perspectives into a collective position. Moreover, the Act recognises the river as one integrated living entity. Together, these constitute steps on the journey to overcome fragmented governance arrangements and denial of the rights of the river.

8.2.3   Pathway 3: Conserving, Restoring, and Maintaining Nature (Ways of Acting)

Our third pathway focuses on ways of acting. We envisage nature-positive cities in which nature conservation and restoration is centred through the implementation of nature-based solutions and ecosystem-based approaches. Nature-based solutions fulfil a range of urban functions and services in addition to their essential role in habitat provision (Cohen-Shacham et al. 2019). They can replace traditional grey infrastructure for water management, cooling cities, and strengthening climate resilience (Frantzeskaki and McPhearson 2021). Shifting from a reliance on grey infrastructure to acknowledging the effectiveness of nature-based solutions in delivering these functions requires transformation in how cities and city infrastructure are planned, funded, and maintained (Bush 2020; Matsler 2019). It also requires ecologists, horticulturalists, and landscape architects and designers to be involved at all stages of urban development, to ensure that nature-based solutions are planned and implemented effectively. Indeed, revegetation programmes in Australian cities have already demonstrated considerable achievements in addressing urban infrastructure needs while also contributing to valuable habitat provision (Bush et al. 2003; McGregor and McGregor 2020; Parris et al. 2020) (Box 8.3).

Box 8.3 Narrap Team: Indigenous Natural Resource Management
The Traditional Owners of much of the Melbourne region, the Wurundjeri Woi-Wurrung peoples, have established a natural resource management team, the Narrap Team,[2] to manage properties owned by the Wurundjeri Council, as well as to work with other landowners and managers, including government agencies[3] and community groups on Country. The Team also works to rediscover and document Wurundjeri Traditional Ecological Knowledge as part of their natural resource management and nature restoration activities. They work to restore and regenerate landscapes, establish biodiversity corridors, manage invasive plants and animals, and carry out ecological cultural burning practices. Since its establishment in 2012 with a small, four-person team, there are now more than 20 members of the Narrap Team, and it is expected to continue to grow as demand for their expertise increases. In 2022 the Narrap Team received a National Landcare Award for Indigenous Land Management.[4]

As living systems, urban landscapes and ecosystems require ongoing maintenance, which needs to be informed by knowledge of how the systems function and thrive (Hansen et al. 2023). As living systems, we need to develop ongoing relationships of care and stewardship with nature-based solutions: for living systems to thrive, Custodianship is essential, and this points to our fourth pathway, which foregrounds local communities and First People's knowledges and practices as a key stepping stone towards nature-positive cities.

---

[2] https://www.wurundjeri.com.au/services/natural-resource-management/narrap-country-team/

[3] https://yoursay.melbournewater.com.au/yan-yean-reservoir/traditional-owners-caring-country-yan-yean

[4] https://nationallandcareconference.org.au/awards/2022-kpmg-indigenous-land-management-award/

8.2.4 Pathway 4: Foreground First Peoples' and Local Communities' Knowledges and Practices (Ways of Knowing)

The potential for urban planning to bridge the gap between aspirations and real-life transformations in urban spaces and infrastructures can be achieved with policies and practices that are informed and guided by evidence, which includes data, accounting for lived experiences of people, and a new appreciation of people-nature relationships in cities (Pineda-Pinto et al. 2023; Potter 2020; Voskamp et al. 2021). The diversity of urban communities means that there are many knowledges and many practices that can contribute to innovation in the planning and management of and with urban nature-based solutions. This diversity, however, needs to be supported, encouraged, and empowered to ensure that it is inclusive of Indigenous and other locally sourced knowledges in decision-making processes and the implementation and maintenance of urban nature-based solutions. Importantly, nature is place-based, as are people's relationships to those places within cities and the biodiversity that constitutes them (Fish et al. 2016; Mattijssen et al. 2020). Therefore, this pathway is underpinned by partnerships among communities and local governments (pathway 2, Sect. 8.2.2) to develop and manage nature-based innovation in cities.

To foreground diverse community knowledges and practices for the planning and management of urban nature-based solutions, we need a shift in which (and whose) knowledge(s) are considered in decision-making processes (e.g. expert, tacit) (Ludwig and Macnaghten 2020; Nagendra et al. 2018; Trisos et al. 2021) and, increasingly importantly, how these knowledges are used in combination (sometimes called knowledge-weaving) (Tengö et al. 2017). Knowledge-weaving, which is understood as those collaborations that respect the integrity of diverse knowledge systems and emerge from knowledge co-production, occurs when Indigenous and other local knowledges are mobilised, as demonstrated in some areas of Australia (Tengö et al. 2017).

As part of the growing recognition that learning from Indigenous ways of knowing, being, and doing is essential to sustainable futures, including overcoming colonial harms, scholars and practitioners are investigating how urban developments can align with these ways (Hill et al. 2020; Porter 2019). Regenerative development approaches weave Indigenous, locally sourced, and Western knowledge systems and practices to align with nature's inherent regenerative processes. Technical and technological knowledges and practices support innovations; however, they do not drive innovation in regenerative development (Hes and du Plessis 2015; Mang and Haggard 2016). Because it sees urban places as living systems, regenerative development considers more than humans to be partners and active agents in regeneration initiatives. Innovations are sourced from the inherent potential of the place and its community; consequently, First Peoples and other human and non-human locals within these communities foreground essential knowledges and practices to innovate with nature (Bellato et al. 2023; Hes and Bush 2020). By foregrounding First Peoples' deep knowledge of Country, urban nature and cultural heritage can be better integrated into Australian cities (Kingsley et al. 2013; Kingsley et al. 2021b; Terare and Rawsthorne 2020) (Box 8.4).

> **Box 8.4 Emu Sky Exhibition**
> "Through art works, storytelling, detailed research and writings, we simultaneously explore our past, our present and our future, as part of concepts that are deeply enmeshed in our ways of being and knowing," writes Barkandji woman Zena Cumpston of the art exhibition she curated, which brought together many first peoples artists and collaborators.[5] the exhibition invites us all to listen and learn from deep cultural knowledge and rich contemporary culture: "Emu sky explores and illuminates indigenous perspectives related to science, innovation, plant use, land management and agricultural practice. Through detailed research, art and storytelling this exhibition is a sustained interrogation of the western lens through which indigenous scientific Endeavour has been historically perceived".[6] the exhibition website provides a permanent record of the artworks after the completion of the physical exhibition, pointing to multimedia approaches to learning, listening, and recording ways of knowing.

8.3 The Way Forwards: Lessons from and Implications for Australian Cities

Planning for nature-positive Australian cities needs to be informed and driven by First Peoples' aspirations and knowledges and by the dual imperatives of nature restoration as encapsulated by the UN's declaration of the 2020s as the Decade of Ecosystem Restoration (UNEP 2021a) and the urgency of just climate change action (IPCC 2023). Actions must be developed and assessed using a justice lens: Who will benefit? Who will be made more vulnerable by the responses to climate change? How do we understand and assess climate justice while designing (or co-designing) resilient solutions? Urban injustices have increased over the past decades for Australian cities due to inadequate social welfare policies, lack of climate policies, and unresolved settler-colonial tensions between its First Peoples and governments regarding sovereignty, land rights, and the resulting ongoing poverty, exclusion, and systemic injustices. Rising housing prices, undefined climate vulnerability zones, and drive for urban development, often with nature's spaces and roles overlooked or deprioritised, all fuel a mix of interconnected drivers for urban injustices in the making.

[5] https://emusky.culturalcommons.edu.au/works/welcome-from-zena-cumpston/
[6] https://emusky.culturalcommons.edu.au/exhibition/

Regenerating urban neighbourhoods is a contested and challenging issue for cities. Many interests, place-based histories, and values need to be considered and expressed. Urban regeneration has always been an opportunity for trialling, learning, and contrasting "old" with "new" urban planning processes. The challenges at the neighborhood or district level resurface in discussion and deliberation at citywide scales: districts/neighbourhoods are microcosms of what urban planning deals with at larger scales, often depicting visions of the past but also presenting potential for navigating the complexities and uncertainties of the present and future. Post-pandemic, many studies have shown the appreciation of urban citizens for urban ecosystems: the multiple benefits of urban parks for social cohesion, stress relief, social encounters, and (the well-researched) cooling effects during hot days, as well as reconnecting people with nature, contributing to biodiversity habitat, and acting overall as urban oases (Ugolini et al. 2020). Research has also reinforced the importance of urban water bodies such as rivers, waterfronts, and lakes both for residents and for biodiversity (Threlfall et al. 2021). As cities now aim to build from these realisations and lessons learned, the COP15 Kunming-Montreal Biodiversity agreement strengthens the position of nature in global agendas, proposing a set of biodiversity targets that cities have an important role in achieving.

References

Almenar JB, Elliot T, Rugani B, Philippe B, Gutierrez TN, Sonnemann G, Geneletti D (2021) Nexus between nature-based solutions, ecosystem services and urban challenges. Land Use Policy 100. https://doi.org/10.1016/j.landusepol.2020.104898

Apfelbeck B, Snep RPH, Hauck TE, Ferguson J, Holy M, Jakoby C, MacIvor SJ, Schär L et al (2020) Designing wildlife-inclusive cities that support human-animal co-existence. Landsc Urban Plan 200:103817. https://doi.org/10.1016/j.landurbplan.2020.103817

Australian Academy of Science (2021) The risks to Australia of a 3°C warmer world. www.science.org.au/warmerworld

Avelino F, Monticelli L, Wittmayer JM (2019) How transformative innovation movements contribute to transitions. In: Howaldt J, Kaletka C, Schröder A, Zirngiebl M (eds) Atlas of social innovation: a world of new practices, vol 2. Oekom Verlag, Munich, pp 70–74. https://www.socialinnovationatlas.net/fileadmin/PDF/volume-2/01_SI-Landscape_Global_Trends/01_15_How-Transformative-Innovation-Movements_Avelino-Monticelli-Wittmayer.pdf

Bellato L, Frantzeskaki N, Tebrakunna Country, Lee E, Cheer JM, Peters A (2023) Transformative epistemologies for regenerative tourism: towards a decolonial paradigm in science and practice? J Sustain Tour 32:1161. https://doi.org/10.1080/09669582.2023.2208310

Bohnet I (2010) Integrating social and ecological knowledge for planning sustainable land- and seascapes: experiences from the great barrier reef region, Australia. Landsc Ecol 25(8):1201–1218. https://doi.org/10.1007/s10980-010-9504-z

Buijs A, Mattijssen TJM, Van der Jagt APN, Ambrose-Oji B, Andersson E, Elands BHM, Steen Møller M (2017) Active citizenship for urban green infrastructure: fostering the diversity and dynamics of citizen contributions through mosaic governance. Curr Opin Environ Sustain 22:1–6. https://doi.org/10.1016/j.cosust.2017.01.002

Buijs A, Hansen R, Van der Jagt S, Ambrose-Oji B, Elands B, Lorance Rall E, Mattijssen T, Pauleit S, Runhaar H, Stahl Olafsson A, Steen Møller M (2019) Mosaic governance for urban green infrastructure: upscaling active citizenship from a local government perspective. Urban For Urban Green 40:53–62. https://doi.org/10.1016/j.ufug.2018.06.011

Bulkeley H, Coenen L, Frantzeskaki N, Hartmann C, Kronsell A, Mai L, Marvin S, McCormick K, van Steenbergen F, Voytenko Palgan Y (2016) Urban living labs: governing urban sustainability transitions. Curr Opin Environ Sustain 22:13–17. https://doi.org/10.1016/j.cosust.2017.02.003

Bush J (2020) The role of local government greening policies in the transition towards nature-based cities. Environ Innov Soc Trans 35:35–44. https://doi.org/10.1016/j.eist.2020.01.015

Bush J, Doyon A (2023) Planning a just nature-based city: listening for the voice of an urban river. Environ Sci Policy 143:55–63. https://doi.org/10.1016/j.envsci.2023.02.023

Bush J, Miles B, Bainbridge B (2003) Merri Creek: managing an urban waterway for people and nature. Ecol Manage Restor 4(3):170–179. https://doi.org/10.1046/j.1442-8903.2003.00153.x

Bush J, Ashley G, Foster B, Hall G (2021) Integrating green infrastructure into urban planning: developing Melbourne's green factor tool. Urban Plan 6(1):20. https://doi.org/10.17645/up.v6i1.3515

Bush J, Frantzeskaki N, Ossola A, Pineda-Pinto M (2023) Priorities for mainstreaming urban nature-based solutions in Australian cities. Nat Based Solutions 3:100065. https://doi.org/10.1016/j.nbsj.2023.100065

Cohen-Shacham E, Andrade A, Dalton J, Dudley N, Jones M, Kumar C, Maginnis S, Maynard S et al (2019) Core principles for successfully implementing and upscaling nature-based solutions. Environ Sci Policy 98:20–29. https://doi.org/10.1016/j.envsci.2019.04.014

Coutts AM, Tapper NJ, Beringer J, Loughnan M, Demuzere M (2013) Watering our cities: the capacity for water sensitive urban design to support urban cooling and improve human thermal comfort in the Australian context. Prog Phys Geogr 37(1):2–28. https://doi.org/10.1177/0309133312461032

Daniels P, El Baghdadi O, Desha C, Matthews T (2020) Evaluating net community benefits of integrating nature within cities. Sustain Earth 3:12. https://doi.org/10.1186/s42055-020-00025-2

Davidson K, Gleeson B (2018) New socio-ecological imperatives for cities: possibilities and dilemmas for Australian metropolitan governance. Urban Policy Res 36(2):230–241. https://doi.org/10.1080/08111146.2017.1354848

Droste N, Alkan Olsson J, Hanson H, Knaggård Å, Lima G, Lundmark L, Thoni T, Zelli F (2022) A global overview of biodiversity offsetting governance. J Environ Manage 316:115231. https://doi.org/10.1016/j.jenvman.2022.115231

Dumitru A, Frantzeskaki N, Collier M (2020) Identifying principles for the design of robust impact evaluation frameworks for nature-based solutions in cities. Environ Sci Policy 112:107–116. https://doi.org/10.1016/j.envsci.2020.05.024

Fish R, Church A, Winter M (2016) Conceptualising cultural ecosystem services: a novel framework for research and critical engagement. Ecosyst Serv 21:208–217. https://doi.org/10.1016/j.ecoser.2016.09.002

Fitzsimons JA, Antos MJ, Palmer GC (2011) When more is less: urban remnants support high bird abundance but diversity varies. Pac Conserv Biol 17:97–109. https://doi.org/10.1071/PC110097

Fors H, Hagemann FA, Sang ÅO, Randrup TB (2021) Striving for inclusion—a systematic review of long-term participation in strategic management of urban green spaces. Front Sustain Cities 3:572423. https://doi.org/10.3389/frsc.2021.572423

Frantzeskaki N, de Haan H (2009) Transitions: two steps from theory to policy. Futures 41(9):593–606. https://doi.org/10.1016/j.futures.2009.04.009

Frantzeskaki N, McPhearson T (2021) Mainstream nature-based solutions for urban climate resilience. BioScience 72(2):113–115. https://doi.org/10.1093/biosci/biab105

Frantzeskaki N, McPhearson T, Collier MJ, Kendal D, Bulkeley H, Dumitru A, Walsh C, Noble K, van Wyk E, Ordóñez C, Oke C, Pintér L (2019) Nature-based solutions for urban climate change adaptation: linking science, policy, and practice communities for evidence-based decision-making. BioScience 69:455. https://doi.org/10.1093/biosci/biz042

Gardens for Wildlife Victoria (2021) Our work. https://gardensforwildlifevictoria.com/our-work/

Garrard GE, Williams NSG, Mata L, Thomas J, Bekessy SA (2018) Biodiversity sensitive urban design. Conserv Lett 11(2):e12411. https://doi.org/10.1111/conl.12411

Gentin S, Herslund LB, Gulsrud NM, Hunt JB (2022) Mosaic governance in Denmark: a systematic investigation of green volunteers in nature management in Denmark. Landsc Ecol 38:4177. https://doi.org/10.1007/s10980-022-01421-z

Giachino C, Pattanaro G, Bertoldi B, Bollani L, Bonadonna A (2021) Nature-based solutions and their potential to attract the young generations. Land Use Policy 101:105176. https://doi.org/10.1016/j.landusepol.2020.105176

Hansen R, Bush J, Pribadi DO, Giannotti E (2023) Planning and maintaining NBS—lessons for foresight and sustainable care from Berlin, Jakarta, Melbourne and Santiago de Chile. In: McPherson T, Kabisch N, Frantzeskaki N (eds) Nature-based solutions for cities. Edward Elgar, Cheltenham

Hanson HI, Olsson JA (2023) Uptake and use of biodiversity offsetting in urban planning—the case of Sweden. Urban For Urban Green 80:127841. https://doi.org/10.1016/j.ufug.2023.127841

Hes D, Bush J (2020) Designing for living environments using regenerative development: a case study of the paddock. In: Pedersen Zari M, Southcombe M, Connolly P (eds) Ecologies design. Routledge, Abingdon, pp 26–33

Hes D, du Plessis C (2015) Designing for hope: pathways to regenerative sustainability. Earthscan, Abingdon

Hill R, Adem Ç, Alangui WV, Molnár Z, Aumeeruddy-Thomas Y, Bridgewater P, Tengö M, Thaman R, Adou Yao CY, Berkes F, Carino J, Carneiro da Cunha M, Diaw MC, Díaz S, Figueroa VE, Fisher J, Hardison P, Ichikawa K, Kariuki P et al (2020) Working with indigenous, local and scientific knowledge in assessments of nature and nature's linkages with people. Curr Opin Environ Sustain 43:8–20. https://doi.org/10.1016/j.cosust.2019.12.006

Hobbie SE, Grimm NB (2020) Nature-based approaches to managing climate change impacts in cities. Philos Trans R Soc B Biol Sci 375:20190124. https://doi.org/10.1098/rstb.2019.0124

Ignatieva M (2010) Design and future of urban biodiversity. In: Müller N, Werner P, Kelcey J (eds) Urban biodiversity and design. Blackwells Publishing, London, pp 118–144

Ignatieva M (2018) Biodiversity-friendly designs in cities and towns: towards a global biodiversinesque style. In: Ossola A, Niemelä J (eds) Urban biodiversity: from research to practice. Francis and Taylor, London, pp 216–235

Ignatieva M, Haase D, Dushkova D, Haase A (2020) Lawns in cities: from a globalised urban green space phenomenon to sustainable nature-based solutions. Land 9:73. https://doi.org/10.3390/land9030073

Intergovernmental Panel on Climate Change (IPCC) (2023) Synthesis report of the IPCC sixth assessment report (AR6). Summary for policymakers. https://report.ipcc.ch/ar6syr/pdf/IPCC_AR6_SYR_SPM.pdf

Intergovernmental Science-Policy Platform on Biodiversity and Ecosystem Services (IPBES) (2022) Summary for policymakers of the methodological assessment of the diverse values and valuation of nature. https://zenodo.org/record/6832427#.YtNXl3ZByUk

Janke T (2021) True tracks: respecting indigenous knowledge and culture. NewSouth Publishing, Montgomery, AL

Jones DS, Choy DL, Tucker R, Heyes S, Revell G, Bird S (2018) Indigenous knowledge in the built environment: a guide for tertiary educators. Department of Education and Training, Sydney, NSW

Kabisch N, van den Bosch M, Lafortezza R (2017) The health benefits of nature-based solutions to urbanization challenges for children and the elderly—a systematic review. Environ Res 159:362–373. https://doi.org/10.1016/j.envres.2017.08.004

Kalliolevo H, Gordon A, Sharma R, Bull JW, Bekessy SA (2021) Biodiversity offsetting can relocate nature away from people: an empirical case study in Western Australia. Conserv Sci Pract 3(10):e512

Keeler BL, Hamel P, McPherson T, Hamann MH, Donahue ML, Meza Prado KA, Arkema KK, Bratman GN et al (2019) Social-ecological and technological factors moderate the value of urban nature. Nat Sustain 2(January):29–38. https://doi.org/10.1038/s41893-018-0202-1

Kingsley J, Townsend M, Henderson-Wilson C, Bolam B (2013) Developing an exploratory framework linking Australian aboriginal peoples' connection to country and concepts of well-being. Int J Environ Res Public Health 10(2):678–698. https://doi.org/10.3390/ijerph10020678

Kingsley J, Egerer M, Nuttman S, Keniger L, Pettitt P, Frantzeskaki N, Gray T, Ossola A et al (2021a) Urban agriculture as a nature-based solution to address socio-ecological challenges in Australian cities. Urban For Urban Green 60:127059. https://doi.org/10.1016/j.ufug.2021.127059

Kingsley J, Munro-Harrison E, Jenkins A, Thorpe A (2021b) Developing a framework identifying the outcomes, principles and enablers of "gathering places": perspectives from aboriginal people in Victoria, Australia. Soc Sci Med 283:114217

Kronsell A, Mukhtar-Landgren D (2018) Experimental governance: the role of municipalities in urban living labs. Eur Plan Stud 26(5):988–1007. https://doi.org/10.1080/0965431 3.2018.1435631

Langton M (2018) Welcome to country: a travel guide to indigenous Australia. Hardie Grant Travel, Melbourne

Lin BB, Ossola A, Ripple WJ, Alberti M, Andersson E, Bai X, Dobbs C, Elmqvist T et al (2021) Integrating solutions to transform cities for climate change. Lancet Planet Health 5(7):e479–e486. https://doi.org/10.1016/S2542-5196(21)00135-2

Lindenmayer DB, Crane M, Evans MC, Maron M, Gibbons P, Bekessy S, Blanchard W (2017) The anatomy of a failed offset. Biol Conserv 210:286–292. https://doi.org/10.1016/j.biocon.2017.04.022

Loorbach D, Wittmayer J, Avelino F, von Wirth T, Frantzeskaki N (2020) Transformational innovation and translocal diffusion. Environ Innov Soc Trans 35:251–260. https://doi.org/10.1016/j.eist.2020.01.009

Ludwig D, Macnaghten P (2020) Traditional ecological knowledge in innovation governance: a framework for responsible and just innovation. J Respons Innov 7(1):26–44. https://doi.org/1 0.1080/23299460.2019.1676686

Malekpour S, Tawfik S, Chesterfield C (2021) Designing collaborative governance for nature-based solutions. Urban For Urban Green 62:127177. https://doi.org/10.1016/j.ufug.2021.127177

Maller C (2021) Re-orienting nature-based solutions with more-than-human thinking. Cities 113:103155. https://doi.org/10.1016/j.cities.2021.103155

Mang P, Haggard B (2016) Regenerative development and design: a framework for evolving sustainability. Wiley, Hoboken, NJ

Matsler MA (2019) Making "green" fit in a "grey" accounting system: the institutional knowledge system challenges of valuing urban nature as infrastructural assets. Environ Sci Policy 99:160–168. https://doi.org/10.1016/j.envsci.2019.05.023

Mattijssen TJM, Ganzevoort W, van den Born RJG, Arts BJM, Breman BC, Buijs AE, van Dam RI, Elands BHM, de Groot WT, Knippenberg LWJ (2020) Relational values of nature: leverage points for nature policy in Europe. Ecosyst People 16(1):402–410

McGregor BA, McGregor AM (2020) Communities caring for land and nature in Victoria. J Outdoor Environ Educ 23(2):153–171. https://doi.org/10.1007/s42322-020-00052-9

Moloney S, Doyon A (2021) The resilient Melbourne experiment: analyzing the conditions for transformative urban resilience implementation. Cities 110:103017. https://doi.org/10.1016/j.cities.2020.103017

Mumaw L (2017) Transforming urban gardeners into land stewards. J Environ Psychol 52:92–103. https://doi.org/10.1016/j.jenvp.2017.05.003

Nagendra H, Bai X, Brondizio ES, Lwasa S (2018) The urban south and the predicament of global sustainability. Nat Sustain 1:341–349. https://doi.org/10.1038/s41893-018-0101-5

Norman B, Newman P, Steffen W (2021) Apocalypse now: Australian bushfires and the future of urban settlements. NPJ Urban Sustain 1:2. https://doi.org/10.1038/s42949-020-00013-7

O'Bryan K (2019) The changing face of river management in Victoria: the Yarra River protection (Wilip-gin Birrarung murron) act 2017 (Vic). Water Int 44(6–7):769–785. https://doi.org/1 0.1080/02508060.2019.1616370

Oke C, Bekessy S, Frantzeskaki N, Bush J, Fitzsimons J, Garrard GE, Grenfell M, Harrison L et al (2021) Cities should respond to the biodiversity extinction crisis. Urban Sustain 1(1):1–4. https://doi.org/10.1038/s42949-020-00010-w

Ossola A, Lin BB (2021) Making nature-based solutions climate-ready for the 50°C world. Environ Sci Policy 123:151–159. https://doi.org/10.1016/j.envsci.2021.05.026

Padma P, Ramakrishna S, Rasoolimanesh SM (2020) Nature-based solutions in tourism: a review of the literature and conceptualization. J Hosp Tourism Res 46(3):442–466. https://doi.org/10.1177/1096348019890052

Palmer GC, Fitzsimons JA, Antos MJ, White JG (2008) Determinants of native avian richness in suburban remnant vegetation: implications for conservation planning. Biol Conserv 141(9):2329–2341. https://doi.org/10.1016/j.biocon.2008.06.025

Parris KM, Barrett BS, Stanley HM, Hurley J (eds) (2020) Cities for people and nature. Clean Air and Urban. Landscapes Hub, Melbourne. https://nespurban.edu.au/wp-content/uploads/2020/11/Cities-for-People-and-Nature.pdf

Pascoe S, Cannard T, Steven A (2019) Offset payments can reduce environmental impacts of urban development. Environ Sci Policy 100:205–210. https://doi.org/10.1016/j.envsci.2019.06.009

Pineda Pinto M, Frantzeskaki N, Nygaard CA (2021) The potential of nature-based solutions to deliver ecologically just cities: lessons for research and urban planning from a systematic literature review. Ambio 16:167. https://doi.org/10.1007/s13280-021-01553-7

Pineda-Pinto M, Kennedy C, Collier M, Cooper C, O'Donnell M, Nulty F, Castañeda NR (2023) Finding justice in wild, novel ecosystems: a review through a multispecies lens. Urban For Urban Green 83:127902. https://doi.org/10.1016/j.ufug.2023.127902

Porter L (2019) Learning to live lawfully on country. In: Maddison S, Nakata S (eds) Questioning indigenous-settler relations: interdisciplinary perspectives, vol 1. Springer, Singapore, pp 137–146. https://doi.org/10.1007/978-981-13-9205-4_9

Potter E (2020) Contesting imaginaries in the Australian city: urban planning, public storytelling and the implications for climate change. Urban Stud 57(7):1536–1552. https://doi.org/10.1177/0042098018821304

Schiller J, Rayner JP, Williams NSG (2023) Guidelines for biodiversity green roofs. Report for the City of Melbourne. https://www.melbourne.vic.gov.au/SiteCollectionDocuments/guidelines-for-biodiversity-green-roofs-2023.pdf

Shade C, Kremer P, Rockwell JS, Henderson KG (2020) The effects of urban development and current green infrastructure policy on future climate change resilience. Ecol Soc 25(4):37. https://doi.org/10.5751/ES-12076-250437

Söderqvist T, Cole S, Franzén F, Hasselström L, Beery TH, Bengtsson F, Björn H, Kjeller E, Lindblom E, Mellin A, Wiberg J, Jönsson KI (2021) Metrics for environmental compensation: a comparative analysis of Swedish municipalities. J Environ Manage 299:113622. https://doi.org/10.1016/j.jenvman.2021.113622

Tabassum S, Manea A, Ossola A, Thomy B, Blackham D, Leishman MR (2021) The angriest summer on record: assessing canopy damage and economic costs of an extreme climatic event. Urban For Urban Green 63:127221. https://doi.org/10.1016/j.ufug.2021.127221

Tebrakunna country, Lee E (2019) "Reset the relationship": Decolonising government to increase indigenous benefit. Cult Geogr 26(4):415–434. https://doi.org/10.1177/1474474019842891

Tengö M, Hill R, Malmer P, Raymond CM, Spierenburg M, Danielsen F, Elmqvist T, Folke C (2017) Weaving knowledge systems in IPBES, CBD and beyond—lessons learned for sustainability. Curr Opin Environ Sustain 26:17–25

Terare M, Rawsthorne M (2020) Country is yarning to me: worldview, health and Well-being amongst Australian first nations people. Br J Soc Work 50(3):944–960. https://doi.org/10.1093/bjsw/bcz072

Threlfall C, Marzinelli EM, Ossola A, Bugnot A, Bishop M, Lowe L, Imberger S, Myers S et al (2021) Towards cross-realm management of coastal urban ecosystems. Front Ecol Environ 19(4):225–233. https://doi.org/10.1002/fee.2323

Tozer L, Hörschelmann K, Anguelovski I, Bulkeley H, Lazova Y (2020) Whose city? Whose nature? Towards inclusive nature-based solution governance. Cities 107:102892. https://doi.org/10.1016/j.cities.2020.102892

Tozer L, Bulkeley H, van der Jagt A, Toxopeus H, Xie L, Runhaar H (2022) Catalyzing sustainability pathways: navigating urban nature based solutions in Europe. Glob Environ Chang 74:102521. https://doi.org/10.1016/j.gloenvcha.2022.102521

Trisos CH, Auerbach J, Katti M (2021) Decoloniality and anti-oppressive practices for a more ethical ecology. Nat Ecol Evol 5:1205–1212. https://doi.org/10.1038/s41559-021-01460-w

Ugolini F, Massetti L, Calaza-Martínez P, Cariñanos P, Dobbs C, Ostoic SK, Marin AM, Pearlmutter D et al (2020) Effects of the COVID-19 pandemic on the use and perceptions of urban green space: an international exploratory study. Urban For Urban Green 56:126888. https://doi.org/10.1016/j.ufug.2020.126888

United Nations Environment Programme (UNEP) (2021a) Becoming #Generation Restoration: Ecosystem restoration for people, nature and climate. https://wedocs.unep.org/bitstream/handle/20.500.11822/36251/ERPNC.pdf

United Nations Environment Programme (UNEP) (2021b) Making peace with nature: a scientific blueprint to tackle the climate, biodiversity and pollution emergencies. Nairobi. https://www.unep.org/resources/making-peace-nature.ck

van der Jagt APN, Kiss B, Hirose S, Takahashi W (2021) Nature-based solutions or debacles? The politics of reflexive governance for sustainable and just cities. Front Sustain Cities 2:583833. https://doi.org/10.3389/frsc.2020.583833

Voskamp IM, de Luca C, Polo-Ballinas MB, Hulsman H, Brolsma R (2021) Nature-based solutions tools for planning urban climate adaptation: state of the art. Sustainability 13(11):6381. https://doi.org/10.3390/su13116381

Wickenberg B, Kiss B, McCormick K, Palgan YV (2022) Seeds of transformative learning: investigating past experiences from implementing nature-based solutions. Front Sustain Cities 4:835511. https://doi.org/10.3389/frsc.2022.835511

Willems JJ, Giezen M (2022) Understanding the institutional work of boundary objects in climate-proofing cities: the case of Amsterdam rainproof. Urban Clim 44:101222. https://doi.org/10.1016/j.uclim.2022.101222

Willems JJ, Kuitert L, Van Buuren A (2022) Policy integration in urban living labs: delivering multi-functional blue-green infrastructure in Antwerp, Dordrecht, and Gothenburg. Environ Policy Gov 33(3):258–271. https://doi.org/10.1002/eet.2028

Williams NSG, Bathgate R, Farrell C, Lee KE, Szota C, Bush J, Johnson KA, Miller RE et al (2021) Ten years of greening a wide brown land: a synthesis of Australian green roof research and roadmap forward. Urban For Urban Green 62:127179. https://doi.org/10.1016/j.ufug.2021.127179

World Economic Forum (2020) The future of nature and business. New Nature Economy Report II. Switzerland.

Woodward E, Hill R, Harkness P, Archer R (eds) (2020) Our knowledge our way in caring for country: indigenous-led approaches to strengthening and sharing our knowledge for land and sea management. Best practice guidelines from Australian experiences, NAILSMA and CSIRO. https://www.csiro.au/en/research/indigenous-science/indigenous-knowledge/our-knowledge-our-way

Niki Frantzeskaki is a Chair Professor of Regional and Metropolitan Governance and Planning, Section Spatial Planning, Geosciences Faculty, Utrecht University, the Netherlands. Her expertise is on urban transitions and transformations, their governance and planning, with a focus on achieving climate change adaptation and mitigation, sustainability, and resilience. Her research also focuses on the governance and planning of nature-based solutions to enhance climate change resilience and promote more just urban futures. She has a rich international research experience with a portfolio of ongoing projects in Australia, Canada, and the USA. She has been a Highly Cited Researcher awardee from Clarivate Analytics in 2020 and 2021, placing her in the top 1% of researchers globally in the cross-field of social sciences and ecology. From 2019 to 2021, she has been a Research Professor and Director of the Centre for Urban Transitions at Swinburne University of Technology, Melbourne, Australia. From 2010 to 2019, she has been an Associate Professor at the Dutch Research Institute for Transitions, affiliated with Erasmus University Rotterdam. She has published over 100 peer-reviewed articles and 18 special issues and released four books on urban sustainability transitions in 2017, 2018, and 2020.

Judy Bush is a Senior Lecturer in Urban Planning at the University of Melbourne, Australia. Her research focuses on urban environmental policy and governance, including governance and policy approaches for nature-based solutions, biodiversity, urban ecology, and climate change perspectives.

Dave Kendal is the Director of Future in Nature and an internationally recognised expert in nature-based solutions, urban forests, biodiversity conservation, climate adaptation, and sustainable people–environment relationship. He co-leads the Oceania node of the global NATURA network, exploring nature-based solutions in cities. Dave aims to make a unique contribution to creating a better world for all species; include the perspectives of local communities, First Nations peoples, and threatened species in decision-making; support the people he works with to achieve their goals; and use evidence to support innovative and creative solutions to address the unprecedented challenges faced by society.

Clare Adams is a PhD candidate at the Centre for Urban Transitions, Swinburne University of Technology, Melbourne, Australia. Her research focuses on the mainstreaming of nature-based solutions in cities from the perspective of urban planning and governance.

Loretta Bellato is a settler woman from Melbourne, Australia. She was recently awarded a PhD from Swinburne University of Technology for examining regenerative tourism approaches to urban development. She has published four peer-reviewed papers on the topic and is co-editing a special issue on regenerative development and tourism. Loretta is a graduate of the Regenesis Regenerative Practitioner Series, holds a Master of International Development from RMIT and Master of International Sustainable Tourism Management, Monash University, and has extensive practitioner experience in the community health and tourism sectors.

Alessandro Ossola is an urban ecologist and environmental scientist, and Assistant Professor at the University of California, Davis, USA. He is honorary faculty at the University of Melbourne, Australia, a 2022 FFAR New Innovator Awardee, and a former NASEM Associate with US-EPA. His research encompasses urban ecology, climate change science, urban sustainability, and governance. Over the years, he has advised several government agencies in various countries on urban nature-based solutions for climate adaptation. He is interested in applied, co-produced research that bridges environmental management, ecological design, and science communication.

Cathy Oke is the Melbourne Enterprise Principal Fellow in Informed Cities within the Faculty of Architecture, Building, and Planning and Deputy Director (Strategy and Operations) for Melbourne Centre for Cities. She has considerable international and local expertise in sustainable, resilient, and liveable cities. Cathy's research interests focus on the interaction between urban nature and climate—research, policy, and practice—for greater impact in cities. She is Senior Advisor to the Innovate4Cities program of the Global Covenant of Mayors for Climate and Energy and is Research Portfolio holder on ICLEI's (Local Governments for Sustainability) Global Executive Committee.

Chapter 9
Regenerative Urban Development Paradigms in a Time of Climate Change and Ecological Crisis

Melissa Pineda Pinto and Wendy Steele

Abstract In this chapter, we draw attention to the need for, and yet radical nature of, embracing regenerative urban futures in the context of the climate and biodiversity crisis. This is a mission-oriented vision that recognises the need to fundamentally reconceptualise cities and urban regions as living entities that must be supported by more regenerative ways of imagining the role of urban nature cultures and multispecies justice. This is an emphasis on the "livingness" of cities and the urgent need to shift *away* from extraction, devaluation, and displacement practices that affect both humans and non-humans. Genuinely addressing a regenerative future vision demands that cities are co-designed with, and for, the flourishing of more-than-human communities. The chapter draws on the Three Horizons approach to put forward a paradigm shift to regenerative futures, which is framed as alternative ways of governing our cities, and illustrates this shift with examples of regenerative practices in Australia. The need for urban regeneration as a transformative mission within the Australian context is highlighted.

Keywords Cities · More-than-human · Multispecies · Regenerative futures · Climate crisis

M. Pineda Pinto (✉)
The University of Melbourne, Melbourne, VIC, Australia
e-mail: melissa.pinedapinto@unimelb.edu.au

W. Steele
RMIT University, Melbourne, VIC, Australia

© The Author(s) 2025
N. Frantzeskaki et al. (eds.), *Future Cities Making*, Theory and Practice of
Urban Sustainability Transitions, https://doi.org/10.1007/978-981-97-7671-9_9

9.1 Hot Cities in Crisis

The need for radical transformations in cities and urban regions to address the climate and biodiversity emergency is already upon us. Emissions are rising fast, and the planet is rapidly heating up, taking global heat temperatures into "uncharted territory", with catastrophic consequences for life on Earth—human and non-human. This crisis requires actively shaping and co-creating more inclusive and sustainable futures as a central societal mission (see Mazzucuto 2021). This chapter focuses on the need to build regenerative urban futures that serve to fundamentally transform the future of "living" cities and wider urban settlements.

As major greenhouse gas emitters, cities—and the activities and practices that sustain them—exacerbate climate change impacts such as urban heat island effects, as well as urban floods and fires (Solecki and Marcotullio 2013). The Sixth Intergovernmental Panel on Climate Change report stressed that the impact of carbon-intensive cities will "intensify human-induced warming locally". Unchecked urbanisation, together with more frequent heat-related extremes, will increase the severity of heatwaves and extreme sea-level events, with rainfall and river flow events exacerbating the likelihood of flooding and landslide disasters (IPCC 2021).

The catastrophic fires that have ravaged human settlements and unique ecosystems around the world are an example of our changing climate and the severity and intensity of events. In Europe, Canada, the United States, India, Pakistan, and Africa, unprecedented fire and heat events in the last decade have also prompted calls for critical changes to urban settlement and development trajectories (Steele et al. 2023). Temperatures as high as 47 °C have caused deaths and prompted large-scale evacuations in Algeria, Croatia, Greece, Italy, Portugal, and Spain, devastating communities in the natural and built environment (Kwai 2023). The number of extreme heat days in Australia is increasing, particularly in cities, with estimations showing a 471% increase in heatwave-related deaths by 2080 in a high-emissions trajectory (Guo et al. 2018).

In Australia, the Black Summer fires of 2020 burnt through millions of hectares, destroyed infrastructure, and killed 400 people and over one billion animals. Alongside ecological world heritage places and rural areas, urban and peri-urban areas also suffered devastating losses, with days that registered the worst air-quality indices in the world. These climate impacts draw attention to the complex interactions between climate change, urban areas, and biodiversity. The fires stopped the nation, inciting deep considerations of the way Australians live, and the implications for urban futures (Norman et al. 2021).

In this chapter, we draw attention to the "livingness" of cities and the urgent need to shift away from extraction, devaluation, and displacement practices that affect both humans and non-humans. Genuinely addressing a regenerative future vision demands that cities are co-designed with, and for, the flourishing of communities—both human and non-human. This is framed as an alternative paradigm that will transform cities into more-than-human regenerative futures. The chapter draws on the Three Horizons (Sharpe et al. 2016; 2020) approach to put forward a new way of governing cities. This framework is a simple approach for navigating complexity that offers insights into the nature of transformation by engaging with patterns of the present and the future (Sharpe et al. 2016; Sharpe 2020). A recent review of futures-thinking literature identified that transformative change is critical to address the challenges of the Anthropocene (Cork et al. 2023). The proposed horizons are tied to existing regenerative practices within the Australian urban context that serve as examples of achieving regenerative futures. The chapter concludes with the implications for regenerative governance policy and planning as a transformative mission within the Australian urban context.

9.2 Alternative Urban Futures

As climate change continues to unfold, multispecies injustices will multiply, with the most vulnerable and marginalised suffering the greatest (Celermajer et al. 2021; Pineda-Pinto et al. 2022; Tschakert et al. 2020). The changes caused by development and growth underpinned by extractive and profit-driven actions lead to displacement, inequity, and deprivations. This is a challenge underpinned by an urgent need for transformation away from the extractive pathways that affect both people and the living planet in increasingly perverse and violent ways and towards more sustainable and regenerative futures.

One way to progress and co-create solutions for more regenerative cities is to develop future-oriented visions, or urban imaginaries. In an urbanising planet, cities as multispecies habitats and spaces for protecting and recovering threatened and vulnerable species—both human and non-human—require a new trajectory in urban policy and planning. A focus on futures involves *looking backwards*, to understand how cities have brought both opportunities and costs; *looking to the present*, to spotlight how cities are finding new ways to work together to create impact; and *looking forwards*, to determine how regenerative practices can be more clearly reflected in urban governance (Sharpe 2023).

The future is not static or separate but inextricably linked to the past and present. A focus on regenerative futures within the context of the climate and biodiversity crisis implies making decisions about what to let go of, what to conserve, and what to radically reimagine or change. This approach to futures thinking is summarised through the *Three Horizons Futures Framework* by Bill (Sharpe et al. 2016; 2020) (Fig. 9.1), which has informed the work of Kate Raworth, author of *Doughnut Economics* and advocate for creating more just and regenerative economies. The Three Horizons model explores different futures coexisting in the seeds of the present. Horizon 1 maps a business-as-usual approach; Horizon 2 focuses on emerging positive changes evident in the present; and Horizon 3 is the reimagining of the present into sustainable and regenerative urban futures.

In *Horizon 1*, cities are key actors in the global effort to address climate change and socio-economic inequalities and to protect biodiversity—but they are also a large part of the problem. Cities are deeply rooted in extractive, colonial practices

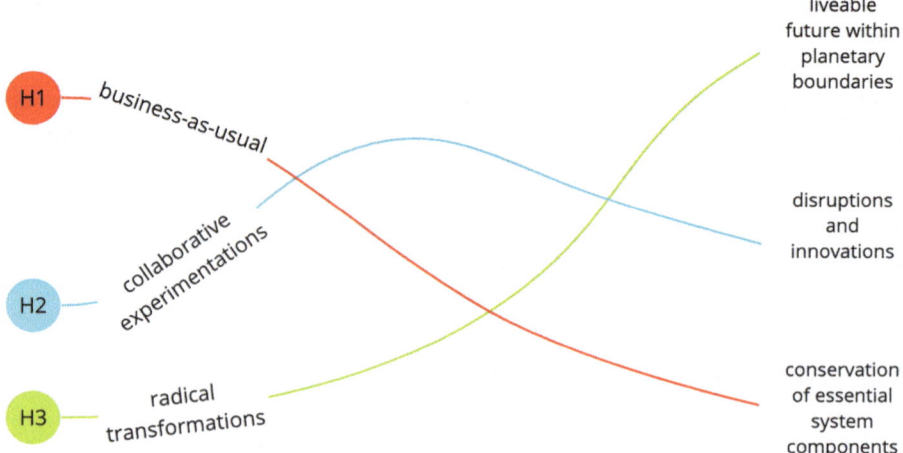

Fig. 9.1 The Three Horizons framework for mapping the current system, including what elements will be lost and retained (H1) and which innovations and disruptions will create new pathways and opportunities (H2) for taking us to Horizon 3 (H3), which involves shared visions of viable futures. (Adapted from Sharpe 2019, 2023)

(Lea 2020; Porter 2020). These have led to the devaluation and displacement of vulnerable human and non-human communities. A central message from research on the role of cities in the climate crisis is the need to radically change the path-dependent patterns of unsustainability (Bai et al. 2018; Frantzeskaki et al. 2017). As the global heat crisis unfolds and the world grapples for ways to move forwards in the climate emergency, there is growing recognition that cities must urgently regenerate for the planet to survive.

This leads to *Horizon 2,* where the mission for cities is increasingly clear: they must meet the urgent need to find creative ways to support more regenerative and more equitable futures in the context of the climate emergency. This is about ethical innovations in thinking and practices. These innovations emerge from alternative paradigms that are usually rooted in indigenous ways of being and doing and/or those that are grounded in an ethics of care and interconnectedness. Buen Vivir, for example, is an Andean way of being that seeks the transformation of post-extractivist futures through leading "a good life": a "vision and a platform for practising alternative futures focused on lived practice, that is connected to global movements that promote economies of sharing and care" (Salazar 2015, p. 1).

Catherine Walsh (2010) highlights how alternative paradigms when applied in practice can be challenging due to the entangled nature of ongoing settler colonial practices, the difficulties in applying these paradigms to diverse cultural contexts, and the inconsistencies and contradictions that emerge in everyday practices where these alternative approaches to sustainability-led transformation are co-opted by the State as a new paradigm for development. The question is, then, how do we enact new urban paradigms and practices that disrupt the status quo? As the Royal Society for Arts (2023) emphasises:

> The future doesn't just happen, it's up to us to create it. As we face the challenges of climate change, inequality and environmental degradation, we know to simply sustain is not enough. We want to see a world where people and communities harness their potential to be sources of health and regeneration for all life on earth. Because people and planetary needs are intertwined, our problem solving should be too. We need to regenerate (p. 1).

Horizon 3 points to the need to radically reimagine more regenerative and more equitable futures, including the need to address in meaningful ways the complex interactions between climate change, cities, and the biodiversity that sustains life on Earth. Lefebvre (2014), for example, called for a planetary-scale urban metamorphosis. If the urban is "the sum of the productive practices and vehicle for new values and an alternate civilization", then, according to Lefebvre, without a major metamorphosis in the urban, society's "hopes are fading". The difference between "change in society" and "metamorphosis in the world" is also outlined by Beck (2016), who argues that a metamorphosis is the overhaul of the social contract in ways that were unthinkable in the past, have become real in the present, and will be critical for the future. This metamorphosis will require leverage points—"…places in the system where a small change could lead to a large shift in behavior"—which are counterintuitive and require deep cultural changes, a shift in society's paradigms (Meadows 2008; Ch. 6 para.1).

All three of these horizons draw attention to the need for greater awareness and recognition of the *intentional* frames for transformation for regenerative urban futures, that is, the intentions needed to repair and heal the already existing more-than-human urban environments. This will require creating opportunities for multi-species life systems to flourish in harsh, new, hot city environments based on recognising and acting on the need for uncertainty and to be out of our current comfort zone; attending carefully to the power of business-as-usual practices and an alternative understanding of what constitutes "progress"; listening deeply to Indigenous communities and cultures that seek to honour and embrace all life on Earth, or on Country; and disrupting the current unsustainable development logics through the creation of new joined-up multispecies and urban-nature imaginaries.

This is not new, as First Nations leadership and knowledge systems demonstrate. Powerful, Earth-centred paradigms and ecological cosmologies already offer alternative understandings of being and seeing in and of the world that differ radically from the ideologies and values that lie beneath the development histories of the world's cities. In Australia, all cities are built, and continue to be developed, on Aboriginal and Torres Strait Islander land, which was never ceded by First Nations people. This is a shared climate crisis—First Nations and colonisers, human, and non-human—but with very differentiated histories and therefore ethical responsibilities. The following section draws on the Three Horizons frame to outline different pathways to regenerative futures, with a focus on the urban governance, policy, and planning implications for multispecies justice and the vital role of our rivers and waterways.

9.3 Pathways for Multispecies Flourishing

The climate and biodiversity crisis is forcing cities and their inhabitants to recognise the shared nature of the more-than-human climate crisis. Cities can change in response to this across the Three Horizons in ways that recognise what values, policies, and practices we need to let go of (and which to keep) (Horizon 1); the new experiments and innovations that are taking place to address the changing climate and impacts that are simultaneously environmental, cultural, social, and economic (Horizon 2); and the radical prospects for transformative change that are needed to achieve sustainable urban settlements in which humans and non-humans are able to survive and flourish (Horizon 3). We propose that the horizon pathways can enable a new set of paradigms for achieving regenerative urban futures (Fig. 9.2). An example of each will be outlined below. The Three Horizons framework, as opposed to other futures-thinking models, creates a "triangle of change" that builds a space to visualise how Horizon 1 pathways start to fail; how and when, and through which innovative actions, change starts to emerge in Horizon 2; and how changes in values and beliefs start to gain influence (Curry and Hodgson 2008). Within this space, it is possible to identify conflicts and power dynamics and visualise divergent views and underpinning values that shape a desired future (Curry and Hodgson 2008).

9.3.1   Horizon 1: The "River with a City Problem" (Brisbane)

In her account of the 2011 Brisbane floods in the Australian state of Queensland, historian Margaret Cook (2019) makes clear that human decisions and actions were the drivers of the disaster, not the "wrath of mother nature" or "weather of mass destruction". And not just any humans, but specifically colonial settlers who built cities on flood-plains. The Brisbane River meanders, and as Cook points out, the riverbed, banks, and floodplains are made of mud, sand, and silt that move and shifts as needed. The floodplains are a living ecosystem, and the overflow of water is part of the cycle that changes with the seasons and, in doing so, supports the cycle of biodiversity that exists symbiotically both in the water and on land. The term "flood", she argues, is a highly anthropocentric term relating to the overflow of water that affects human settlements.

Since colonisation, Australian cities have traditionally relied on big engineering solutions for their water security. The impact of property rights on water access and use and the privatisation of water and water authorities has reinforced a maximum of consumption and profits and increases in both access and supply. As Troy (2008, p. 1) highlights, "now the cities must cope with the stresses these policies have imposed on the eco-systems from which they harvest water, into which they discharge wastes, and on which they are located. Residents are having to pay more for their water, while the cities themselves are becoming less sustainable". More broadly, the approach to urban water has reflected a presumption that urban

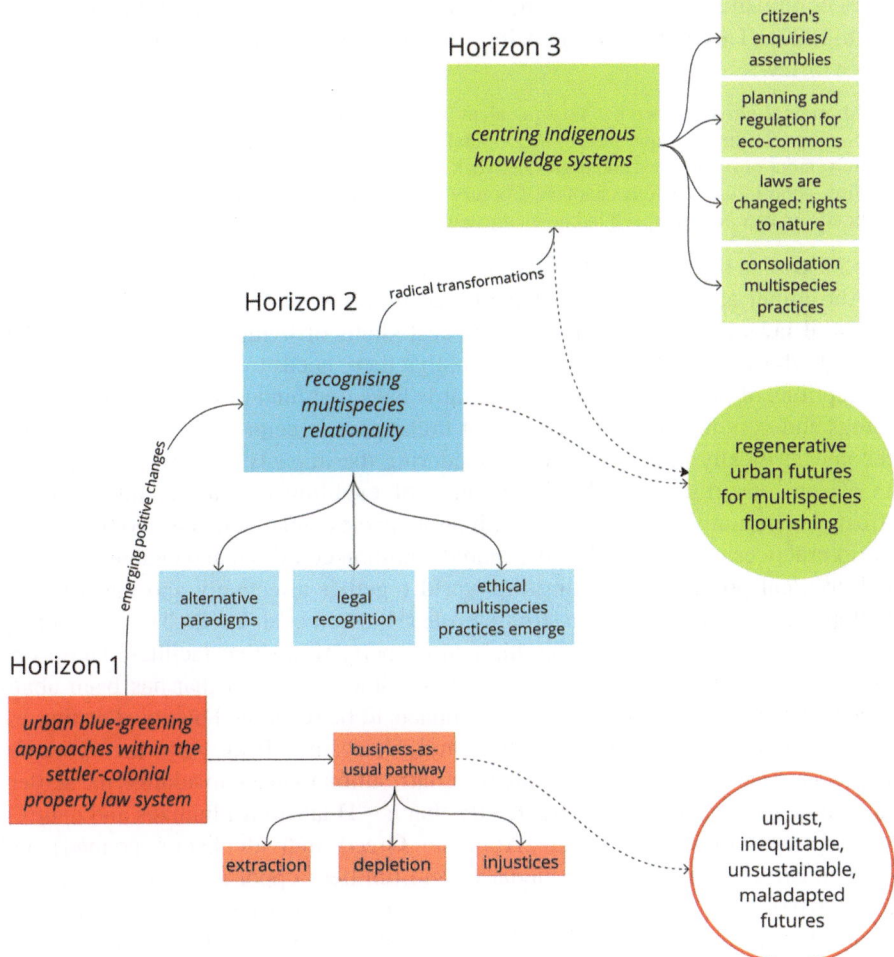

Fig. 9.2 The Three Horizon pathways for achieving urban regeneration for multispecies flourishing

development will be able to control "nature"—the case of water, through the building of dams, desalination plants, seawalls, and the like.

The settler-colonial property law system in Australia emerges from and serves neoliberal markets and economies. This causes two key problems: first, markets do not capture social-ecological dynamics that emerge and are entangled in the land and thus fail to recognise the interests of the more-than-human—they are focused on the individual; and second, individual private-property rights supersede the rights of the public or the commons (Schuijers and Bush 2022). In trying to find ways to green the city to "bring nature back", adapt to or mitigate climate change, or enhance biodiversity, urban blue-greening approaches within the private-property system do not accomplish social-ecological justice in cities (Cooke et al. 2019; Schuijers and Bush 2022). This is also in opposition to Indigenous and First Nations jurisprudence or law:

> Aboriginal jurisprudence has no equivalent notion of private property. Being bound –not binding—is the sine qua non of Aboriginal jurisprudence: being bound is a reciprocal movement of obligations and duties between both humans and non-humans. In light of the discussion in the previous chapters, Country is not land and neither is it soil. It is law, as bonds, matter and life fused in knowledge and relational practice (Lay 2016, pp. 255–256).

The Australian approach to *Aqua Nullius*, or "no-one's water", denies the existence of Aboriginal and Torres Strait Islander sovereignty, connection to Country, and ancestral laws that position them as the custodians of living waters (see Marshall 2017; Taylor et al. 2022). It follows, then, that there needs to be a shift in the liberal conceptions of property and a move towards laws that bind or bring together living beings and systems through practices of multispecies reciprocity (Lay 2016). This becomes critically important when considering the impacts of climate change and its effects post-disasters. The importance of enabling displaced and otherwise affected communities by engaging with multispecies entanglements and becoming-world practices that acknowledge a shared common condition and enable an ethical and political process of creating the world together affectively and ensuring the well-being of all need to play a stronger role (Houston et al. 2018). This would help create new models of land ownership and property rights that facilitate land to be "donated", or returned to the interspecies commons, or land that has been abandoned or decimated by climate change impacts to be reclaimed by novel ecologies.

Indigenous perspectives, developed on Country in holistic ways incorporating lore and law, have a particularly valuable contribution to make to address the settler-colonial legacy and capitalist DAMAGE: that is, "**D**ualism (of humans and nature), **A**nthropocentricism, **M**aterialism, **A**tomism, **G**reed (individualism gone mad), and **E**conomism (the myth of no boundaries and limitless opportunities)" (Bosselman et al. 2010). While this is a relatively new concept in Australia, similar "rights of nature" laws, which change the legal status of nature, exist in New Zealand, Ecuador, Bolivia, Colombia, India, and Uganda, to name a few. A shift to more regenerative futures requires far better legal recognition of the role of traditional owners, which includes cultural and environmental heritage protection, in the governance, policy, and planning of cities and urban regions. As Cook (2019) describes in the Brisbane context:

> For millennia the Brisbane River followed its own hydrological rhythms with floods replenishing the Estuarine environment and regenerating the floodplains. For 60,000 years the Turrbal and Jagera people had a spiritual connection with the country, respecting and accommodating the river's life cycles. British colonisation in 1824 brought a problem for the river: settlement of the society on a floodplain by a people imbued with notions of human superiority over nature, a mindset that viewed nature as bounty for progress. To the colonists, riverine floods brought a moment of disorder as the river left its "proper place" with catastrophic results, shattering the ideal of the linear path of progress (pp. ix–x).

9.3.2 Horizon 2: A Swimmable Birrarung in Naarm (Melbourne)

The pathways to regenerative urban futures begin with embedding new thinking and practices in cities, which include tangible actions to shift cities from business as usual towards addressing climate change and biodiversity loss in creative and collaborative ways. These include the creation of climate and biodiversity-responsive cities where an ethic of care is central to building community, sharing responsibilities, and bringing together spaces and opportunities that facilitate multispecies flourishing. Finding ways to develop eco-social commons is a critical agenda for regenerative futures. This includes developing socially innovative forms of governance and planning that place ecological or multispecies rights at the forefront of justice and decision-making (Metzger 2016).

Current examples of this include supporting different forms of community and nature land trusts that model providing affordable shelter, alternative sources of food, and protect ecosystem functions through stewardship, community participation, and multistakeholder ownership. These types of more-than-human communing practices, based on shared responsibilities, duties, and custodianship, can be acts of "quiet activism" such as reclaiming vacant and abandoned land or creating gardens in street verges. These efforts are already present in Australian urban regions in the shape of urban wildlife habitats in private gardens, long-standing community gardens, and informal tactical urbanism blue-green spaces that are tended and used by local communities. These experimental earth-centred practices can be amplified to increase their impact, replicability, and acceptance through community engagement and local council support (Steele et al. 2021).

An example of this can be found in the innovative projects of Regen Melbourne (RM), a not-for-profit movement committed to addressing systemic problems by building alliances between unusual actors from the business, non-profit, government, and university sectors and the general public in Naarm (Melbourne, Australia). RM's (2023) argument is that our urban systems are not currently built for this type of collaboration. They describe the need to "break out of siloes and increase collective ambition, create new structures that reactivate and reorganise systems, and initiate ambitious, tangible projects that chart a collective course to a regenerative and resilient future" (p. 1).

The Regen Melbourne "Swimmable Birrarung" project, for example, is a novel adaptation of a broader global movement on how cities undertake transitions towards environmental sustainability, by "daylighting", or working to bring back and restore, urban rivers, waterways, and ecological habitats that have been built over or hidden underground (Lerner 2019). Their aim is much more than restoring the possibility of swimming for the purposes of human leisure; instead, they aim to establish a collective way to address climate change in Melbourne by "reorienting our city to recognise our waterway as a living entity, as a place inextricably linked to health and biodiversity, and working as a coalition of action supporting the holistic regeneration of the lifeforce that is the Birrarung" (p. 1).

Aboriginal author Tony Birch writes that in the context of the Birrarung, there is the great privilege of being on Wurundjeri land and paying respects not only to their elders, past, present, and emerging, but to all of the Wurundjeri people and their sacred Country including the rivers, waterways, lands, air, and other living species. Tony grew up next to the Merri Creek and has written powerfully about the impact of the south-east freeway development on local waterways, biodiversity, and community:

> Before the outsiders arrived in Wurundjeri country the billabong enjoyed a vital ecological connection with other waterways on country. Many of them have since been suffocated by occupation and development. The vast network of wetlands surrounding the Birrarung, from its birth in the mountains to its mouth at what we now call Port Phillip Bay, previously acted as both a repository of life and a sponge, absorbing and distributing water across large tracts of land. These days the river is governed, held in place, against its will. The same could be said for the billabong. If our river and creek valleys are "the lungs of the city", historically we have forced them to breathe toxins. Over the following 40 years, many more freeways and extensions have been built, crisscrossing and extending the infamous Melbourne sprawl—a city that has undergone more than one quadruple bypass which is yet to save the patient (pp. 18–19).

In response, communities are starting to advocate for the rights of nature to exist, thrive, and evolve. In Naarm (Melbourne), under the *Yarra River (Wilip-gin Birrarung murron) Act 2017*, there is legal recognition of the connection between the traditional owners and the river. (In the Woi-wurrung language of the traditional owners, Wilip-gin Birrarung murron means "keep the Birrarung alive".) As the 2020 preamble to the act states, "This Act recognises the intrinsic connection of the traditional owners to the Yarra River and its Country and further recognises them as the custodians of the land and waterway which they call Birrarung". The following statement (in the Woi-wurrung language and in English) is from the Woi-wurrung people:

> Woiwurrungbaluk ba Birrarung wanganyinu biikpil Yarrayarrapil, manyi biik ba Birrarung, ganbu marram-nganyinu Manyi Birrarung murrondjak, durrung ba murrup warrongguny, ngargunin twarnpil Birrarungwa nhanbu wilamnganyinu Nhanbu ngarn.ganhanganyinu manyi Birrarung Bunjil munggany biik, wurru-wurru, warriny ba yaluk, ba ngargunin twarn Biiku kuliny munggany Bunjil Waa marrnakith-nganyinBalliyang, barnumbinyu Bundjilal, banyu bagurrk munggany Ngarn.gunganyinu nhanbu nyilam biik, nyilam kuliny—balit biik, balit kuliny: balitmanhanganyin manyi biik ba Birrarung. Balitmanhanganyin dur-rungu ba murrupu, ba nhanbu murrondjak!

> We, the Woi-wurrung, the First People, and the Birrarung, belong to this Country. This Country, and the Birrarung are part of us. The Birrarung is alive, has a heart, a spirit and is part of our Dreaming. We have lived with and known the Birrarung since the beginning. We will always know the Birrarung. Bunjil, the great Eagle, the creator spirit, made the land, the sky, the sea, the rivers, flora and fauna, the lore. He made Kulin from the earth. Bunjil gave Waa, the crow, the responsibility of Protector. Bunjil's brother, Palliyang, the Bat, cre-ated Bagarook, women, from the water. Since our beginning it has been known that we have an obligation to keep the Birrarung alive and healthy—for all generations to come (Wilip-gin Birrarung murron Act 2017, pp. 1–2).

9.3.3   Horizon 3: Centring Indigenous Knowledge Systems in City-Regions

Horizon 1 practices outlined here demonstrated how settler-colonial actions of control and domination over living systems, such as rivers and other social-ecological systems, have resulted in the displacement and destruction of people and nature. Currently, these practices are maladapted to a changing climate and have only aggravated the culture-nature divide and deepened injustices. From this, however, complementary forms of nature-based actions and alternative forms of governance have emerged as practices that have led the way to innovative changes in the Horizon 2 pathways (Newton and Bai 2008). These, in turn, are starting to give way to, and are a foundation for, more radical paradigms that envision regenerative cities and regions through a pluralistic, planetary politics grounded in the achievement of eco-social well-being on Earth. First Nations and Indigenous cosmologies have long recognised the interconnected nature of a living planet that includes the sea, sky, weather, and species including flora, fauna, algae, and other microorganisms. For vital, thriving cities and regions to occur, a more expansive understanding of urban nature is needed within the context of the climate crisis that is driven by cities that are still being developed against, not with, Nature.

Aboriginal and Torres Strait Islander approaches to Country offer an ethics of intergenerational, multispecies care and repair. Care for Country in the context known as Australia is a deeply felt ethic and sacred alternative to the borders and boundaries of settler-colonialism that separate out civilised versus wild, human versus non-human, past versus future, and cities versus nature so cruelly and crudely. Indigenous spaces and places within cities and regions do not exist as a separate entity from the urban land, sea, sky or weather. Through the sharing of stories, sculpture, and yarning, Uncle Bud Marshall, a Waambung man of the Baga baga bari, in collaborative research with Fabri Blacklock and non-Indigenous geographers Lara Daley and Sarah Wright, describes the infinite ways in which "no place, no matter how colonised or urbanised, exists outside of, or separate to, Aboriginal relational ontologies and more-than-human sovereignties" (Marshall et al. 2022).

A working example of this type of resistance can be seen in *The Australian Peoples' Tribunal (APT) for Community and Nature's Rights*. This is a unique forum for ecological and social justice that has emerged in Australia, inspired by the first International Rights of Nature Tribunal in Quito, Ecuador, in 2014. Through citizen enquiry, the APT hears Ecological Justice Cases brought on behalf of flora, fauna, ecosystems, bioregions, and local communities around Australia, involving First Nations Peoples, lawyers, community representatives, and scientists (Australian People's Tribunal 2022). Three citizen enquiries have been undertaken (in 2016, 2018, and 2019) on industrial-scale agriculture and bioregional impacts at the community scale, including the Murray-Darling River Basin, Australia's largest river system, cultural homelands to Indigenous nations, and internationally protected wetlands, all which are lifelines for Australian cities.

The 2022–2023 citizen enquiry focuses on the healthy regeneration and flourishing of Australia's biodiversity and ecosystems. The enquiry is also focused on how policy and planning laws can be transformed to ensure that people live within their ecological limits and balance so as to restore and regenerate the living world upon which they depend. The Tribunal has a strong focus on enabling Indigenous people to share their concerns and solutions about land, water, and culture with the global community. This is underpinned by a commitment to the Rights of Nature to support living ecosystems and communities in existing and flourishing and to imagine a future that creates the conditions for multispecies flourishing in the climate crisis through an expanded vision of cities as shared commons advanced by acts of solidarity-building (Fitz-Henry 2022).

Cities have been built on discriminatory, exclusionary, and displacement practices. An important action towards multispecies inclusivity and integration is to unpack, identify, and make visible those exclusionary and violent precepts that are present in planning laws and regulations. Reimagining zoning, for example, can serve as a mechanism to protect nature-cultures and vulnerable communities, particularly in areas predicted to be affected by climate change or that have been extremely extracted and degraded and are at risk of more impacts. Rezoning for ecological recovery and reparation can help reconstruct degraded urban habitats to build social-ecological resilience, create habitats for climate-displaced species, and provide temporary shelters from climate and other anthropogenic-related impacts. Planning for an eco-social commons must be attuned to how ecosystems and diverse communities function, including the sovereign rights of Indigenous groups to reclaim, reformulate, and reconstitute their "right to the city" (O'Malley 1996; Yates et al. 2022), particularly in the climate crisis. Horizon 1 practices are focused on technological, engineering solutions that continue to be extractive but begin to solve at small scales our grand urban challenges. They provide the socio-economic foundation for new technological and social-ecological innovations to emerge, experiment, and put forward alternative ways of being and doing. However, Horizon 3—regenerating cities and regions by transforming planning and governance paradigms—is not achieved through technology and engineering solutions. It requires a radical change in human behavior at the individual and collective scales (Newton and Bai 2008). Accordingly, individual- and system-level behavior change should be positioned as a critical element in research and practice, allowing resistance, innovation, experimentation, shock, loss, and recovery to be part of the systems transformation.

9.4 Regenerative Urban Futures: Mission (Im)Possible?

Within the context of the climate emergency, there is a radical need for change in the way cities and settlements are understood and experienced. As a result of the climate crisis, global catastrophes will increase in number and frequency and cities will become increasingly uninhabitable (United Nations Environment Programme

2022). Achieinge regenerative urban change requires creating opportunities for multispecies life systems to flourish in harsh, new, hot city environments; listening deeply to Indigenous communities and cultures that seek to honour and embrace all life on Earth, or on Country; and disrupting current unsustainable development logics through the creation of new joined-up multispecies and urban-nature imaginaries.

In a regenerative future, shock- and loss-driven communities will be able to form new identities, collectives, and ways of being in and with a landscape (Schlosberg et al. 2020). Identifying and creating mechanisms that can confront powerful structures that try to undermine established or emerging collective eco-social interests is critical. These counterhegemonic social-ecological practices include the recognition that other life forms affected by global catastrophes will need protection and assistance in recovery and adaptation if they are to survive. Recognising that other species are climate or ecological refugees leads to very different urban futures and prospects (Christmas 2017).

Guided by the Three Horizons future visions, this chapter has focused on specific examples of new paradigms to activate a multispecies city flourishing in climate change. The Three Horizons pathways are presented as a roadmap to see into the future through multiple horizons. By identifying our current failures and the inadequacies of our "worth-sustaining" governance and planning structures, we can bring a paradigm of multispecies, regenerative, Indigenous-centred governance. The future of urban regions demands a mission-oriented vision that addresses the urgent need to transition *away* from the extraction, devaluation, and displacement of business-as-usual city practices and *towards* imagining and enacting urban sustainability. Regenerative urban practices are a transformative response to short-term thinking and profit-driven urban development. Achieving regenerative futures will require transformative changes that spring from a co-collective understanding of the present to a co-imagining of the pathways for achieving these futures (Cork et al. 2023). In the Australian context, the mission of achieving regenerative futures recognises the need to reimagine "living cities and regions" in close collaboration with Indigenous leadership and knowledge systems. This is the critical recognition that the past is already present in the future of cities and regions.

References

Australian People's Tribunal (2022) Community and Nature's Rights—Australian Hearings. https://tribunal.org.au/sessions/

Bai X, Dawson RJ, Ürge-Vorsatz D, Delgado GC, Barau AS, Dhakal S, Dodman D, Leonardsen L, Masson-Delmotte V, Roberts D, Schultz S (2018) Six research priorities for cities and climate change. Nature 555(7694):23–25

Beck U (2016) The metamorphosis of the world: how climate change is transforming our concept of the world. Polity, London

Bosselman F, Eisen J, Rossi J, Spence D, Weaver J (2010) Energy, economics and the environment: cases and materials. Foundation Press, Goleta, CA

Celermajer D, Schlosberg D, Rickards L, Stewart-Harawira M, Thaler M, Tschakert P, Verlie B, Winter C (2021) Multispecies justice: theories, challenges, and a research agenda for environmental politics. Environ Polit 30(1–2):119–140. https://doi.org/10.1080/0964401 6.2020.1827608

Christmas M (2017) How the warming world could turn many plants and animals into climate refugees. The Conversation. https://theconversation.com/how-the-warming-world-could-turn-many-plants-and-animals-into-climate-refugees-72722

Cook M (2019) A river with a city problem: a history of Brisbane floods. University of Queensland Press, Brisbane

Cooke B, Landau-Ward A, Rickards L (2019) Urban greening, property and more-than-human commoning. Aust Geogr 51(2):169–188. https://doi.org/10.1080/00049182.2019.1655828

Cork S, Alexandra C, Alvarez-Romero JG, Bennett EM, Berbés-Blázquez M, Bohensky E, Bok B, Costanza R, Hashimoto S, Hill R, Inayatullah D, Kaper Kok K, Jn Kuiper J, Moglia M, Pereira L, Peterson G, Weeks R, Wyborn C (2023) Alternative futures in the Anthropocene. Annu Rev Env Resour 48:25. https://doi.org/10.1146/annurev-environ-112321-095011

Curry A, Hodgson A (2008) Seeing in multiple horizons: connecting futures to strategy. J Fut Stud 13(1):1–20

Fitz-Henry E (2022) Multi-species justice: a view from the rights of nature movement. Environ Polit 31(2):338–359. https://doi.org/10.1080/09644016.2021.1957615

Frantzeskaki N, Broto VC, Coenen L, Loorbach D (2017) Urban sustainability transitions: the dynamics and opportunities of sustainability transitions in cities. In: Frantzeskaki N, Broto VC, Coenen L, Loorbach D (eds) Urban sustainability transitions. Routledge, London, pp 1–20

Guo Y, Gasparrini A, Li S, Sera F, Vicedo-Cabrera AM, et al. (2018) Quantifying excess deaths related to heatwaves under climate change scenarios: A multicountry time series modelling study. PLOS Medicine 15(7): e1002629. https://doi.org/10.1371/journal.pmed.1002629

Houston D, Hillier J, MacCallum D et al (2018) Make kin, not cities! Multispecies entanglements and "becoming-world" in planning theory. Plan Theory 17(2):190–212

IPCC (2021) Summary for policymakers. In: Masson-Delmotte V, Zhai P, Pirani A et al (eds) Climate change 2021: the physical science basis. Contribution of working group I to the sixth assessment report of the intergovernmental panel on climate change. Cambridge University Press, Cambridge

Kwai I (2023) Extreme weather hits Europe, and it's not over yet. New York Times. https://www.nytimes.com/article/europe-heat-wave-forecast.html

Lay B (2016) Juris Materiarum: empires of earth, soil and dirt. Atropos Press, New York

Lea T (2020) Wild policy: indigeneity and the unruly logics of intervention. Stanford University Press, Redwood City, CA

Lefebvre H (2014) Dissolving city, planetary metamorphosis. Environ Plann D Soc Space 32(2):203–205. https://doi.org/10.1068/d3202tra

Lerner D (2019) Many urban rivers are hidden underground—"daylighting" them would bring nature back to cities. The Conversation, 10 December. https://theconversation.com/many-urban-rivers-are-hidden-underground-daylighting-them-would-bring-nature-back-to-cities-128441

Marshall V (2017) Overturning aqua nullius: securing aboriginal water rights. Aboriginal Studies Press, Canberra

Marshall B, Daley L, Blacklock F, Wright S (2022) Re-membering weather relations: urban environments in and as country. Urban Policy Res 40(3):223–235. https://doi.org/10.1080/0811114 6.2022.2108394

Mazzucuto M (2021) Mission economy: a moonshot guide to changing capitalism. Penguin Books, London

Meadows DH (2008) Thinking in systems: a primer. Chelsea Green Publishing, Vermont

Metzger J (2016) Cultivating torment: the cosmopolitics of more-than-human urban planning. City 20(4):581–601

Newton PW, Bai X (2008) Transitioning to sustainable urban development. In: Newton PW (ed) Transitions: pathways towards sustainable urban development in Australia. Springer Science & Business Media, Dordrecht, pp 3–19

Norman B, Newman P, Steffen W (2021) Apocalypse now: Australian bushfires and the future of urban settlements. NPJ Urban Sustain 1:2. https://doi.org/10.1038/s42949-020-00013-7

O'Malley P (1996) Indigenous governance. Econ Soc 25(3):310–326. https://doi.org/10.1080/03085149600000017

Pineda-Pinto M, Frantzeskaki N, Chandrabose M, Herreros-Cantis P, McPhearson T, Nygaard CA, Raymond C (2022) Planning ecologically just cities: a framework to assess ecological injustice hotspots for targeted urban design and planning of nature-based solutions. Urban Policy Res 40(3):1–17. https://doi.org/10.1080/08111146.2022.2093184

Porter L (2020) Indigenous Cities. In: Rogers D, Keane A, Alizadeh T, Nelson J (eds) Understanding Urbanism. Palgrave Macmillan, Singapore. https://doi.org/10.1007/978-981-15-4386-9_2

Regen Melbourne (2023) Swimmable Birrarung. https://www.regen.melbourne/swimmable-birrarung

Royal Society of Arts (2023) Join the re-generation. https://www.thersa.org/regenerative-futures

Salazar JF (2015) Buen Vivir: South America's rethinking of the future we want. The Conversation, 24 July. https://theconversation.com/buen-vivir-south-americas-rethinking-of-the-future-we-want-44507

Schlosberg D, Della Bosca H, Craven L (2020) Disaster, place, and justice: experiencing the disruption of shock events. In: Lukasiewicz A, Baldwin C (eds) Natural hazards and disaster justice. Palgrave Macmillan, pp 239–259. https://doi.org/10.1007/978-981-15-0466-2_13

Schuijers L, Bush J (2022) Stewardship: retrofitting private property with the public interest in ecology. In: Graham N, Davies M, Godden L (eds) The Routledge handbook of property. Law and Society, Routledge, pp 312–324

Sharpe B (2019) Three horizons mapping, facilitation guide, *H3Uni.*, 2 August. https://www.h3uni.org/facilitation-guide/three-horizon-mapping-guide/

Sharpe B (2020) Three horizons: the patterning of hope. International Futures Forum. Triarchy Press, Axminster

Sharpe B (2023) Seeing and thinking in three horizons. H3Uni, 2 August. https://www.h3uni.org/foundational-insights/seeing-and-thinking-in-three-horizons/

Sharpe B, Hodgson A, Leicester G, Lyon A, Fazey I (2016) Three horizons: a pathways practice for transformation. Ecol Soc 21(2):47. http://www.jstor.org/stable/26270405

Solecki W, Marcotullio PJ (2013) Climate change and urban biodiversity vulnerability. In: Elmqvist T, Marcotullio PJ (eds) Urbanization, biodiversity and ecosystem services: challenges and opportunities. Springer, Dordrecht. https://doi.org/10.1007/978-94-007-7088-1_25

Steele W, Hillier J, MacCallum D, Byrne J, Houston D (2021) Quiet activism: climate action at the local scale. Palgrave, New York

Steele W, Handmer J, McShane I (2023) Hot cities: a transdisciplinary agenda. Edward Elgar City Series, London

Taylor K, Poelina A, Grafton Q (2022) The lie of aqua nullius, "nobody's water", prevails in Australia Indigenous water reserves are not enough to deliver justice. The Conversation, December 23. https://theconversation.com/the-lie-of-aqua-nullius-nobodys-water-prevails-in-australia-indigenous-water-reserves-are-not-enough-to-deliver-justice-195557

Troy P (ed) (2008) Troubled waters: confronting the water crisis in Australia's cities. ANU Press, Canberra

Tschakert P, Schlosberg D, Celermajer D, Rickards L, Winter C, Thaler M, Stewart-Harawira M, Verlie B (2020) Multispecies justice: climate-just futures with, for and beyond humans. Wiley Interdiscip Rev Clim Chang 12(2):e699

United Nations Environment Programme (2022) Emissions gap report 2022: the closing window—climate crisis calls for rapid transformation of societies. UNEP, Nairobi. https://www.unep.org/emissions-gap-report-2022

Walsh C (2010) Development as Buen Vivir: Institutional arrangements and (de) colonial entangle-
    ments. Development 53(1):15–21. https://doi.org/10.1057/dev.2009.93
Yates A, Dombroski K, Dionisio R (2022) Dialogues for wellbeing in an ecological emergency:
    wellbeing led governance frameworks and transformative indigenous tools. Dialog Hum Geogr
    13(2). https://doi.org/10.1177/204382062211029

Melissa Pineda Pinto 's research examines urban nature through diverse justice
lenses for achieving sustainable futures. She is currently a Mckenzie Postdoctoral
Research Fellow at the Melbourne Centre for Cities at the University of Melbourne.
Before this she worked on the project NovelEco, at Trinnity College Dubin, Ireland,
which examined wild ecosystems in cities through forecasting methodologies and
policy analysis. Prior to this, she completed her PhD at Swinburne University of
Technology and a Master of Environment from the University of Melbourne. This
work draws on her previous experience in the architectural and planning industries
and not-for-profit sectors. Her academic experience and interests cut across social
research methods, inter-transdisciplinary collaboration, and systems thinking in the
context of urban ecosystems, justice, and ethics.

Wendy Steele is a Professor of Sustainability and Urban Governance and Interim
Director of the Urban Futures Enabling Impact Platform at RMIT University. She
leads research on the regenerative future of cities and settlements in the context of
climate change, with a particular emphasis on human–nature relationships, urban
governance, critical infrastructure, climate justice, and re-imagining the contemporary role of the university in society. She is the Co-chair of the Future Earth Australia
National Steering committee and President of the Australasian Cities Research
Network and sits on the Australian Research Council (ARC) College of Experts.
Her recent books include Wild Cities: Human–Nature Relationships in the Urban
Age (Routledge 2020), Quiet Activism: Climate Action at the Local Scale (Palgrave
2021), and Hot Cities: A Transdisciplinary Agenda (Elgar 2023).

Part IV
Smart Cities

Chapter 10
Digital Innovations for City Sustainability Analysis and Decision-Making

Peter Newton, Chris Pettit, Stuart Barr, and Loren Bruns Jr

Abstract This chapter examines the potential for accelerating a convergence between fields of research associated with digitalisation (incorporating disciplines of sensing, data science, data analytics, and information technology) and urban-sustainability transition (relating to disciplines such as planning, design, environmental science, economics, transport, and politics), acknowledging that the threads of such linkages have been emerging over decades. Convergence research is a fundamental underlying principle of scientific progress that assembles and integrates all relevant capabilities to answer contemporary grand challenges. It is a frontier area for applied research that is critical for an accelerated transition to a green economy and sustainable urban development—two mission-scale challenges of the twenty-first century. The framework for this chapter comprises four interconnected innovation arenas associated with digitalisation that together constitute a basis for more rapidly advancing urban-sustainability research and development. These are information and communications technologies and digital infrastructure platforms that enable stakeholder engagement across distributed collaborative networks, advances in data science, and advanced urban analytics that support integrated urban analyses and decision-making.

Keywords Digitalisation · Smart sustainable urban development · Collaborative decision-making · Digital infrastructure platforms · Data science · Urban analytics · Urban experimentation

P. Newton (✉)
Swinburne University of Technology, Hawthorn, VIC, Australia
e-mail: pnewton@swin.edu.au

C. Pettit
City Futures Research Centre, University of NSW, Sydney, NSW, Australia

S. Barr
School of Engineering, Newcastle University, Newcastle upon Tyne, UK

L. Bruns Jr
Australian Urban Research Infrastructure Network, Carlton, VIC, Australia

© The Author(s) 2025   215
N. Frantzeskaki et al. (eds.), *Future Cities Making*, Theory and Practice of
Urban Sustainability Transitions, https://doi.org/10.1007/978-981-97-7671-9_10

10.1   Introduction

Sustainability pressures have been accelerating since the mid-twentieth century, and we are now in a unique era (the Anthropocene) where human activity has become the dominant influence on the earth's climate and environment (Steffen et al. 2015; Lewis and Maslin 2015). Rapidly increasing population, consumption, urbanisation, and industrialisation are principal contributors and are all concentrating in cities. In both developed and developing societies, current ecological, carbon, and urban footprints and trajectories are unsustainable and need to be radically and rapidly wound back (Newton 2012; IRP 2018; Guterres 2021). The shrinking window of opportunity for achieving this transition requires accelerated action on multiple fronts, including the application of digital information technologies and infrastructures.

The body of research focused on sustainable urban development has been growing rapidly over the decades since the first UN conference was held in 1970 (Ward and Dubos 1972), with the most recent synthesis emerging in 2016 with the United Nation's Sustainable Development Goals (UNSDGs, https://unsdg.un.org/) providing an important global focus for applied research. The latest progress report to the UN indicates that none of the 17 SDGs will be met by the self-imposed 2030 deadline, reflecting the complexity of the challenge (Tollefson 2023).

The dimensions of urban performance on which cities need to be assessed are now well established and reflect the complexity of the topic: environmental quality; ecological justice (balance between human and non-human needs); liveability-amenity; human well-being-health-quality of life; equity-inclusion; economic competitiveness-productivity; resilience to both endogenous and exogenous shocks; and climate neutrality. These mission-scale challenges now feature in the strategic plans of most cities and regions, but are often poorly supported by linked sets of indicators, targets, and benchmarks. International standards for defining and measuring urban sustainability that can help drive the comparative performance, and the associated potential for adaptation and learning, of cities nationally and globally are emerging via ISO (2022) as core sets of indicators. However, at present, they are mostly tied to individual sectors and more commonly report on "traditional" statistics compared to those reflective of the dynamics of twenty-first-century urban forces such as evolving types and places of work, work practices and work-life preferences, flows in multiple types of networks (physical and digital), changing patterns of urban metabolism, and the intensity of different exogenous as well as endogenous threats to settlement systems (e.g. Florida 2003; Hall and Pain 2006; Newton and Doherty 2014; Moglia et al. 2021). OECD (2020) is calling for innovative international studies involving a territorial approach to the UNSDGs by focusing on indicators that operate at multiple spatial scales: national, state, city/region, municipal, and precinct. Contributions in these areas are lagging for Australia: in 2021 the federal government ranked 35th of 190 countries, with 75% of its SDG targets achieved (https://dashboards.sdgindex.org/profiles/australia); and in 2021 it abandoned its National Cities Performance Framework (https://www.bitre.gov.au/

national-cities-performance-framework), leaving academic urban observatories to fill the gap (e.g. https://auo.org.au/measure/scorecards/). Attempts at developing precinct−/neighbourhood-scale performance assessments to guide local development are in their early stages. Here it is argued that "If cities are to achieve the international performance goals and objectives outlined by the United Nation's Sustainable Development Goals and the New Urban Agenda as well as those identified at a national level then it will be necessary for their constituent precincts to demonstrate performance outcomes that align with and add to, rather than subtract from, these objectives" (Newton 2019, p. 359).

Developing conceptual models of cities as complex integrated systems continues to be a focus for urban research, and significant advances have been made. Important contributions have been based on metabolic representations of cities (Musango et al. 2017), land-use-transport-environment models (Hunt et al. 2005), nature-based blue-green conceptualisations (Newton and Rogers 2020), and building-precinct-city information models (Newton et al. 2017), to name a few. Developing applied operational models and a measure of standardisation in these areas remains challenging, however, as do processes for integrating urban analytics and modelling into city governance and participatory decision-making (Biermann et al. 2022) accompanied by an effective supply of relevant data. Internationally, capacity for digital transformation in the built environment sector continues to lag compared to other sectors (Bello and Galindo-Rueda 2020).

10.2   A Critical Nexus

A key twenty-first-century grand challenge involves successfully meshing information technology and digitalisation, the fifth long wave of global innovation (Batty 2018), with sustainable development—the emerging sixth long wave (Hargroves and Smith 2005) to deliver a smart, sustainable urban-development transition. Regenerative urban development is an emerging field that constitutes an even greater challenge for applied research than sustainable urban development (Girardet 2015; Newton et al. 2022).

This chapter begins to explore the potential for increased integration and an accelerated convergence between fields of research associated with digitalisation and sustainable urban development. The challenges are formidable, but need to be articulated as a focus for a mission-oriented response. Convergence research has been identified as a fundamental underlying principle of scientific progress that assembles and integrates all relevant capabilities to answer contemporary grand challenges (Bainbridge and Roco 2016). It is a critical arena for applied research central to a much-needed accelerated transition to smart and sustainable urban development.

The principal mission is to accelerate a convergence of the extensive body of scattered research on urban sustainability with the similarly fragmented digital urban information infrastructures (data, analytics, and platforms) necessary to effectively tackle the accumulating volume of urban problems catalogued in multiple reports and daily media. It has become clear that the window of opportunity for transition to more sustainable forms of urban development without significant environmental, social, and economic disruption is closing at a faster rate than previously expected.

The extent to which an urban-sustainability transition can be realised more rapidly will depend on the speed with which digital transformation and sustainability transformation can merge into one coherent process of change. The critical connections between the two are briefly outlined below as a prelude to a more focused discussion of the key digital domains capable of accelerating sustainable development:

1. *Stakeholder collaboration and engagement.* Capacity to more effectively assemble representatives for more collaborative "joined-up", top-down, and bottom-up decision-making is central to successful urban-development projects, whether at neighbourhood or metropolitan scale. Diffusion of networked urban labs and urban rooms (discussed in more detail later) that are purpose-designed for the exposure and discussion of future urban-development scenarios and projects among stakeholder groups can be expected to spread from specialist academic settings to planning and design practices as well as state and local governments. Providing an arena for both face-to-face and virtual input (e.g. from geographically remote experts) represents a new mode of urban governance capable of achieving greater alignment of multi-actor intentions and practices associated with urban regeneration.

2. *Urban data.* There is an exponential growth of economic, social, and environmental data—the three domains of sustainable development. The complexity of urban and environmental systems requires assembling multiple indicators for analysis, an area where considerable time is lost in accessing and harmonising data for evidence-based discussion and decision-making. Pathways for accelerating the emergence of a data commons covering built, environmental, and population data from both public and private sectors are also key to sustainable development.

3. *Urban analytics, integrated modelling and AI.* The UN 2030 Agenda for Sustainable Development highlights the fact that all 17 Sustainable Development Goals are complex and integrated, and these interlinkages need to underpin analyses that guide planning frameworks, strategies, and plans. The level of integration needs to incorporate multi-criteria, multi-scale, and multi-stakeholder connections, embodying key trade-offs that are a common feature in urban decision-making.

4. *Digital infrastructure platform.* The complexity of urban systems needs to be matched by capabilities now offered by digital infrastructure platforms. These platforms enable interoperability of both urban data and analytical software assembled for major urban projects where new urban technologies such as digital twins are increasingly being employed to simulate the behavior of built

environment systems—buildings, transport system, and precincts—at different stages of their life cycles.

The sections that follow present the state of progress in each of these areas offering the most potential to accelerate a transition to smarter and more sustainable cities.

10.3   Frontiers in Multi-Stakeholder Engagement, Networked Collaboration, Experimentation, and Decision-Making

A recent review of the national ecology of urban innovation in Australia revealed the absence of any effective national collaboration and experimentation platform relevant to urban systems at all scales. Current urban research is poorly co-ordinated and ill-suited to the forms and scale of collaboration required for addressing the grand urban challenges we now face. What is needed is a national collaboratory (Fig. 10.1): one capable of enabling enhanced research synthesis, policy and governance innovation, and sociotechnical design innovation (e.g. quantitative and qualitative evaluation of alternative urban-development project options, urban-infrastructure technologies, alternative land-use arrangements, or mobility

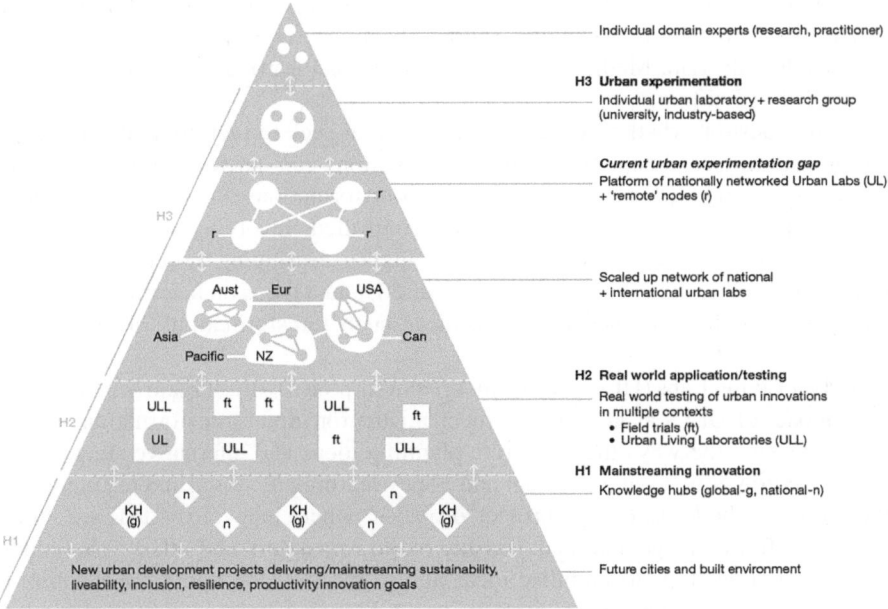

Fig. 10.1 The ecology of urban innovation, knowledge production, collaboration, and information dissemination in the Australian context

options) and citizen engagement, results from which hold the promise of more rapid evidence-based policy and implementation and transformational change (Newton and Frantzeskaki 2021).

The process of effectively assembling and managing geographically distributed transdisciplinary teams to pursue new concepts and processes for major urban-development problem-solving challenges remains. This is despite significant repositories of domain knowledge and urban-transition blueprints (Webb et al. 2018, 2023). Achieving smart, sustainable cities requires a shift from the siloed approach of much scholarship in the built environment, social science, environmental science, engineering, and information science disciplines to integrated thinking and analysis. Creating research groups within public- and private-sector organisations reflects a search for greater critical mass and new synergies, but they often represent only slightly larger silos. The comparative lack of research funding in the built-environment sector accentuates this situation: its level of research intensity remains low compared to most other sectors (Innovation and Science Australia 2020), and its funding tends to be piecemeal and small-scale, scattered across many activities and with too little co-ordination. Until relatively recently, there has been an absence of research laboratories in the built-environment sector capable of acting as a focus for innovation and experimentation. This stands in contrast to the STEM disciplines, where purpose-built laboratories have been at the heart of major advances in the physical, chemical, biological, materials, and medical sciences, as well as engineering, for decades. Diffusion of ICT, the rapid growth of urban spatial data, and the gradual growth of computational expertise and capacity among researchers and practitioners in the built-environment field is beginning to be reflected in the emergence of urban labs in Australia, following in the footsteps of leaders at MIT (Senseable City Lab, Media Lab, etc.), UCL (Urban Lab), and Singapore (ETH's Future Cities Lab).

Some question whether urban experiments that seek to identify pathways capable of outlining roadmaps of urban transformations can emerge from a traditional laboratory setting. They suggest that such transformations "need to take place in real-world settings that cannot be tightly controlled, involving societal actors in initiating and carrying out the experiments (necessitating co-creation or co-production, rather than sole reliance on experts), and a focus on learning about what the system ought to be and how to achieve such transformation" (Torrens et al. 2019, p. 212).

Urban Living Labs (ULLs) have emerged in response to this (again see Fig. 10.1). The impact of ULLs as governance mechanisms for urban transformations can be traced through the ways they generate planning ideas and alternatives that lead to the adaptation of plans and to social and policy learning by doing and testing within the setting of the ULL. Being "protected spaces" where innovative ideas are safe to fail and safe to change and adapt further empowers actors and other stakeholders and their partnerships to mobilise ideas and to transfer concepts from ideation to implementation. As governance mechanisms, ULLs are advanced as fruitful grounds for policy innovation and planning adaptation (Pereira et al. 2020).

ULLs also have their weaknesses in relation to their role in driving urban innovation. Many ULL experiments remain as iconic and stand-alone projects with no direct or implicit connection to wider urban agendas and urban programmes, often making the case for experimentation to occur in the shadow of unsustainable or conventional urban planning and policy processes and programmes. Specific weaknesses include an inability to substitute alternative (new) scenarios or develop options that encompass different land-use and building mixes, densities, streetscapes, mobilities, blue-green infrastructures, resident socio-demographics, and other factors and to assess variations in performance outcomes. ULLs are typically "bespoke", where urban designs, fabrics, and technologies are tailored to a particular political and geographic context, and where project boundaries are predefined, limiting examination of project-scale issues and limiting or preventing opportunities for comparative spatial and temporal analyses. This extends to issues of socio-technological experimentation, project governance, management, and levels of stakeholder engagement and roles (e.g. involving the extent of co-design and co-production possible). In a recent book that reviews ULLs, Marvin et al. (2018, p. iii) concluded that "despite the experimentation taking place on the ground, we lack systematic learning and international comparison across urban and national contexts about their impacts and effectiveness. We have limited knowledge on how good practice can be scaled up to achieve the transformative change desired".

The question this raises is whether it is possible to envisage smart processes and platforms capable of more realistically examining scenarios that represent potential future urban systems and living-environment options. These need to be represented in a sufficient variety of geographical, environmental, social, built-form, and urban-fabric contexts that lend themselves to experimentation and comparative performance assessment. A gap in the national urban innovation ecology has been identified as a networked collaboration platform (supported by data, analytics, and digital infrastructure capable of linking geographically distributed researchers, their urban laboratories and "urban rooms" (Dixon and Farrally 2018). The objective is to enable synchronous distributed collaboration between experts and key stakeholders (including citizens) on planning and design issues ranging from local to global (as shown in the indented section of Fig. 10.1). This platform will facilitate linking leading university, government, and industry research centres and ULLs with the skill mix and innovation intensity necessary to initiate and support a step-change in urban planning and design. Over time it can be expected to be scaled up and mainstreamed to what the Australian government has envisaged in its National Research Infrastructure Roadmap (Department of Education, Skills and Employment (2021; pp. 16–17).

The first five nodes in a national collaborative urban-research network (iHUB; Newton and Burry 2018) have been established in Australia's four largest capital cities, located in universities with leading urban-research centres and urban laboratories (Fig. 10.2). The basic infrastructure employed to establish these labs is now becoming commonplace and easily replicable: an integrated suite of information and communication technologies encompassing computing, high-speed communications, high-definition graphics, and tools for synchronous distributed

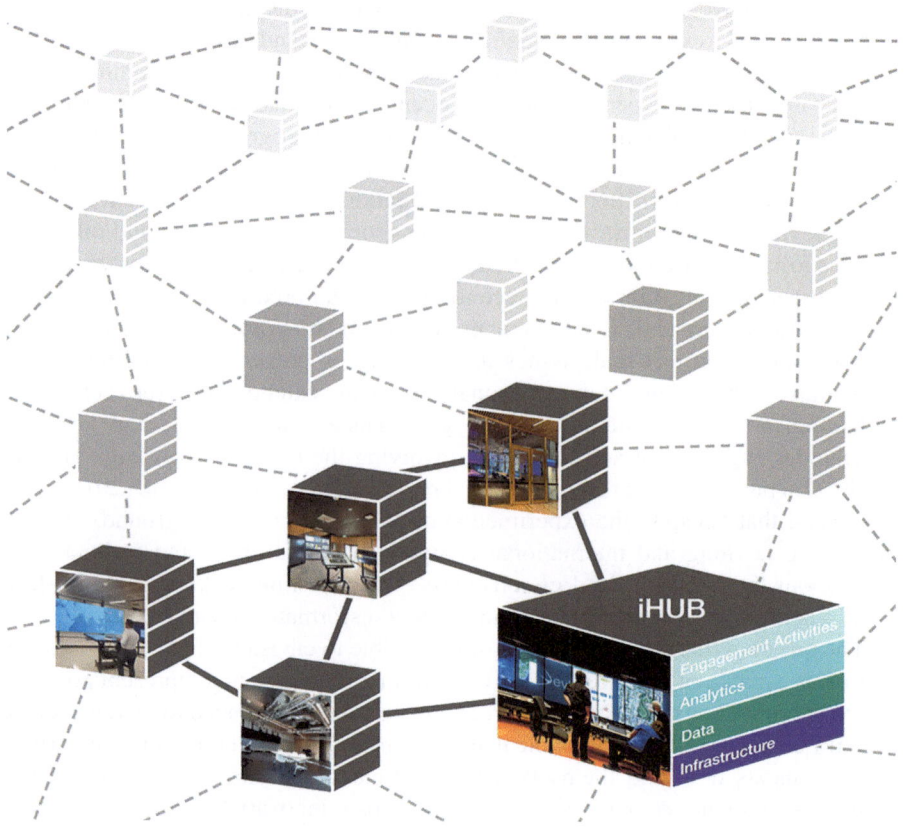

Fig. 10.2 Networked urban collaboratories, each featuring four integrated layers of innovation: information and communications infrastructure, data, analytics, and hybrid forms of engagement. Note: images feature each iHUB Urban Laboratory counterclockwise from Swinburne [RHS], University of Queensland, Curtin University, University of New South Wales, and Monash University

collaboration, most of which are now available off the shelf (these are specified in more detail in Newton and Frantzeskaki 2021). Figure 10.3 depicts a section of the Urban Lab established at Swinburne University of Technology. These nodes are positioned to support a wide range of collaborative and applied research activities in highly reconfigurable spaces capable of being arranged for research group discussions, lectures, boardroom meetings, etc. (see https://www.swinburne.edu.au/research/platforms-initiatives/ihub-network/). Foremost among these are:

- A national collaboratory (see Muff 2017) that maximises the input of all stakeholder groups in responding to the federal government's call for the development of a vision of the future settlement system of Australia (House of Representatives 2018); enhanced national state of the environment reporting across multiple themes (SoE 2021); national infrastructure planning and prioritising

Fig. 10.3 Urban Lab at Swinburne University of Technology, depicting two of the four pods of visualisation panels; part of the five-node national iHUB network

(Infrastructure Australia 2021); and Future Earth Australia's transdisciplinary engagement and national urban strategy development for Australia (Webb et al. 2018, 2023).

- A platform that can more effectively assemble stakeholder groups locally or nationally on issues that require a broad alignment of actor intentions and practices, and a shift in thinking, which in turn requires clear visions to be articulated, discussed, and agreed to. A case in point is more than 30 years of vacillating interest in planning for a very fast train linking capital cities along Australia's east coast (Newton 2016).
- A platform that supports research planning, project management, and knowledge dissemination and training for national collaborative research centres where expertise, partners, and locations of projects are geographically distributed (Co-operative Research Centres, Centres of Excellence, AURIN, etc.)
- A platform for supporting strategic metropolitan planning, where levels of vertical integration between the three tiers of government and community have traditionally been lacking, as have horizontal connections between government departments and major urban utilities.
- Support for local governments in their strategic and statutory planning processes (especially development assessment) where citizen engagement is poor and expert input needs to augment that available within councils, where there is often a lack of capacity (Productivity Commission 2012; Newton et al. 2022).

- Support for virtual project teams involved in major infrastructure projects where key nodes are represented by the head office, site office, and collaborating project partners (e.g. designers, quantity surveyors, constructors, building fit-out; Newton 1995).

To be fully effective, however, the urban labs platforming such networks need to be able to seamlessly access the required data and analytics necessary for their effective operation via a digital research infrastructure platform.

10.4 Frontiers in Spatial Data Science

High-quality, trusted data lies at the heart of enabling analytical innovation and improved collective decision-making concerning the economic, social, engineered, and environmental dimensions of the built environment (Kharrazi et al. 2016; Talebkhah et al. 2021). The built environment is undergoing a revolution in terms of the volume of digital data it generates and consumes; approximately 2.5 quintillion bytes (2500 petabytes) of data is generated globally on a daily basis, much of which relates to the status, condition, activities, movements, flows, and interactions taking place within the world's urban conurbations (Manjunatha and Annappa 2018). Access to such data, compiled as structured longitudinal spatio-temporal data collections, will provide the evidence base for improved decision-making and planning to deliver long-term sustainable, livable cities today and into the future (Kharrazi et al. 2016).

Over the next decade, we can expect the volume of urban digital data to continue to increase, driven by new technologies such as new IoT (Internet of Things) devices, intelligent location-based services, and smart health devices. For example, by 2025 it is estimated that there will be 75 billion IoT (Internet of Things) devices in use, serving a global population of 8.5 billion by 2030, of which 90% over the age of five will be "regular" consumers of Internet-delivered digital data. For future generations, "urban big data" will form the foundation upon which major decisions will be made in relation to the anthroposphere (Shah et al. 2019; Talebkhah et al. 2021).

The potential benefits of the "era of urban big data" have been well documented. These include the ability to make better-integrated decisions regarding the built environment; improved real-time, system-wide situational awareness; increased productivity and efficiency across commercial and public services; cost savings and optimisation of flows, movements, and logistics; improved societal outcomes; and improved short- and long-term environmental policy and planning (Chauhan et al. 2016; Kharrazi et al. 2016; Manjunatha and Annappa 2018; Shah et al. 2019; Talebkhah et al. 2021).

The increasingly widespread availability of urban data has been a fundamental driving force in the development of the smart-city concept from a technological perspective, where "traditional networks and services are made more efficient with the use of digital solutions for the benefit of its inhabitants and business" (European Union 2022). The "era of urban big data" has also been critical in driving the development of operational system-scale urban digital twins: the digital representation of the physical (and increasingly social) fabric and the dynamics of the built environment, such as transport networks, buildings, and critical infrastructure, where direct feedback to the physical built environment is provided to enable improved situational awareness, decision-making, and planning (Schrotter and Hurzeler 2020; Deng et al. 2021; Shahat et al. 2021).

To accommodate the huge volumes of urban data available today and in the future, we have seen the development of a new generation of data ETL (Extract, Transform, and Load) management and delivery of digital infrastructure that uses cloud storage and computing (e.g. Amazon Web Service, Azure, etc.), modern containerisation and software orchestration, and open-source, advanced, big-data software tools such as those supported by the APACHE ecosystem (Lee and Kang 2015; Pan et al. 2016). Such capability has driven the delivery globally of a suite of urban-data access platforms at multiple levels of governance: from the city (e.g. the mySMARTLife Helsinki Urban Platform; https://mysmartlife.eu/interventions/helsinki-urban-platform/) and the regional/state level (e.g. New York State NYC Open Data; https://opendata.cityofnewyork.us/) to national capability (e.g. the UK Urban Big Data Centre; https://www.ubdc.ac.uk/ and the Australian Urban Research Infrastructure Network (AURIN) Data Provider; https://aurin.org.au/resources/einfrastructure/).

However, significant challenges remain in relation to providing high-quality data for large-scale economic, social, engineered, and environmental system-wide collaborative analysis and decision-making for sustainable urban futures and long-term liveable cities (Kharrazi et al. 2016). These can be broadly grouped as technical, governance, and training challenges that encompass the 5Vs of big data: volume, variety, velocity, veracity, and value (Lim et al. 2018).

From a technical perspective, data volume is perhaps the least significant future challenge, as storage is becoming more cost-effective, and database technology and associated management systems are now highly optimised for large-scale storage and retrieval (query) of huge data repositories. However, the highly distributed nature of data storage across many different data platforms and providers is a significant limiting factor to the use of urban big data (Granell et al. 2014). There is an urgent need for data providers to adopt a far more integrated approach that involves federated meta-data management, where data APIs (application programming interfaces) are openly accessible and a suite of data "shopping" tools can be developed that allow easy melding, integration, and delivery of data from multiple sources relating to a common geographical area, time period, and/or set of themes or domains (Pan et al. 2016). Without such capability, we will find analytical innovation being quashed by the overhead of data search and acquisition.

A related technical challenge lies in the significant variety of data that is now routinely collected in relation to the built environment (Liu et al. 2016); today, digital urban data is recorded in a wide range of formats and data types. Increasingly, decision-making in relation to urban areas looks to employ not only traditional zonal administrative geographies, such as census data, but also critical infrastructure networks; transport networks; environmental conditions recorded by satellite, airborne, and sensor networks; 2D and 3D city models (and related Building Information Models); CCTV and video streams; and transaction data such as touch-on/touch-off travel cards, mobile phone data, social media, and IoT data feeds. However, few urban-data platforms are well adapted to handle such data variety (Baumann et al. 2021). The development of hybrid data-management systems that employ tailored database systems optimised for specific types of data is now recognised as a key future requirement (Badidi et al. 2020). For example, array databases have been employed for the storage and management of large volumes of gridded/raster data such as image data (Baumann et al. 2021), while graph databases have been shown to be more efficient for the retrieval and analysis of critical infrastructure networks such as transport, electricity, gas, and water, where the topology (connectivity) between assets is important (Płuciennik and Płuciennik-Psota 2014; Robinson et al. 2015).

The past decade has seen a dramatic increase in the deployment of real-time sensor networks within the built environment (e.g. see the individual Australian State Digital Twin (DT) initiatives such as Victoria (https://www.land.vic.gov.au/maps-and-spatial/digital-twin-victoria), New South Wales (https://nsw.digitaltwin.terria.io/) and Queensland (https://qld.digitaltwin.terria.io/), and the UK UKCRIC Urban Observatory programme (e.g. https://www.ukcric.com/facilities/newcastle-urban-observatory/ and https://www.ukcric.com/facilities/cranfield-urban-observatory/). Such data offers enormous potential for real-time mitigation, as well as invaluable highly granular temporal data for long-term adaption and resilience planning. However, to fully realise this potential, new data-delivery platforms are required that can provide users with data that may be from multiple sensors, at multiple locations and recorded at different frequencies and times. Such variable *velocity* data requires new event store and data streaming-processing capability (e.g. such as APACHE Kafka (https://kafka.apache.org/)) to provide high-throughput, low-latency platforms for handling real-time data feeds (Gilbert et al. 2018).

Urban big-data *veracity* relates to the consistency, accuracy, quality, and trustworthiness of the data to be employed within the urban decision-making and planning process. While technical developments may help us measure and quantify the confidence that we can have in the data collected, governance, including standards, plays a key role in ensuring that we can have trust in the huge volume of data that is now routinely collected. Work by the Open Geospatial Consortium (OGC) is pioneering not only the development and adoption of accepted standards for 2D and 3D spatial data but also standards regarding metadata, IoT (e.g. OGC SensorThings API), and analytics and simulation modelling (e.g. OpenMI), and standards that are increasingly recognised as critical for the delivery of internationally leading initiatives such as the OGC Urban Digital Twins summit(s). However, while standards can help develop strong governance models in the era of urban big data, major challenges remain in this regard, particularly in relation to ensuring FAIR (findable, accessible, interoperable, and reusable) data and software principles while maintaining strong models of data privacy and security, particularly in relation to sensitive unit (individual) level data pertaining to individuals (e.g. health-data records) (Sta 2017; Deng et al. 2021; Singh et al. 2021).

As the *volume* and complexity of urban data increase over the next decade, training will be critical if the full societal benefits of these new forms of data are to be realised. Increasingly, API endpoints are being provided by data providers and "open" urban-data platforms to ensure access to the most authoritative, up-to-date versions of data sets. However, using these often requires programming skills (e.g. Python or R) and/or "advanced" skills in API configuration within GIS packages. As the complexity and diversity of urban data increase, there will be a need for new programmatic-based tools that allow users to discover, access, query, refactor, meld, integrate, and retrieve data; training in the use of such tools, aligned with the advances being made within the analytical and statistical domains, will be key to delivering the best science-led data-centric evidence for urban policy and planning. Indeed, training of a highly digitally enabled workforce and an advanced data-science capability has been recognised as a national priority for Australia (Australian Academy of Science 2021, 2022) and critical in the successful future delivery of location-based services (Geospatial Commission).

Ultimately, the value of urban big data lies in its ability to address many of the pressing societal, economic, engineering, and environmental challenges facing our urban conurbations and cities. However, to meet these challenges, we need technological solutions aligned with strong governance models (which themselves need to be linked to the analytical and decision-making dimensions) and training across academia, government, and industry to ensure that the potential of the data leads to genuine benefit for all and does not inadvertently result in increasing inequalities within settlement systems (Nugent and Suhail 2021).

10.5   Frontiers in Urban Analytics

Urban analytics has been defined by Singleton et al. (2018, p. xv) as "the multi-disciplinary area of research concerned with using new and emerging forms of data, alongside computational and statistical techniques, to study cities". Synonymous with urban analytics are the terms city analytics, urban informatics, urban science, and city science, all of which came about in the 2010s. As noted by Batty (2019), all of these disciplines essentially encompass the frontiers of data and technology for the purposes of better city-shaping. It was also around this time that the journal *Environment Planning B—Planning and Design* was renamed *Environment and Planning B—Urban Analytics and City Science* (Batty 2017), signalling the importance and maturing of this field of research.

While urban analytics is a relatively new frontier, it is important to note that it builds upon previous endeavours to use data and technology to provide evidence to support future city-making. The use of data, technology, and computers to support planning and city-shaping dates back over 50 years. In the 1960s, the Harvard Laboratory for Computer Graphics saw the likes of Peter Rogers and Carl Steinitz running early urban models for a better understanding of urbanisation and land-use change in Boston (Steinitz and Rogers 1970). As noted by Chrisman (2006), such early mapping work was the precursor to the formalisation of geographical information systems (GIS). It was also around that time that the first generation of large-scale urban models was being built and applied to cities in an attempt to predict their future state. These models were essentially driven by a series of equations aimed at optimising the future city-state. Such models were, unsurprisingly, critiqued as unwieldy, black box-like, and essentially impenetrable to city planners and those charged with shaping the future of our cities (Lee 1973).

The rise of urban analytics has emerged as somewhat of a response to large and unwieldy top-down models. Urban analytics comprise a toolkit of methods and techniques, where geospatial analysis and visualisation interfaces can be written as code and shared in lightweight ways using Jupyter notebooks. For example, Reades and Rey (2021) have developed GEOPYTER, a hub comprising a number of note-books, for conducting geographical analysis. Whether via notebooks, or plugins into a GIS, or the programming of new lightweight digital planning tools or easy to use city dashboards, such tools are becoming increasingly useful for envisioning possible or probable future cities (Newton and Taylor 1985; Dixon and Tewdwr-Jones 2021). Urban analytics represents a significant maturing of how data and technology can be more easily used to assist planners, decision-makers, and communities in shaping the future city.

Urban analytics provides a toolkit to enable the future design of our cities to be more liveable, resilient, productive, inclusive, and sustainable. For example, many cities throughout the world are focused on reducing travel times between home and work destinations through land-use-transport planning objectives such as the 15-minute neighbourhood in Paris (Moreno et al. 2021) or the 30-minute city in Sydney (Leao et al. 2021). Urban analytics is assisting planners and decision-makers in assessing these policies and communicating the performance of the city against such metrics through the use of city dashboards, as illustrated in Fig. 10.4.

While big data can be seen as the engine for creating enhanced metrics of our current and future cities (Pettit et al. 2022), urban analytics also plays a critical role in enabling planners and designers envision and explore future *what if?* scenarios for our cities. Frameworks such as geodesign (Steinitz 2012) incorporate systems-thinking methodologies for bringing together urban residents, domain experts, data scientists, and decision-makers to collectively create and explore alternative city futures. Again, if we take the example of Sydney, a geodesign approach undertaken for exploring future resilient-city scenarios drew upon a suite of urban analytic tools including GIS, GeoJson.io, and Kepler.GL (Debnath et al. 2021). These tools were made available to participants undertaking a geodesign studio in a purpose-built digital-planning and decision-support urban laboratory, known as the City Analytics Lab (a partner in the national iHUB network described earlier in this chapter; also, see Punt et al. 2020). This physical space comprised several multi-touch screens that enabled participants to work together in a collaboratory of small groups using the digital tools as instruments for sharing and discussing scenarios for future urban development.

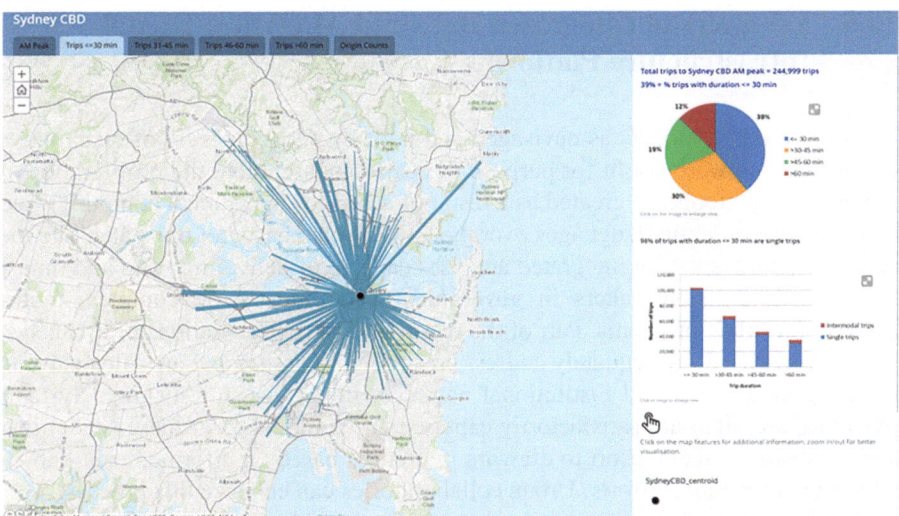

Fig. 10.4 30-minute city dashboard for Sydney, https://cityfutures.ada.unsw.edu.au/
cityviz/30-min-city/

Urban analytics is currently undergoing a revolutionary transformation due to the increasing utilisation of artificial intelligence (AI) methods and in particular unsupervised and supervised machine learning (ML). Indeed, the emergence of AI methods has led to the coining of the term UrbanAI as a discipline in its own right (Luusua and Ylipulli 2020) with researchers and practitioners considering it the next logical step in the development of the "autonomous" smart city of the future (Cugurullo 2020). Across all aspects of the built environment: transport, construction, engineered infrastructure, retail, finance, and social infrastructure, AI has the potential to deliver significant benefits via improved situational awareness and decision-making (Luusua et al. 2023). Already, many researchers and planning practitioners are exploring the opportunity of AI approaches to aid city planning (Yigitcanlar et al. 2023). More recently, we have seen large language models (LLMs) such as ChatGPT arrive. Such AI methods, and their next-generation derivatives, will undoubtedly form the basis of urban analytics over the coming decade(s).

However, as with many aspects of the digitisation of the "city", careful consideration needs to be given to the ethical, governance, and privacy issues involved in the use of such technology (Sanchez 2023). In particular, Cornwell (2023) points to concerns associated with what questions to ask and what data to use to inform AI algorithms, given that AI is only as good as the data it is trained on. Also, urban planning involves a wide range of social, cultural, and political considerations that go beyond the purely technical aspects of AI. As such, human involvement and judgement need to remain core to decision-making.

10.6 Frontiers in Integrated Data and Analytics: Digital Infrastructure Platforms

Monolithic urban models as envisaged and built by early pioneers of quantitative urban planning were not fit for purpose. It is also apparent that the proliferation of urban "analytic toolkits" created as a result of advances in personal computing and software programming languages over the last 20 years have not translated into the sought-after advances in integrated analyses capable of delivering solid and timely evidence to decision-makers in government and industry on complex, multi-stakeholder urban problems. Part of the reason is that representations of urban systems, as modelled, inadequately address cross-sectoral issues and are often naïve to the integrated social and institutional nature of urban decision-making. Systems dynamics are often not satisfactorily captured in the models. For these sets of reasons, decision-makers resort to drawing inferences based on informed tacit knowledge and deliberative efforts. Urban collaboratories can enhance this process.

Another reason for this—and one that is central to this section of the chapter—is that the urban-research community lacks a standardised approach for the creation, curation, and execution of analytical models. This deficiency causes the translation of research software into decision-support tools to become an artisanal enterprise. The research-infrastructure specialists building these tools are required to write bespoke software for each new tool and to have extensive knowledge in both modern software engineering and the specific urban-research domain the tools are built for. Coupled with the short funding periods and long ramp-up times associated with the creation of decision-support tools, the conversion of urban research into infrastructure is slowed dramatically and delivered in an unsustainable way. Most decision-support tools resulting from research projects remain as prototypes, where their contribution to advancing urban analytics is lost.

What is needed is a strategic transition towards interoperable analytics in the urban-research community, driven by the digital research infrastructure specialists who serve that community. Other, more traditionally data-intensive research fields have adopted this approach as a natural response to the increased complexity of their domain areas, with bioinformatics being a leading example. The scale of the data within biological sciences has necessitated the creation of workflow-management systems and workflow languages to standardise repetitive data tasks and give researchers the cognitive space to interpret the results rather than be bogged down in data processing. These same tools are now being used in research-infrastructure facilities and platforms targeting other disciplines that are becoming overwhelmed with data, with the notable examples of the Data & Analytics Facility for National Infrastructure (DAFNI: https://www.dafni.ac.uk/) in the UK and the EcoCommons platform (EcoCommons: https://doi.org/10.47486/PL108) in Australia. Both projects look to create standardised and independent units of work out of research models that can be composed into more complex workflows.

A catalyst to this needed change within the urban context will be the creation of both a new standard for containerising urban-research analytics [1] and a digital infrastructure platform capable of running these analytics containers in an interoperable and scalable environment. This approach, capable of providing "Urban Analytics as a Service" (UAaaS), defines a clean separation of concerns as urban-research analytics are progressed up the technology-readiness (TR) scale (Levels 1–9):

- *Urban* Researchers (TRL 1–3): Use their expert domain knowledge in urban research to create analytical models that underpin new decision-support tools. They do not have to concern themselves with current software-engineering best practices.
- Research Software Engineers (TRL 4–6): Use their highly specialised understanding of software engineering and the urban research domain [2] to uplift the urban analytical models using software-engineering best practices and to containerise them so that they are capable of running on a UAaaS platform.
- Research Infrastructure Specialists (TRL 7–9): Use their expert knowledge in digital research infrastructure (NCRIS 2022) to create platforms that apply the capabilities of the containerised analytics using standard application programming interfaces (APIs). They do not need to interact with the modelling code directly or to be an expert in the specific urban-research domain.

This separation allows each group to work rapidly and more autonomously in their areas of expertise to turn urban research into real-world applications and impact and lessen the dependence of a project on a handful of "unicorns" who are comfortable across the entire spectrum of work.

For this approach to gain traction, three components to be prototyped and made available at scale and to have their usage championed by a national research-infrastructure facility are required (Fig. 10.5):

- A standard Urban Analytics Container format that defines how urban analytics should be containerised.
- An open Urban Analytics Exchange that provides access to uploaded containers on a cloud-based platform—hosted on a national research infrastructure—and makes them findable and accessible following the FAIR Principles for Research Software (FAIR4RS Principles; see Hong 2022).
- An Urban Analytics Engine that can execute the analytics within these containers on demand and at scale on cloud-based computational infrastructure, providing UAaaS to both individual researchers and large, persistent decision-support platforms through a common API following modern software-architecture best practices.

The technology required for each of these components already exists and has been commodified into software-, platform-, and infrastructure-as-a-service offerings. The only thing to be done technically is to compose these together as required. The challenges are in community engagement and striking a balance between the needs and capabilities of the researchers and the research infrastructure specialists in the Urban Analytics Container format. Going through the additional steps of having their model containerised should be something with a very high return on time

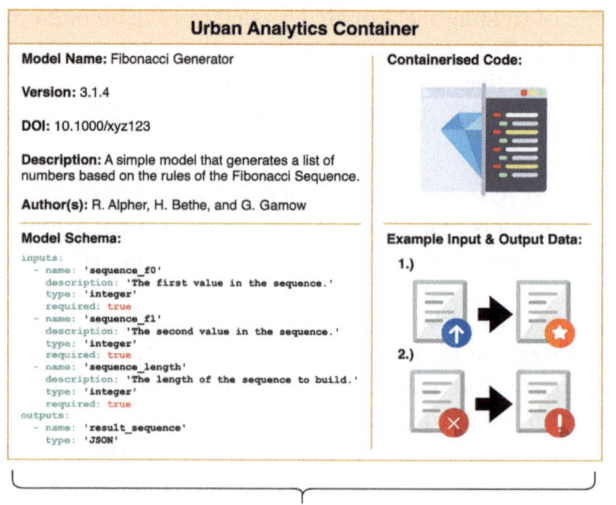

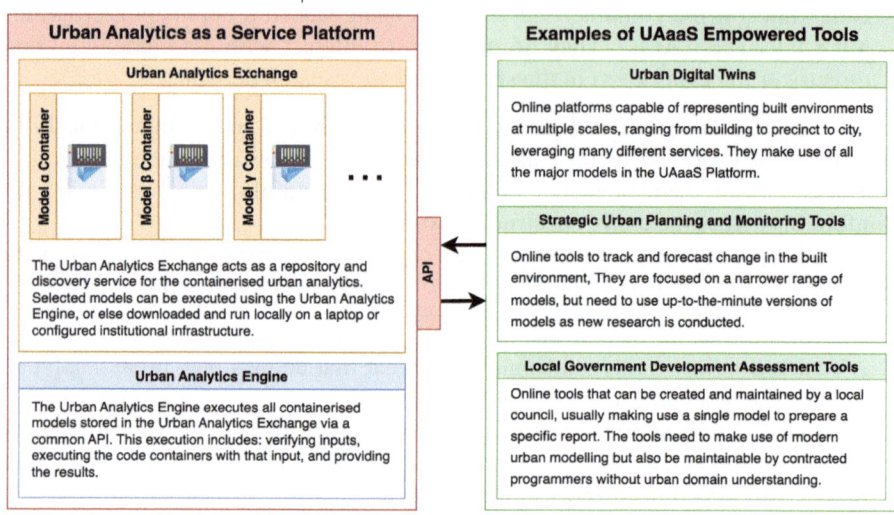

Fig. 10.5 Ecosystem for future delivery of urban analytics as a service

invested for an urban researcher, and so there is a balance to be struck in what goes into the analytics container and how complex it is to create. Broadly, each Urban Analytics Container stored on the Urban Analytics Exchange would incorporate the following (Fig. 10.5):

- A unique persistent identifier for rapid and consistent discoverability.
- A manifest of rich human- and machine-readable metadata, fully describing the contained model's operations.
- The containerised code representing the analytical model, packaged so that the Urban Analytics Engine (or any other compatible platform that can execute the contained analytics) can interact with it in a standardised way.

- Examples of inputs and corresponding outputs that can be used to test and validate the model and to give researchers in the broader community a hands-on example of its usage.

This containerisation process should be a simple task for a research infrastructure specialist, given that it only requires that the original analytics code follows a few basic rules to be compatible with the UAaaS approach and common workflow paradigms (see https://www.commonwl.org/).

Reworking existing models and modes of work to follow these rules can be a tedious process and in some cases requires a completely new analytical approach. A large part of the transition of urban researchers to this new paradigm, however, will be to make explicit the rewards gained from expending the time and energy required to adopt the framework outlined above. Ideally, this would start with the creation of an initial curated collection of high-value containerised models that would form the basis for a suite of small demonstration platforms (Fig. 10.5). These demonstrators would showcase the benefits of integrating with the UAaaS: the shortened time to an initial platform, the ability to painlessly patch or upgrade an analytical model on a live system without major code rewrites, and simpler, more standardised code that can be understood and modified by contracted programmers who do not require an intimate knowledge of the research domain.

For an urban practitioner, there are numerous benefits that come from the UAaaS platform:

- Lower costs in both time and money to create new decision-support tools based on cutting-edge analytics stored in Urban Analytics Containers.
- Decision-support tools that will have longer useful lifetimes, as the models they are based on can be easily extended and upgraded centrally in the Urban Analytics Exchange.
- The ability to create decision-support tools that can solve more complex issues by easily composing multiple analytical tools and running them remotely in the cloud on the Urban Analytics Engine platform.

In the not-too-distant future, leading analytical tools will be delivered as modular, web-based services and composed together to create rich and dynamic decision-support platforms. These platforms will be straightforward to build and maintain while simultaneously being complex and detailed enough to cover a wide range of urban projects and planning processes that require greater efficiencies in integrated assessment and delivery time than is presently the case. Leading examples include (Fig. 10.5) the following:

- *Urban Digital Twins*: online platforms capable of representing built environments and their resident populations at multiple scales, ranging from building to precinct to city, where *what if* questions can be posed during any project lifecycle stage from planning, design, construction, and operation.
- *Strategic Urban Planning and Monitoring*: given the rapid growth of cities, a facility to track and forecast change in built and natural environments across

multiple time periods and sectors, represents a step-change advance in urban-planning practice.

• *Local Government Development Assessment*: providing an accessible toolkit for integrated triple bottom line performance assessment of a large pipeline of proposed urban-development projects.

10.7 Conclusion

Commentary associated with the release of the IPCC (2022) Sixth Assessment Report on Climate Change has reinforced an increasing number of scientific claims that the next decade will prove a crucial test of whether the world can move fast enough to implement plans to adapt to this increasingly complex existential challenge: "Any further delay in concerted anticipatory global action on adaptation and mitigation will miss a brief and rapidly closing window of opportunity to secure a liveable and sustainable future for all". Climate change has been identified as a clear threat multiplier for Australian cities, their populations, and the systems that support them (Newton et al. 2018). As such, an accelerated transition to smart, sustainable urban development represents one of the most critical pathways for avoiding such threats.

The new frontiers of data science, urban analytics, digital platforms, and real-time collaboration networks are rapidly advancing and converging (see also Batty and Yang 2022). They offer the prospect of creating an unprecedented digital research platform capable of meshing with the knowledge, data, and models of domain experts and leading urban practitioners across all built-environment sectors. This represents a critically important contribution to sustainability science and sustainable urban development: a capacity for rapid integrated assessment of complex urban problems, hitherto limited—dramatically reducing time to deliver evidence-based insights for decision-making and governance associated with the mission-scale challenges of this century (Mazzucato 2021).

Notes

1. Code containerisation was standardised and championed by the Docker corporation in the early 2010s and has since changed the way digital technology is written, deployed, and operated at scale. Docker's standard allowed any code, and the computer environment it relies on (i.e. operating system, dependant software and libraries, etc.), to be "containerised" and stored in a static file called an "image". These images can then be distributed and run on any computer that can execute them into live "containers" of code, regardless of their underlying hardware or operating system.

2. The term "research software engineering", first coined in the early 2010s, is an attempt to describe the unique work that goes into actualising analytics- or data-heavy research through the creation of research software (Prause et al. 2010). It requires the practitioner, a "research software engineer", to both be fluent in

modern software engineering and have a strong passion for and understanding of research. While the term is not universally accepted within the academic and professional communities, it is gaining international recognition (https://researchsoftware.org/).

Acknowledgements This chapter has been a collaboration of the authors across four frontier areas of national digital infrastructure research. All have held senior positions within AURIN: the Australian Urban Research Infrastructure Network (https://aurin.org.au/), a federally funded (NCRIS) facility. Its charter is to develop the next generation of national digital research infrastructure capable of being meshed with leading urban and environmental systems analytics. The objective is to significantly boost R&D productivity among urban system domain experts and practitioners engaged in the mission-scale challenge of delivering smarter, more sustainable, liveable, resilient, competitive, and inclusive cities—more rapidly. The chapter has also benefited from connection with the National Urban Research Platform (iHUB), established as an Australian Research Council-funded research infrastructure project (LE1900100201) involving five universities: Swinburne, University of NSW, Monash University, University of Queensland, and Curtin University—all ranked "above to well above world standard" in the latest Excellence in Research Australia rankings for Built Environment and Design field of research (ARC 2016).

References

ARC (2016) State of Australian university research 2015–2016. Australian Government, Canberra

Australian Academy of Science (2021) Advancing data intensive research in Australia. AAS, Canberra

Australian Academy of Science (2022) Australia's data-enabled research future: science. AAS, Canberra

Badidi E, Mahrez Z, Sabir E (2020) Fog computing for smart cities' big data management and analytics: a review. Future Internet 12:190

Bainbridge W, Roco M (2016) Handbook of science and technology convergence. Springer, Berlin

Batty M (2017) The future journal. Environ Plan B Urban Anal City Sci 44(1):6–9

Batty M (2018) Inventing future cities. MIT Press, Cambridge, MA

Batty M (2019) Urban analytics defined. Environ Plan B Urban Anal City Sci 46(3):403–405

Batty M, Yang W (2022) A digital future for planning: Spatial planning reimagined. https://digital4planning.com/wp-content/uploads/2022/02/A-Digital-Future-for-Planning-Full-Report-Web.pdf

Baumann P, Misev D, Merticariu V, Huu BP (2021) Array databases: concepts, standards, implementations. J Big Data 8:28

Bello M, Galindo-Rueda F (2020) Charting the digital transformation of science. OECD, Paris

Biermann F, Hickmann T, Senit C-A (2022) The political impact of the SDGs: transforming governance through global goals. Cambridge University Press, Cambridge

Chauhan S, Agarwal N, Kar AK (2016) Addressing big data challenges in smart cities: a systematic literature review. Info 18(4):73–90

Chrisman N (2006) Charting the unknown: how computer mapping at Harvard became GIS. ESRI Press, Redlands, CA

Cornwell S (2023) Growing discourse in AI and data's impact on urban planning and placemaking. FastCompanycom. Accessed 5 Jun

Cugurullo F (2020) Urban artificial intelligence: from automation to autonomy in the smart city. Front Sustain Cities 2:38. https://doi.org/10.3389/frsc.2020.00038

Debnath R, Pettit C, Leao SZ, Lock O (2021) The role of technology tools to support geodesign in resilience planning. Urban Inform Fut Cities 2021:447–463

Deng T, Zhang K, Shen Z-J (2021) A systematic review of a digital twin city: a new pattern of urban governance toward smart cities. J Manag Sci Eng 6(2):125–134

Department of Education, Skills and Employment (2021) National Research Infrastructure Roadmap (exposure draft). Australian Government, Canberra

Dixon T, Farrally L (2018) Urban rooms: where people get to design their city's future. The Conversation, January

Dixon T, Tewdwr-Jones M (2021) Urban futures: planning for city foresight and city visions. Bristol University Press, Bristol

European Union (2022) Smart cities. https://ec.europa.eu/info/eu-regional-and-urban-development/topics/cities-and-urban-development/city-initiatives/smart-cities_en

Florida R (2003) Cities and the creative class, vol 2. Wiley, New York, p 3

Gilbert T, Barr S, James P, Morley J, Ji Q (2018) Software systems approach to multi-scale GIS-BIM utility infrastructure network integration and resource flow simulation. ISPRS Int J Geo Inf 7(8):310

Girardet H (2015) Creating regenerative cities. Routledge, New York

Granell C, Fernandez OB, Diaz L (2014) Geospatial information infrastructures to address spatial needs in health: collaboration, challenges and opportunities. Future Gener Comput Syst 31:213–222

Guterres A (2021) The IPCC Report is a code red for humanity. https://unric.org/en/guterres-the-ipcc-report-is-a-code-red-for-humanity/

Hall P, Pain K (2006) The polycentric metropolis. Routledge, London

Hargroves K, Smith M (eds) (2005) The natural advantage of nations: business opportunities, innovation and governance in the 21st century. Routledge, London

Hong NC (2022) FAIR principles for research software (FAIR4RS Principles). https://doi.org/10.15497/RDA00068

House of Representatives Standing Committee on Infrastructure, Transport and Cities (2018) Building up or moving out. Inquiry into the Australian Government's role in the development of cities. Canberra

Hunt JD, Kriger DS, Miller J (2005) Current operational urban land-use—transport modelling frameworks: a review. Transplant Rev 25:329–376

Infrastructure Australia (2021) Australian Infrastructure Plan. https://www.infrastructureaustralia.gov.au/2021-australian-infrastructure-plan-implementation-and-progress

Innovation and Science Australia (2020) Stimulating business investment in innovation. DISER, Australian Government, Canberra

IPCC (2022) Climate change: a threat to human wellbeing and health of the planet. Taking action now can secure our future. https://www.ipcc.ch/2022/02/28/pr-wgii-ar6/

IRP (2018) The weight of cities: resource requirements of future urbanization. In: Swilling M, Hajer M, Baynes T, Bergesen J, Labbé F, Musango JK, Ramaswami A, Robinson B, Salat S, Suh S et al (eds) A report by the international resource panel. United Nations Environment Programme, Nairobi

ISO (2022) ISO and sustainable cities. https://www.iso.org/files/live/sites/isoorg/files/store/en/PUB100423.pdf

Kharrazi A, Qin H, Zhang Y (2016) Urban big data and sustainability goals: challenges and opportunities. Sustainability 8:1293

Leao S, Hassan M, Rashidi T, Pettit C (2021) Is Sydney a 30-minute city? Big data analytics assisting to bring political rhetoric into practice. In: Birkin M, Clarke G, Corcoran J, Stimson R (eds) Big data applications in geography and planning. Edward Elgar Publishing, Cheltenham, UK, pp 169–188

Lee B (1973) Requiem for large-scale models. J Am Inst Plann 39(3):163–178

Lee J-G, Kang M (2015) Geospatial big data: challenges and opportunities. Big Data Res 2(2):74–81

Lewis S, Maslin M (2015) Defining the Anthropocene. Nature 519:171–180

Lim C, Kim K-J, Maglio PP (2018) Smart cities with big data: reference models, challenges and considerations. Cities 82:86–99

Liu J, Li J, Li W, Wu J (2016) Rethinking big data: a review on the data quality and usage issues. ISPRS J Photogramm Remote Sens 115:134–142

Luusua A, Ylipulli J (2020) Urban AI: formulating an agenda for the interdisciplinary research of artificial intelligence in cities. In: Companion publication of the 2020 ACM designing interactive systems conference, pp 373–376. https://doi.org/10.1145/3393914.3395905

Luusua A, Ylipulli J, Foth M et al (2023) Urban AI: understanding the emerging role of artificial intelligence in smart cities. AI Soc 38:1039–1044. https://doi.org/10.1007/s00146-022-01537-5

Manjunatha M, Annappa B (2018) Real time big data analytics in Smart City applications. In: Proceedings of the International conference on communication, computing and internet of things (IC3IoT), 15-17 February

Marvin S, Bulkeley H, Mai L, McCormick K, Palgan Y (eds) (2018) Urban living labs: experimenting with city futures. Routledge, London

Mazzucato M (2021) Mission economy. Allen Lane (Penguin), London

Moglia M, Frantzeskaki N, Newton P, Pineda Pinto M, Witheridge J, Cook S, Glackin S (2021) Accelerating a green recovery of cities: lessons from a scoping review and a proposal for mission-oriented recovery towards post-pandemic urban resilience. Dev Built Environ 7:100052. https://doi.org/10.1016/j.dibe.2021.100052

Moreno C, Allam Z, Chabaud D, Gall C, Pratlong F (2021) Introducing the "15-Minute City": sustainability, resilience and place identity in future post-pandemic cities. Smart Cities 4:93–111

Muff K (2017) The collaboratory. A co-creative stakeholder engagement process for solving complex problems. Routledge, London

Musango J, Currie P, Robinson B (2017) Urban metabolism for resource efficient cities: from theory to implementation. UN Environment, Paris

NCRIS (2022) Research infrastructure specialist: position paper. National Collaborative Research Infrastructure Strategy, Canberra

Newton PW (1995) Virtual project teams. In: Fisher MA, Law KH, Luiten B (eds) Modeling of buildings through their life cycle. CIB Publication 180. Stanford University, Stanford, CA

Newton PW (2012) Liveable and sustainable? Socio-technical challenges for 21st century cities. J Urban Technol 19(1):81–102

Newton P (2016) Delays at Canberra: why Australia should have built fast rail decades ago. The Conversation, 13 April

Newton P (2019) The performance of urban precincts: towards integrated assessment. In: Newton P et al (eds) Decarbonising the built environment: charting the transition. Palgrave Macmillan, Singapore, pp 357–386

Newton P, Burry M (2018) Game changing new digital platform to drive innovation in urban design. The Fifth Estate, 11 December. https://thefifthestate.com.au/urbanism/planning/game-changing-new-digital-platform-to-drive-innovation-in-urban-design/

Newton P, Doherty P (2014) The challenges to urban sustainability and resilience. In: Pearson L et al (eds) Resilient sustainable cities. Routledge, London

Newton P, Frantzeskaki N (2021) Creating a national urban research and development platform for advancing urban experimentation. Sustainability 13:530. https://doi.org/10.3390/su13020530

Newton P, Rogers B (2020) Transforming built environments: towards carbon neutral and blue-green cities. Sustainability 12:4745. https://doi.org/10.3390/su12114745

Newton P, Taylor V (1985) Probable urban futures. In: Brotchie J, Newton P, Hall P, Nijkamp P (eds) The future of urban form: the impact of new technology. Croom Helm, London

Newton P, Plume J, Marchant D, Mitchell J, Ngo T (2017) Precinct information modelling: a new digital platform for integrated design, assessment and management of the built environment. In: Sanchez A, Hampson K, London G (eds) Integrating information in built environments. Routledge, London

Newton P, Bertram N, Handmer J, Tapper N, Thornton R, Whetton P (2018) Australian cities and the governance of climate change. In: Tomlinson R, Spiller M (eds) Australia's metropolitan imperative. An agenda for governance reform. CSIRO, Melbourne

Newton PW, Newman PW, Glackin S, Thomson G (2022) Greening the greyfields: new models for regenerating the middle suburbs of low-density cities. Palgrave Macmillan, Singapore. https://link.springer.com/book/10.1007/978-981-16-6238-6

Nugent D, Suhail A (2021) Crisis, disorder and management: smart cities and contemporary urban inequality. In: Pardo I, Prato G (eds) Urban inequalities. Palgrave studies in Urban Anthropology. Palgrave Macmillan, Cham

OECD (2020) A territorial approach to the sustainable development goals: synthesis report. OECD, Paris

Pan Y, Tian Y, Liu X, Gu D, Hua G (2016) Urban big data and the development of city intelligence. Engineering 2(2):171–178

Pereira L, Frantzeskak N, Hebinck A, Charli-Joseph L, Drimie S, Dyer M, Eakin H, Galafassi D, Karpouzoglou T, Marshall F et al (2020) Transformative spaces in the making: key lessons from nine cases in the global south. Sustain Sci 15:161–178

Pettit CJ, Zarpelon Leao S, Lock O, Ng M, Reades J (2022) Big data: the engine to future cities—a reflective case study in urban transport. Sustainability 14(3):1727

Płuciennik T, Płuciennik-Psota E (2014) Using graph database in spatial data generation. In: Gruca D, Czachórski T, Kozielski S (eds) Man-machine interactions 3. Advances in intelligent systems and computing, vol 242. Springer, Cham

Prause C, Reiners R, Dencheva S (2010) Empirical study of tool support in highly distributed research projects. In: 5th IEEE International Conference on Global Software Engineering, Princeton, 23-26 August

Productivity Commission (2012) Performance benchmarking of Australian business regulation: the role of local government. Australian Government, Canberra

Punt EP, Geertman SC, Afrooz AE, Witte PA, Pettit CJ (2020) Life is a scene and we are the actors: assessing the usefulness of planning support theatres for smart city planning. Comput Environ Urban Syst 82:101485

Reades J, Rey SJ (2021) Geographical python teaching resources: Geopyter. J Geogr Syst 23(4):579–597

Robinson I, Webber J, Eifrem E (2015) Graph databases: new opportunities for connected data. O'Reilly Media, Newton, MA

Sanchez TW (2023) Planning on the verge of AI, or AI on the verge of planning. Urban Sci 7(3):70

Schrotter G, Hurzeler C (2020) The digital twin of the city of Zurich for urban planning. J Photogramm Remote Sens Geoinf Sci 88:99–112

Shah SA, Seker DZ, Rathore MM, Hameed S, Yahia SB, Draheim D (2019) Towards disaster resilient smart cities: can internet of things and big data analytics be the game changers? IEEE Access 7:91885–91903

Shahat E, Hyun CT, Yeom C (2021) City digital twin potentials: a review and research agenda. Sustainability 13:3386

Singh RK, Agrawal S, Sahu A, Kazancoglu Y (2021) Strategic issues of big data analytics applications for managing health-care sector: a systematic literature review and future research agenda. Big data analytics applications. TQM J 35(1):262–291

Singleton AD, Spielman S, Folch D (2018) Urban analytics. Sage, London

SoE (2021) Australia State of the Environment 2021. https://soe.dcceew.gov.au/

Sta BH (2017) Quality and the efficiency of data in smart cities. Future Gener Comput Syst 74:409–416

Steffen W, Broadgate W, Deutsch L, Gaffney O, Ludwig C (2015) The trajectory of the Anthropocene: the great acceleration. Anthropocene Rev 2(1):81. https://doi.org/10.1177/2053019614564785

Steinitz C (2012) A framework for geodesign: changing geography by design. https://www.esri.com/en-us/esri-press/browse/a-framework-for-geodesign-changing-geography-by-design

Steinitz C, Rogers P (1970) A systems analysis model of urbanization and change: an experiment in interdisciplinary education. MIT Press, Cambridge, MA

Talebkhah M, Sali A, Marjani M, Gordon M, Hashim SJ, Rokhani FZ (2021) IoT and big data applications in smart cities: recent advances, challenges, and critical issues. IEEE Access 9:55465–55484

Tollefson J (2023) World recommits to 2030 plan to save humanity—despite falling short so far. Nature, 20 September. https://doi.org/10.1038/d41586-023-02970-2

Torrens J, Schot J, Raven R, Johnstone P (2019) Seedbeds, harbours and battlegrounds: on the origins of favourable environments for urban experimentation with sustainability. Environmental innovation and societal transitions 31:211–232

Ward B, Dubos R (1972) Only one earth. WWNorton, NY

Webb R, Bai X, Stafford Smith M, Costanza R, Griggs D, Moglia M, Neuman M, Newman P, Newton P, Norman B, Ryan C, Schandl H, Steffen W, Tapper N, Thomson G (2018) Sustainable urban systems: co-design and framing for transformation. Ambio 47(1):57–77. https://doi.org/10.1007/s13280-017-0934-6

Webb R, O'Donnell T, Auty K, Bai X, Barnett G, Costanza R, Dodson J, Newman P, Newton P, Robson E, Ryan C, Stafford Smith M (2023) Enabling urban systems transformations: co-developing national and local strategies. Urban Transform 5(1):1–31

Yigitcanlar T, Agdas D, Degirmenci K (2023) Artificial intelligence in local governments: perceptions of city managers on prospects, constraints and choices. AI Soc 38:1135–1150. https://doi.org/10.1007/s00146-022-01450-x

Peter Newton is an Emeritus Professor in the Centre for Urban Transitions at Swinburne University of Technology in Melbourne, Australia. From 2007 to 2021, he held the position of Research Professor, following 15 years as Chief Research Scientist with the Commonwealth Scientific and Industrial Research Organisation (CSIRO). He has had a distinguished career in built-environment research and was elected a Fellow of the Academy of Social Sciences in Australia in 2014. His principal fields of research have focused on the technology of planning, sustainability science, and urban transitions. He has published over 25 books, including *Greening the Greyfields* and *Migration and Urban Transitions in Australia* in 2022.

Chris Pettit is the inaugural Professor of Urban Science and Director of the City Futures Research Centre at the University of NSW, Sydney. He serves as Chair of the Board of Directors for Computational Urban Planning and Urban Management. He is a member of the Planning Institute of Australia's National Plantech Working Group, the NSW Government's Expert Advisory Group for Planning Evidence and Insights, and the Australian Government's Urban Policy Consultation Network. Over the past 25 years, he has been conducting research and development in the use of digital tools to support what if? analysis and scenario planning of city futures. He has edited five books and published more than 200 academic papers.

Stuart Barr is Professor of Geospatial Systems Engineering at the School of Engineering Newcastle University, UK (2017–2020 and 2022–2023). His research focuses on the use of geospatial data, analytics, simulation modelling, and decision support for integrated urban-infrastructure systems analysis. Between 2020 and 2022, he was Director of the Australian Urban Research Infrastructure Network (AURIN) and Professor of Urban Data Science at the University of Melbourne.

Loren Bruns Jr is the Digital Research Infrastructure Manager at the Australian Urban Research Infrastructure Network (AURIN), headquartered in Melbourne. He directs a team of research infrastructure engineers who build modern, cloud-first digital infrastructure for the urban research community. Prior to joining AURIN, Loren completed a BA in physics from Reed College, USA, and a PhD in astrophysics from the University of Melbourne. He then went on to spend a decade as a full-stack research software engineer in the School of Computing and Information Systems at the University of Melbourne, where he created research platforms and co-authored papers across multiple fields before pivoting to national research infrastructure.

Chapter 11
Embedding Transformative Innovation into Mission-Oriented Policy and Innovation Districts: The Case of Melbourne

Thi Minh Phuong Nguyen, Kathryn Davidson, and Megan Farrelly

Abstract This chapter presents key interventions to provide a step-change in the understanding of innovation, from orthodox to transformative, in the development of local government public policy in the Melbourne Innovation District (MID) City North, Australia. Innovation refers to the introduction of new technological, organisational, and social solutions in response to problems and challenges that arise in existing social, economic, and environmental settings. A step-change is necessary as traditional innovation practice is not currently fit for purpose to deliver a more inclusive and sustainable society. To illustrate this step-change, we reference the MID City North with the key actors including the City of Melbourne (CoM) and two universities (University of Melbourne and RMIT). The MID City North, established in 2017, is a maturing initiative that has evolved over a period of time punctuated by considerable disruptions, particularly the Australian bushfire (Black Summer 2019–2020) catastrophe and the COVID-19 pandemic. We argue these disruptions are a catalyst for change and therefore an important context for the social shaping of the evolving policy discourse relevant to the MID City North. Both disruptions were powerful pressures, as well as windows of opportunity, to effect policy reform mainly by the key actor: the CoM. We propose four key interventions to build on the policy-reform momentum to better develop pathways to deliver on transformative innovation in policy, specifically for the MID City North and the CoM. These inter-

T. M. P. Nguyen (✉)
Fenner School of Environment and Society, Australian National University,
Canberra, ACT, Australia
e-mail: thiminhphuong.nguyen@anu.edu.au

M. Farrelly
Faculty of Architecture, Building and Planning, University of Melbourne, Parkville, VIC,
Australia

K. Davidson
Faculty of Arts, School of Social Sciences, Monash University, Clayton, VIC, Australia

© The Author(s) 2025   241
N. Frantzeskaki et al. (eds.), *Future Cities Making*, Theory and Practice of
Urban Sustainability Transitions, https://doi.org/10.1007/978-981-97-7671-9_11

ventions are key because they foster the establishment of favourable policy mixes supporting innovation for the transformative change required for cities transitioning towards sustainability.

Keywords Innovation districts · Mission-oriented innovation · Transformative change · Melbourne

11.1 Introduction

Innovation now features prominently on policy agendas across governance levels in Australia, for example, the national plan "Australian 2030 – Prosperity through Innovation" (Australian Government 2017) or the Victorian Government's "Innovation Statement" (Victoria State Government 2021). However, many key thinkers question the implied assumption that all innovation is inherently positive and universally desirable (Røpke 2012; Uyarra et al. 2019). For example, a traditional understanding of innovation that emphasises economic growth and competitiveness is generally not aligned with wider societal and environmental needs. It is clear to most policy-makers across all levels of government that the status quo is no longer an option. For example, the OECD (2023), in its "Science, Technology and Innovation Outlook 2023", indicates that governments' business-as-usual approaches to low-carbon innovation are insufficient to achieve net-zero emissions by 2050. Thus, there is an urgent need to better respond to the multiple social, economic, and environmental crises that now confront society. Moreover, we know time is running out to respond to the climate emergency (IPCC 2022); it is, therefore, essential that we now accelerate our understanding of transformative innovation to deliver on sustainable pathways.

Delivering on sustainable futures for cities requires transformative change to established sociotechnical-environmental systems. Transformative change refers to "fundamental change that is distinguished from minor and marginal adjustments" (Heikkinen et al. 2019, p. 94). Such structural change seeks to address the deep-rooted causes of vulnerability, which requires attending not only to technical aspects but also to social and environmental components. Examples include fundamental changes in world-views, rules of business-making, social networks, ecosystems, political and power relations, citizen lifestyles, physical infrastructure, and technology (Heikkinen et al. 2019). The achievement of city-level transformations requires transformative change across all city domains, including economic, social, political, and environmental dimensions (Wolfram 2016; Heikkinen et al. 2019). Sustainable transformations are therefore complex, long-term, and multidimensional processes that require policy-makers to employ innovative and transformative mechanisms to bring about practical transformations.

Transformative innovation means fostering transformative change via innovation to address "wicked" social and environmental challenges and ensuring that innovation processes are open, inclusive, reflexive, and experimental in trialling new things to generate social learning (Schot and Steinmueller 2018). Adopting transformative innovation can trigger interconnected systematic changes that have positive influences on social, environmental, and economic outcomes. In searching for potential pathways for transformative innovation, countries and cities have grappled with a move from "business-as-usual" policy approaches to mission-oriented innovation policies that address grand societal challenges and achieve the global Sustainable Development Goals (SDGs) (United Nations 2015). Mission-oriented policies are slowly developing internationally, mainly with a narrow focus on well-designed and planned experimentation and pilot programmes to achieve defined objectives. The OECD's systemic review of the global policy landscape in 2019–2020 revealed that governments had implemented at least 40 mission-oriented policy initiatives that aimed to reduce greenhouse gas emissions as part of tackling climate change and achieving the SDGs, among other objectives (Larrue 2022). While the majority of these mission-oriented initiatives remain in their infancy, they offer valuable lessons for countries and cities, particularly in terms of "how to co-develop a strategic agenda to the different ways to integrate policy instruments across administrative silos" (Larrue 2022, p. 5).

Delivering on mission-oriented policies is even more critical in light of the current unpredictable state of the world, where unanticipated events such as natural disasters (such as fire, flood, drought, and heatwaves) or global pandemics can create havoc. Yet in some instances, disruptions are a trigger for change, opening a window of opportunity to catalyse new pathways. For example, the Black Summer bushfire in 2019–2020 and the COVID-19 pandemic caused significant disruptions and imposed huge costs for major Australian cities, which raises questions regarding how to proactively respond to and recover from these events. Therefore, in times of crisis, if we can act quickly and effectively with relevant policy pathways, such as mission-oriented innovation policies, we can potentially accelerate our transition towards more sustainable futures.

Promoting city-level sustainable transformations also requires a clearly articulated vision for transformative cities. A transformative city is one that seeks to generate fundamental change in all aspects of urban life (including economy, socio-political structures, physical infrastructure, and natural environment), aiming to create a more sustainable, resilient, and equitable future for all. It is a city that recognises the "wicked" challenges it faces and proactively works to tackle these interconnected issues through transformative innovation policies and initiatives. In doing so, a transformative city embraces and sustains processes of experimentation and learning via which diverse possibilities and pathways for sustainable urban development can emerge and thrive. Such a city also involves multiple actors in processes of structural change, including civil society and users who can play an important role in supporting system innovation. In general, a transformative city is a future-proof city that can adapt to the challenges and opportunities of the twenty-first century and contribute to the ambitious SDG agenda.

This chapter presents four key interventions aimed at driving a step-change in understanding innovation, from "orthodox" to transformative, within policy reform in the City of Melbourne (CoM). Our step-change has a specific focus on innovation districts; therefore, we will reference the Melbourne Innovation Districts (MID) City North. The key actors involved with the MID are the local government (CoM) and two universities (University of Melbourne and RMIT). Established in 2017, the MID is still maturing and has evolved through periods of significant disruption (Black Summer and COVID-19), which have played a key role in shaping this initiative. The disruptive forces were powerful pressures that created windows of opportunity to influence policy reform driven mainly by the key actor: the CoM. We explore how the dominant actor, the CoM, could use these windows of opportunity to further reform policy in a way that better embeds transformative innovation.

In this chapter, we first summarise contemporary understandings of innovation districts and transformative innovation. Second, we review the innovation objectives and progress of the MID towards transformative change. Third, we explore the mission-oriented innovation policy approaches for transformative change. Finally, we suggest four interventions for driving a step-change in understanding innovation in policy.

11.2   Learnings from Melbourne Innovation District City North

The MID City North has an explicit vision as follows:

> City North will become a world class urban district and environment that supports and develops next generation Melbourne, a place designed to leverage emerging technologies and innovation and build on our city's unique characteristics to enhance education and economic outcomes, create new knowledge and city experiences and enrich inclusion and public amenity (CoM 2020, p. 4).

11.2.1   What Are Innovation Districts?

Innovation districts are typically understood as place-based urban-development strategies that seek to regenerate under-performing neighbourhoods into attractive and desirable places for innovative workers and businesses (Morisson 2020). The development and operation of innovation districts are built on the ideas that innovation arises from vibrant and collaborative environments supported by different forms of proximity that encourage people to share ideas and knowledge as they meet and socialise together (Boschma 2005). Such districts may share several characteristics of urban test sites, but are normally larger in scale and integrated with more conventional framings of innovation that promote economic growth and competitiveness. Conventional framings of innovation districts face criticisms in their limited potential to deliver social transformation. For example, some critiques have specifically pointed to their gentrifying effects, noting that many innovation districts have become solid real-estate businesses, failed to connect with their surrounding environments, and exacerbated social inequality and exclusion (Massey et al. 2003; Zukin 2020). Despite these criticisms, innovation districts have proven to be practical policy approaches for cities seeking to modernise their economies and accelerate technological innovation processes (Morisson and Bevilacqua 2019).

Several living instances of innovation districts exist around the world. 22@ Barcelona presents the first example of a planned innovation district for long-term urban transformation, operating as a mixed-used technological district with a compact industrial fabric, high-density, and convenient transport connections (Bottero et al. 2020). Another example is San Francisco's SOMA district, which represents a new spatial arrangement centred around innovation, productivity, and creativity. In Melbourne, Australia, the MID City North is the first attempt to create an "ecosystem" for innovation and is our case study for this chapter.

11.2.2 What Is Transformative Innovation?

Transformative innovation differs from traditional conceptualisations of innovation that prioritise economic growth by articulating a more "capacious" understanding of innovation that addresses "wicked" social and environmental challenges. Transformative innovation refers to the comprehensive mobilisation of innovation, technology, and science for meeting societal needs and achieving the SDGs (Schot and Steinmueller 2018). In particular, transformative innovation emphasises the need to create opportunities for transformative change via innovation processes that are open, inclusive, reflexive, and experimental in trialling new things and practices to generate social learning (Schot and Steinmueller 2018). In the longer term, transformative innovations will contribute to shape the composition and directionality of innovation to ultimately facilitate a systems-wide transformation of sustainability. For example, transformative innovation policies can focus on supporting the development of new mobility systems that discourage private car ownership and encourage other (new) sustainable mobility modalities such as public transportation, bicycling, walking, and electric vehicles. This new system should also promote sustainable mobility planning (including the reduction of non-sustainable mobility modalities) as a standard of modern behavior to tackle urban challenges. Additionally, with the multiplicity of the city, a shift in sustainable mobility can potentially lead to a complex multilayered response, in the sense that reduced car dependency can encourage the establishment of larger greening corridors or water-sensitive urban design features for stormwater (e.g. see Nielsen and Farrelly 2019). The development of new mobility systems may contribute to sociotechnical-environmental system transformation as it fosters the co-production of social, technological, environmental, and behavioral change in an interconnected way (Schot and Steinmueller 2018).

Schot and Steinmueller (2018) present three frames for understanding innovation: (1) innovation for growth, (2) national systems of innovation, and (3) transformative change. The first two frames present "traditional" understandings of innovation as a means for economic growth and city competitiveness, focusing on advancing science and technology for mass production and consumption, as well as for knowledge creation and commercialisation. The third frame presents a more "capacious" understanding of innovation that integrates innovation objectives with pressing environmental and societal challenges, promoting experimentation and learning for broader societal transformations. Table 11.1 summarises the three frames' key points.

Frame 3 in Table 11.1 acknowledges the shortcomings of innovation, technology, and science in solving the sustainability puzzle and driving broader socio-environmental transformations, in that it explicitly focuses on "open, inclusive, reflexive" practices. These processes are often considered external to innovation policies built on Frames 1 and 2 (Schot and Steinmueller 2018). This reflects the tensions between these three frames, making them partially incompatible. Yet a shift towards Frame 3 for transformative innovation does not necessarily imply that policy-makers should completely forgo the first two frames. For example, in real-world policy contexts, investment in science and research for knowledge creation, as promoted in Frames 1 and 2, remains a crucial foundation of any innovation policy. However, Frame 3 suggests that to effectively deliver transformative outcomes, policy-makers should invest in aligned processes of experimentation and learning to develop more sustainable pathways. While real-world policy practices may reflect a mixture of frames, we suggest that Frame 3 should inform and shape the directionality and composition of innovation policies in the longer term.

11.2.3 MID City North: State of Play

Launched in 2017, MID City North is a collaborative partnership between the CoM and two major universities in Victoria—The University of Melbourne and RMIT University—which have a footprint in the central business district (Fig. 11.1).

Table 11.1 Three frames of understanding innovation by Schot and Steinmueller (2018)

Frame	Key points
Frame 1—Innovation for growth	• Promoting innovation as a means for economic growth • Focusing on science and technology for mass production and on consumption to foster prosperity and productivity
Frame 2—National systems of innovation	• Promoting innovation as means for economic growth and fostering city competitiveness to attract talent and investments • Focusing on science and technology to facilitate knowledge creation, transfer, diffusion, and commercialisation
Frame 3—Transformative change	• Calling for transformative change in innovation to effectively address social and environmental challenges • Promoting experimentation and innovation processes that are open, inclusive, reflexive, and experimental in trialling new things to generate social learning

As detailed in the CoM's "MID City North Opportunity Plan" (2020), the MID City North focuses on five key work streams: (1) enabling innovation activities in the urban realm, (2) facilitating enterprise activation, (3) advancing technology, (4) promoting social innovation, and (5) creating supportive institutional design that facilitates learning, collaboration, and creativity (CoM 2020). As the CoM is the key actor within the MID City North, we focus on numerous CoM policies that shape these work streams and innovation objectives. Table 11.2 lists key influential policies for the MID City North between 2017 and 2023 organised according to whether they were enacted pre-pandemic or during and post-pandemic (the post-pandemic period is defined as beginning in November 2021, at the end of State-mandated lockdowns in Melbourne).

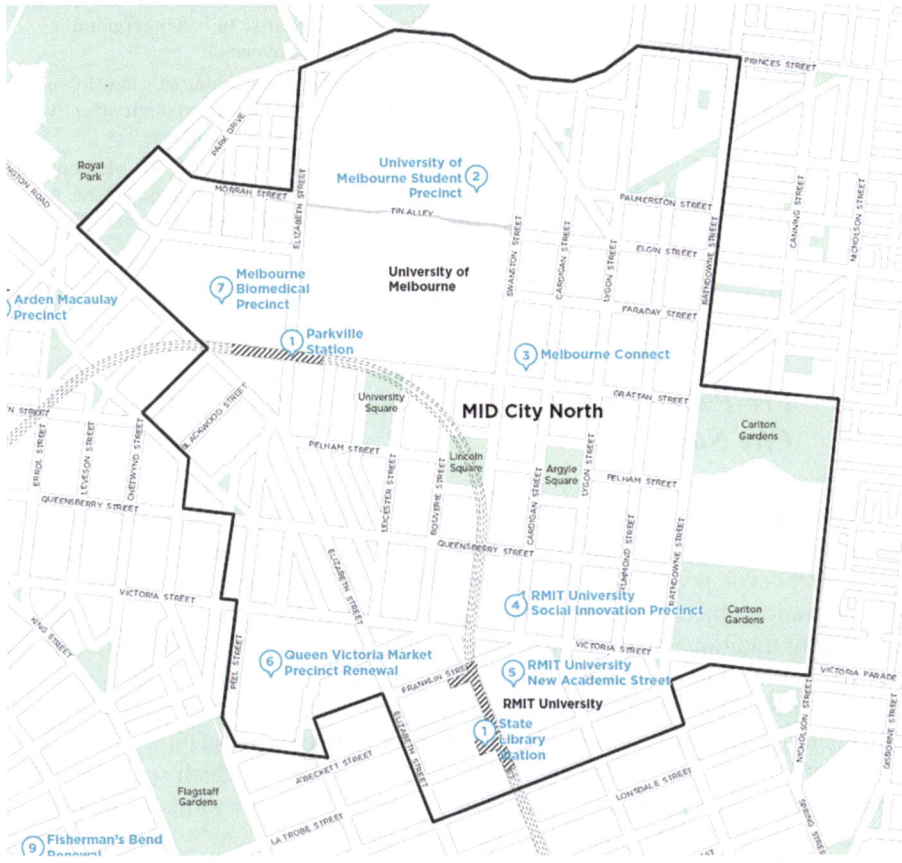

Fig. 11.1 MID City North's location. (Source: Map data © OpenStreetMap and contributors, CC-BY-SA, as used in CoM 2020, p. 15)

Table 11.2 Key CoM influential policies for the MID City North between 2017 and 2023

Time-frame	Policy	How does the policy influence the MID City North?
Pre-pandemic (2017–2020)	CoM Council Plan 2017–2021	Provided guiding directions for Melbourne's development between 2017 and 2021 including objectives for innovation and, specifically, the development of the MID City north as a major initiative for the city's growth and technological advancement
	MID City North Opportunity Plan 2020	Outlines core guiding principles for the design and use of the MID City north
	Urban Realm Action Plan: Melbourne Innovation Districts City North 2018–2023	Outlines key steps to enhance the urban realm around the MID City north area
During and post-pandemic (2021–2023)	Melbourne's Thriving Economic Future—Economic Development Strategy 2031 (released in 2021)	Outlines key priorities and actions for guiding Melbourne's economic development post-pandemic, with reference to the MID City north as an initiative to foster economic growth and competitiveness
	CoM Council Plan 2021–2025	Sets out directions for Melbourne's development during 2021–2025, including explicit objectives for innovation and experimentation
	Inclusive Melbourne Strategy 2022–2032	Presents directions and priorities for Melbourne to become an inclusive city, with objectives for innovation as a key driver of social inclusion

11.3 Innovation Objectives and Progress of the MID Towards Transformative Change

11.3.1 Pre-Pandemic Innovation Objectives of the MID City North

To understand the pre-pandemic objectives of the MID City North, we refer to a study by Davidson et al. (2023) that drew upon the three frames of understanding innovation developed by Schot and Steinmueller (2018). The study concluded that the innovation objectives of the MID City North are entrenched within an understanding of traditional innovation, aiming to support Melbourne's economic growth and international competitiveness (i.e. the first and second frames). However, the study also highlighted how the MID City North carries promise with respect to moving towards more "capacious" innovation objectives (i.e. the third frame). This is reflected within the "MID City North Opportunity Plan", which specifies directions for place-based, socially, and environmentally oriented forms of innovation and experimentation that facilitate visibility and engagement within the public realm (CoM 2020). The "Urban Realm Action Plan: Melbourne Innovation Districts City North 2018–2023" also provides design principles that guide the district's design, planning, and implementation to "improve quality of the social and urban environments" (CoM 2018, p. 11). In particular, "Design principle 5 – sustainable and healthy environments" explicitly focuses on addressing environmental sustainability and creating healthy environments, signalling the potential for the MID to generate positive environmental outcomes (CoM 2018, p. 21).

11.3.2 Disruptions as Catalysts for Transformative Change: Implications of the Black Summer 2019–2020 and the COVID-19 Pandemic for the MID City North

Innovation districts often embrace location-specific forms of innovation. These initiatives are contextually sensitive and the fact that "locally specific diagnoses of a societal deficiency and equally specific understandings of acceptable remedies" are key to their success or failure (Pfotenhauer and Jasanoff 2017, p. 1). Simply put, contextual pressures can influence the vision and design of innovation districts and their associated urban test sites. In the case study of MID City North, the Black Summer 2019–2020 bushfire and the COVID-19 pandemic were powerful contextual pressures that have affected the goals and objectives of the MID City North, requiring a shift towards more transformational innovation pathways to effectively address "wicked" societal challenges.

During the 2019–2020 Australian bushfire season, Melbourne was heavily affected by the hazardous smoke, sparking debate and activism in the city about the impact of climate change and the need for more progressive climate actions. In July 2019, the CoM declared a climate and biodiversity emergency, officially making a commitment to "take urgent action to reduce emissions and waste in order to protect public health, strengthen the economy and create a city that mitigates and adapts to climate change" (CoM 2022a, p. 1). This event clearly opened up opportunities for reallocating resources to plan and develop future green, climate-, or biodiversity-focused test sites within the MID City North; such opportunities were also aligned with design principles set out in the "Urban Realm Action Plan: Melbourne Innovation Districts City North 2018–2023" (Davidson et al. 2023).

In 2020 and 2021, the COVID-19 pandemic hit the city hard, as Melbourne experienced one of the world's longest and most severe State-mandated, citywide lockdowns. This crisis created significant uncertainty for the CoM, resulting in empty CBD streets due to tight movement restrictions. Such a situation presented new opportunities for improving the MID City North's physical environment, focusing on creating more creative, more flexible, and safer public spaces to attract people back to Melbourne's CBD (Davidson et al. 2023).

In sum, the 2019–2020 bushfire catastrophe (with the associated "climate emergency" activism in Melbourne) and the COVID-19 crisis represented powerful contextual pressures that could trigger and accelerate more innovative sustainability shifts for Melbourne and thus could influence the MID City North as a major innovation initiative of the city. These disruptions specifically presented windows of opportunities for developing new innovative urban test sites within the MID City North and trialling new approaches in the public realm, as originally recognised within the "MID City North Opportunity Plan" (CoM 2020).

11.3.3   Pandemic Recovery and Opportunities for Moving Towards Transformative Innovation for the MID City North

Despite severe setbacks with the COVID-19 crisis, innovation remains high on the policy agenda for the CoM and is featured among key drivers of pandemic recovery and Melbourne's future development. In this context, Melbourne's first innovation district is still considered a major project, having already secured $one million in the 2021–2022 annual CoM budget, and a further $500,000 that was allocated in the 2022–2023 financial year (CoM 2021a, 2022b).

During and post-pandemic, there was a shift in the CoM's policy discourse regarding their understanding of innovation, experimentation, and transformation. The change in the CoM's way of thinking towards transformative change was signalled in the recently released "Inclusive Melbourne Strategy 2022–2032". This new strategy, alongside the "Economic Development Strategy 2031", falls under the "Council Plan 2021–2025". This is the first time that the concept of inclusivity has been presented with equal prominence to economic development within the CoM policy hierarchy. This signals an important shift towards more transformative innovation in the CoM's objectives for driving post-pandemic recovery and development—a mission to foster not only economic growth and competitiveness but also social inclusion. Specifically, we identify key points of change within these three policy documents:

- The "Council Plan 2021–2025" focuses on Melbourne's future development post-COVID, aiming to foster not only "long-term economic growth" but also "social cohesion and environmental resilience" (CoM 2021b, p. 3). Such goals demonstrate positive change towards transformation in the policy agenda of the CoM and still align with the Council's official "Climate and Biodiversity Emergency" declaration in 2019. For example, the plan sets out a priority to "lead innovative responses to climate change", aiming to "protect public health, strengthen the economy and create a city that mitigates and adapts to climate change" and achieve the goal of being "a leading city globally that sets the standard in climate action" (CoM 2021b, p. 26). In addition, this plan explicitly promotes the notion of a "City of Possibility"—"a place where anything is possible" (CoM 2021b, p. 3). Such a narrative continues to signal the vision of the CoM in supporting innovation and experimental activities to help the city "recover from the pandemic, [...and] prepare for the challenges and opportunities that will impact the city into the future" (CoM 2021b, p. 3).

- By way of example, the Council's plan lists the following major initiative (CoM 2021b, p. 17):

  Drive economic growth and resilience by implementing the Economic Development Strategy, focusing support on existing and emerging industry sectors. This will include close collaboration with industry and universities, development of globally competitive innovation districts (particularly in our renewal areas), strengthening of the creative sector, facilitation of digital and technology innovation, support for re-establishment of international education and efforts to unlock climate capital.

- While the plan does not explain explicitly what "climate capital" means in practice, the above statements signal that CoM's innovative environmental actions post-COVID will focus on supporting economic recovery and strengthening the city's global reputation.
- The "Economic Development Strategy 2031" also hints at some movements towards transformative innovation, but the traditional understanding of innovation remains dominant. For example, in reference to the MID City North and other emerging innovation districts, this strategy sets out a long-term action that supports "the growth and development of existing, emerging and new globally competitive innovation districts in City North, Arden and Fishermans Bend", as these "innovation precincts drive economic growth and create high-value jobs" (CoM 2021c, pp. 24–25). This statement reinforces innovation districts as a means to ensure economic growth. Nevertheless, a key goal within this strategy is for Melbourne to become "an inclusive city" and highlights innovation as key to enabling "innovative social and commercial opportunities that benefit Aboriginal and non-Aboriginal Melburnians and Victorians alike" (CoM 2021c, p. 28). Thus, there remains a mismatch between goals and long-term actions for innovation and transformation within this particular strategy.
- The strategy also clearly outlines a priority of "unlocking climate capital" as listed in the "Council Plan 2021–2025". This strategy notes that the idea of "climate capital" suggests that "responding to climate change represents a significant economic opportunity" and that "City of Melbourne will demonstrate ongoing leadership and innovation on climate change mitigation and adaptation solutions to help create and grow markets in Melbourne and beyond" (CoM 2021c, p. 26). From these statements, it is clear that the goal of innovation for accelerating "climate capital" focuses squarely on the city's economic growth and competitiveness, to "establish Melbourne as a leading market for innovative climate change solutions" (CoM 2021c, p. 27).
- The "Inclusive Melbourne Strategy 2022–2032" can be regarded as the most progressive strategy for fostering social transformation in the CoM, reinforcing the goal of Melbourne's becoming "an inclusive city". While the "Inclusive Melbourne Strategy 2022–2032" does not explicitly mention the word "innovation", the document sets out a vision that "[o]ur city economy is inclusive – all people can contribute to and benefit from our shared prosperity" and that "inclusion is built into the governance, implementation, and advocacy of our economic recovery initiatives" (CoM 2022c, p. 24). Additionally, under "Priority 2:

Sustainable and fair recovery", the strategy states that the CoM will "continue to adapt to our approach based on what is working: be prepared to try and fail" (CoM 2022c, p. 22). This statement hints towards a crucial role for innovation and experimentation, particularly trialling new things and accepting failure, in supporting the goals of social inclusion and social transformation post-pandemic.

Collectively, the three newly released strategies of the CoM create an enabling environment for shaping the policy discourse to accelerate sustainability shifts that are more innovative and transformative for Melbourne. These approaches are expected to encourage the development and future success of major innovation initiatives across the city, including the MID.

11.3.4   Assessing the Ongoing Progress of the MID City North Towards Its Objectives

Melbourne's first innovation district is still a maturing project, and thus it is not possible to assess its final impacts at this point. However, in tracing the development of the MID City North over time, Davidson et al. (2023) reported that the transformative impacts of MID have been largely discursive, while organisational and material changes have been relatively limited, despite the ambitious objectives set by MID City North's leading partners. Davidson et al. (2023) also suggested that MID City North is inherently a solution looking for a problem. While it outlines a broader objective of generating "significant social, economic and environmental benefits" (CoM 2020, p. 3), Davidson et al. (2023 p. 9) pointed out that "precisely what it is and what its objectives are have never been clearly explicated", with the result that each of the three founding institutions "can interpret the initiative to suit its own purpose". While a key advantage of this interpretive flexibility is that the MID City North can be conveniently activated and mobilised by its stakeholders to address any strategically pressing challenge—be it COVID-19, climate change, or economic recovery—the lack of focus and challenge of mobilising resources means this solution-led pathway for innovation (Davidson et al. 2023) also risks underrepresenting trade-offs between societal values (Wanzenbock et al. 2020). This reflects a tangible risk of weakening the directionality, legitimacy, and potential of MID City North as an innovative solution for solving Melbourne's societal and environmental challenges and achieving the SDGs.

11.4   Moving Towards Mission-Oriented Innovation Policies for Transformative Change: What Does It Mean for the MID City North?

From the discussion above, it is clear that more can be done to support innovation initiatives to generate more transformative impacts within the MID City North. While the recent strategies demonstrate progressive movements towards transformation paradigms, it remains to be seen whether such new post-pandemic objectives can be fully translated to other policies and successfully implemented in practice. To achieve desired outcomes, local policy-makers and planners need to adopt a more systemic policy instrument that provides a long-term, strategic, and coherent orientation to support transformative innovation.

In moving towards the third frame of understanding innovation, conceptualisations such as "mission-led" and "challenge-led" innovation have emerged in policy thinking as potential approaches (Schot and Steinmueller 2018). Mission-oriented innovation policies refer to systemic, challenge-led policies that draw on frontier knowledge to provide solutions for addressing complex, multidimensional societal problems (Mazzucato 2018; Wanzenbock et al. 2020; Bellinson 2022). Such policies have the potential to support transformative system change, especially when they are directed to correct transformational system failures, which include a lack of directionality, a lack of demand articulation, limited reflexivity, and a lack of coordination between policy actors across domains and levels (Weber and Rohracher 2012; see also Bugge et al. 2017). In relation to innovation districts, we recommend that policy-makers apply a mission-oriented innovation policy approach to ensure the directionality and coherence of the policy mix aimed at supporting the long-term sustainability of these initiatives. Policy-makers should attend to key features of policy statements to ensure that they foster legitimacy and facilitate broad engagement and collaboration between multiple actors to govern complex societal problems (Kuhlmann and Rip 2018; Miedzinski et al. 2019).

To trigger and accelerate changes in the understanding of innovation in policy through the mission-oriented approach, we propose four key interventions. We argue that these interventions are key to creating a favourable policy mix that supports the achievement of shared mission objectives with more "capacious" understandings of innovation required for cities transitioning towards sustainability.

11.4.1   Intervention 1: Establishing Relevant Policy Directionality

A crucial feature of supporting missions with policy mixes is how the policies provide incentives for different stakeholders to become involved in innovation activities that contribute to a desirable direction of change in the long term. Policy directionality implies a shared vision and guiding principles that lead policy interventions towards "a particular direction of transformative change" (Weber and Rohracher 2012, p. 1042).

We recommend introducing long-term directionality into the policy mix. Such long-term "missions" should target concrete challenges and translate them into broad societal goals (Mazzucato 2018). Developing mission-led innovation policies should include the question of how to determine, define, and target a convoluted and intractable societal problem, for which solutions cannot be predefined (Wanzenbock et al. 2020). Therefore, if the aim of mission-oriented policies is to address high-impact sustainability missions, it is those sustainability challenges that display the trade-offs, contestation, uncertainty, and complexity that should be at the heart of innovation directionality.

To deliver on policy directionality, we suggest that policy-makers ensure there are no mismatches between goals and actions of policies *within* a policy mix. For example, in the "Economic Development Strategy 2031" of the CoM, we observed a mismatch between the new goal of promoting innovation for social inclusion while simultaneously recognising innovation districts as means of economic growth and competitiveness. Additionally, while CoM declared a "Climate and Biodiversity Emergency" in 2019 to show commitment towards accelerating environmental outcomes, both its "Council Plan 2021–2025" and "Economic Development Strategy 2031" support the goal of innovation for "unlocking climate capital" that reflects an economically oriented paradigm. Such tensions risk weakening the overall policy directionality of the CoM to achieve a more inclusive Melbourne. To address this issue, we recommend that policy-makers within the CoM establish clear policy directionality from the beginning by identifying major societal challenges in policy visions and forming specific policy goals, targets, and milestones, as well as turning those goals into tangible guiding principles for designing and implementing policy interventions.

11.4.2 Intervention 2: Improving Policy Consistency and Coherence

Policy consistency and coherence are key characteristics of policy mixes supporting innovation for transformative change and sustainable development (Rogge and Reichardt 2016; Miedzinski et al. 2019). Policy consistency means ensuring that multiple policies within a policy mix are not internally contradictory, but work together to achieve shared policy goals (Kern and Howlett 2009). Policy coherence goes beyond consistency, referring to the "ability of multiple policy goals to co-exist with each other" to generate and maintain synergies towards achieving shared mission objectives (Howlett and Rayner 2013, p. 174). Improving policy consistency and coherence, therefore, contributes to accelerating the effectiveness and efficiency of a policy mix (Miedzinski et al. 2019). Policy coherence is also particularly important for policy mixes aiming to facilitate multi-actor, cross-scale, and cross-sectoral innovation for sustainable development (Rogge and Reichardt 2016). Failure to corroborate policy consistency and coherence in a policy mix can result in low effectiveness or bring about unintended negative effects or conflicts that limit the achievement of systemic change for sustainability transitions (Miedzinski et al. 2019).

In "real-world" policy mixes at the local government level, different actors and instruments involved in policy processes often have different (or even contradictory) objectives that hinder opportunities for improving overall policy consistency and coherence. Local policy-makers need to recognise these complexities and develop relevant strategies to ensure the achievement of defined mission objectives (Miedzinski et al. 2019). Two key approaches can be applied to policy design to increase the consistency and coherence of a policy mix: policy packaging and policy patching (Howlett and Rayner 2013)

- *Policy packaging* is a policy design process where all preceding policies are abandoned, and a new policy package (with new tools and objectives) is developed and implemented. This straightforward approach explicitly contributes to promoting consistency and coherence in policy designs across different policy domains (e.g. economic, planning, and development policies). However, we consider that such a blunt-force shift in policy is difficult to implement in practice because it can be costly and there are likely numerous structural, organisational, capacity, knowledge, and institutional-legacy issues that would need to be addressed before completely removing old policies for newer ones. Additionally, this approach may lead to conflicts and be challenged by actors and stakeholders who have vested interests in prior policy packages, which can impede implementation. Therefore, we suggest a layering and patching policy approach as a way of moving forwards.
- *Policy patching*, generally considered a more practical strategy for enhancing policy consistency and coherence, refers to the process of gradually changing and adapting policies. This approach functions "in the same way as software designers issue 'patches' for their operating systems and programmes in order to correct flaws or allow them to adapt to changing circumstances" (Howlett and Rayner 2013, p. 177). To ensure the effectiveness and efficiency of the policy-patching approach, stronger political support (i.e. in the form of funding, favourable regulations, and coordination mechanisms for introducing new "patches") is required to lead the gradual adaptation of policy mixes for sustainable development.

Moving beyond the local government level, we recommend the importance of connecting local policy mixes with others across governance levels to ensure a more robust policy coherence for system-wide transformation. For example, in the case of the CoM, there is a need to connect CoM's policies with other local governments' policies and State-wide policies—such that all policies across all levels of government are directed towards shared, defined objectives for sustainable development. The CoM—as the central, most powerful, and most well-resourced local government in Victoria—can first implement the policy-patching design within its own remit and then share experience and lessons learnt with other local governments (and the State government) to facilitate system-wide policy coherence. For example, in Victoria, local governments tend to create formal or informal regional networks to work together on particular topics and/or shared regional issues, such as regional alliances for greenhouse action.[1] Such forums offer great opportunities for local councils such as the CoM to share knowledge and experience (including "policy-patching" experience) to facilitate system change. However, it is important to note that best practices learnt from one local government do not necessarily represent a one-size-fits-all approach for others. With the CoM being the most well-resourced council of Victoria, the capacity for applying solutions that are suitable for CoM in other councils may be limited. Thus, along with learning and sharing knowledge, local policy-makers need to also consider place-based conditions to develop relevant policy-patching approaches that can contribute to accelerating system-wide policy coherence.

11.4.3   Intervention 3: Intermediary Activities to Support Innovation for Transformative Change

Innovation districts often have the intended purpose of facilitating place-based collaboration, innovation, and public engagement (Davidson et al. 2023). To generate transformative impacts and ensure the desired direction of change, creative and fruitful collaboration needs to be fostered between different stakeholder groups (Haddad et al. 2022). Here we strongly recommend the use of transition intermediaries, which can connect multiple actors and activities to support experimentation and facilitate transformative systemic change (see, e.g. Matschoss and Heiskanen 2017; Kivimaaa et al. 2019; Sovacool et al. 2020).

According to Kivimaa et al. (2019, p. 1072), transition intermediaries are defined as:

> actors and platforms that positively influence sustainability transition processes by linking actors and activities, and their related skills and resources, or by connecting transition visions and demands of networks of actors with existing regimes in order to create momentum for socio-technical system change, to create new collaborations within and across niche technologies, ideas and markets, and to disrupt dominant unsustainable socio-technical configurations.

Intermediaries are key catalysts that can speed transitions towards more sustainable sociotechnical systems through performing functions of visioning, networking, brokering, political advocacy, and innovation management (Kivimaaa et al. 2019; Kanda et al. 2020; Sovacool et al. 2020; Kundurpi et al. 2021). Intermediaries hold

---

[1] In metropolitan Melbourne, there are four regional alliances for greenhouse actions driven by local governments,: Western Alliance for Greenhouse Action [WAGA], Northern Alliance for Greenhouse Action [NAGA], Eastern Alliance for Greenhouse Action [EAGA], and South East Councils Climate Change Alliance [SECCCA].

a unique space in being able to drive and facilitate stakeholder involvement in the innovation process, which contributes to building trust, aligning interests, handling conflicts, and fostering collaboration between and among stakeholders (Haddad et al. 2022). In working with intermediaries, local policy-makers and planners should use role-based intermediary typologies. In this context, Kivimaa et al. (2019, pp. 1068–1070) developed a comprehensive intermediary typology that distinguishes five categories of intermediaries driving transitions (Table 11.3). Adopting this typology is highly relevant for capturing the wide range of interfaces where intermediating emerges as an essential activity for facilitating innovation processes and accelerating transformative impacts.

Using this typology will be important for policy-makers seeking to identify the different styles of intermediation required. For example, in supporting local experimentation, niche intermediaries will be important for enabling systemic innovations and facilitating knowledge translation to support the uptake of niche innovations, whereas user intermediaries will be critical for translating these practices within and beyond the innovation district. Alongside understanding the different types and functions of intermediaries, policy-makers must also develop relevant incentives and mechanisms to encourage and facilitate intermediating activities. A key incentive is to provide funding for intermediaries, as this is an important determinant shaping intermediary actions and spaces (van Veelen 2019). In addition, establishing favourable structures in which "the policy principal delegates the choice of support activity and external accountability to the intermediary" (Talmar et al. 2022, p. 1) is key. This incentive will potentially give intermediaries both flexibility and

Table 11.3 Five categories of intermediaries for transformative innovation

Category	Definition	Examples
Systemic intermediaries	Taking the lead for change, operating on all levels, and promoting an explicit transition agenda	• City councillors who operate across policy scales (i.e. local, regional, international) and policy domains
Regime-based intermediaries	Being tied to contemporary institutional arrangements or interests, but have a goal to progress transformational changes for transitions	• Government agencies • Business networks
Niche intermediaries	Working to support experimentation and advance a specific niche	• Local research communities and/or knowledge organisations (in this case, the University of Melbourne, RMIT)
Process intermediaries	Facilitating a change process or a niche project based on context-specific and/or external priorities set by other actors	• Project manager • Sustainability consultant
User intermediaries	Promoting the value of new niche technologies and translating those to users and other actors	• Community members • Representatives of local not-for-profit groups

Based on Kivimaaa et al. 2019, pp. 1068–1070

responsibility to implement and adjust supporting activities for innovation and experimentation.

11.4.4 Intervention 4: Supporting Experimentation Culture and Learning

Experimentation and learning are important elements of a policy mix that supports innovation for transformative change (Schot and Steinmueller 2018; Miedzinski et al. 2019). Experiments typically denote social and technical interventions implemented in real-world settings to test new ideas and methods that can enable the restructuring of existing sociotechnical systems (Bos and Brown 2012; Matschoss and Repo 2018; Fastenrath and Coenen 2020). Such experiments represent "seeds of change that may eventually lead to profound shifts in the way societal functions such as the provision of energy or mobility are met" (Sengers et al. 2016, p. 15). Across metropolitan Melbourne, experiments can be identified within the flagship urban-resilience action—"Living Melbourne: Our urban metropolitan forest", designed and implemented by Resilient Melbourne and 32 local councils. This action is a useful test for new ways of governance through urban innovation and collaboration that attend to the nexus of socio-economic and environmental benefits for the public (Fastenrath et al. 2019, p. 7; see also Fastenrath and Coenen 2020). Hence, experimentation and technical and policy learning may contribute to articulating new shared visions, building new networks of actors, and shaping new markets, which eventually will accelerate changes in existing structures, practices, and cultures of sociotechnical systems (Laakso et al. 2017; Schot and Steinmueller 2018).

We recommend that policy-makers adopt an experimentation approach to innovation practice and policy-making whereby a broad suite of stakeholders, such as governments, universities, businesses, and civil society, can experience iterative processes of "learning-by-doing and doing-by-learning" (Fuenfschilling et al. 2019, p. 224). Inclusive and open experimentation allows a range of relevant actors to not only gain new knowledge and skills but also learn to accept uncertainty and failure as part of the experimentation process (Schot and Steinmueller 2018). While experimentation plays an important role in fostering diversity, learning, and network development, policy-makers need to avoid incidental, isolated experiments and identify coordinative mechanisms to promote and share learning experiences. In this sense, the policy process itself should become an ongoing process of experimentation, learning, and embracing uncertainty and failure, rather than relying on strict monitoring and evaluation frameworks.

Overall, our four recommended interventions play a crucial role in driving a step-change in understanding innovation in policy, thus contributing to the achievement of "transformative cities" for sustainable urban development. Table 11.4

Table 11.4 Connections between four key interventions and the vision for transformative cities

Intervention	How the intervention contributes to the vision for transformative cities
1. Establishing relevant policy directionality	Policy directionality contributes to informing and shaping policy interventions towards a shared direction of transformative change by recognising "wicked" sustainability challenges in cities and translating them into broad societal and environmental goals
2. Improving policy consistency and coherence	Policy consistency and coherence contribute to improving the effectiveness of policy mixes for transformative change and fostering multi-actor, cross-scale, and cross-sectoral innovation for sustainable development
3. Supporting intermediary activities to facilitate innovation for transformative change	Intermediaries play an important role in connecting multiple actors and activities to support innovation and experimentation for transformative systemic change
4. Supporting experimentation culture and learning	Experimentation and learning play a crucial role in fostering diversity and network development and thus support systemic innovation for transformative change

outlines the connections between the four interventions and our vision for transformative cities, as discussed in Sect. 11.1.

11.5 Summary

Transformative innovation policy is increasingly recognised as a mission-oriented approach to delivering on social and environmental objectives by embracing open, inclusive, and reflexive experimentation. Using this framing, the chapter analysed the contemporary and nascent MID City North, within the CoM, Australia, and the relevant policy and strategy documents shaping practices within the innovation district before, during, and after recent external disruptions (bushfires and COVID-19) and proposed future directions for accelerating city-level sustainable transformations. Our analysis revealed promising signs of mission-oriented policies; however, we suggest that inherent tensions remain across the existing policy mix. While many of the strategic documents speak of innovation for societal and environmental advancement, this continues to be largely shaped by deeply entrenched economic norms and language, which may constrain the potential for true innovation and transformation. Responding to this, the chapter identifies four key interventions for policy-makers to act upon when designing future policy initiatives. By attending to policy directionality, policy coherence, and consistency, supporting careful use of different intermediary types, and using experimentation as a mechanism for inclusive, open, and reflexive innovation, we seek to provide policy-makers with the tools to drive sustainable transformations.

References

Australian Government (2017) Australian 2030—Prosperity through innovation

Bellinson R (2022) Mobilising local action to address 21st century challenges: considerations for mission-oriented innovation in cities. Policy brief series (IIPP PB 19)

Bos JJ, Brown RR (2012) Governance experimentation and factors of success in socio-technical transitions in the urban water sector. Technol Forecast Soc Chang 79:1340–1353

Boschma R (2005) Proximity and innovation: a critical assessment. Reg Stud 39(1):61–74

Bottero M, Bragaglia F, Caruso N et al (2020) Experimenting community impact evaluation (CIE) for assessing urban regeneration programmes: the case study of the area 22@ Barcelona. Cities 99:102464

Bugge M, Coenen L, Marques P et al (2017) Governing system innovation: assisted living experiments in the UK and Norway. Eur Plan Stud 25(12):2138–2156

CoM (2018) Urban Realm Action Plan: Melbourne Innovation Districts City North 2018–2023

CoM (2020) MID City North Opportunities Plan 2020

CoM (2021a) Budget 2021–22

CoM (2021b) Council Plan 2021–2025

CoM (2021c) Economic Development Strategy 2031

CoM (2022a) Climate and biodiversity emergency. CoM

CoM (2022b) Budget 2022–23

CoM (2022c) Inclusive Melbourne Strategy 2022–32

Davidson K, Håkansson I, Coenen L et al (2023) Municipal experimentation in times of crises: (re-)defining Melbourne's innovation district. Cities 132:104042

Fastenrath S, Coenen L (2020) Future-proof cities through governance experiments? Insights from the Resilient Melbourne Strategy (RMS). Reg Stud 55:138–149

Fastenrath S, Coenen L, Davidson K (2019) Urban resilience in action: the resilient Melbourne strategy as transformative urban innovation policy? Sustainability 11(693):1–10

Fuenfschilling L, Frantzeskaki N, Coenen L (2019) Urban experimentation & sustainability transitions. Eur Plan Stud 27(2):219–228

Haddad CR, Naki V, Bergek A et al (2022) Transformative innovation policy: a systematic review. Environ Innov Soc Trans 43:14–40

Heikkinen M, Ylä-Anttila T, Juhola S (2019) Incremental, reformistic or transformational: what kind of change do C40 cities advocate to deal with climate change? J Environ Policy Plann 21(1):90–103

Howlett M, Rayner J (2013) Patching vs packaging in policy formulation: assessing policy portfolio design. Pol Governance 1(2):170–182

IPCC (2022) Climate change 2022: impacts, adaptation, and vulnerability. In: Contribution of working group II to the sixth assessment report of the intergovernmental panel on climate change. Cambridge University Press, Cambridge

Kanda W, Kuisma M, Kivimaa P et al (2020) Conceptualising the systemic activities of intermediaries in T sustainability transitions. Environ Innov Soc Trans 36:449–465

Kern F, Howlett M (2009) Implementing transition management as policy reforms: a case study of the Dutch energy sector. Pol Sci 42(4):391–408

Kivimaaa P, Boon W, Hyysalo S et al (2019) Towards a typology of intermediaries in sustainability transitions: a T systematic review and a research agenda. Res Policy 48:1062–1075

Kuhlmann S, Rip A (2018) Next-generation innovation policy and grand challenges. Sci Public Policy 45:448–454

Kundurpi A, Westman L, Luederitz C et al (2021) Navigating between adaptation and transformation: how intermediaries support businesses in sustainability transitions. J Clean Prod 283:125366

Laakso S, Berg A, Annala M (2017) Dynamics of experimental governance: a meta-study of functions and uses of climate governance experiments. J Clean Prod 169:8–16

Larrue P (2022) Do mission-oriented policies for net zero deliver on their many promises? Lessons for tackling complex and systemic societal challenges. In: 2022 GGSD Forum. OECD, Paris

Massey D, Quintas D, Wield D (2003) High-tech fantasies: science parks in society, science and space. Routledge, London

Matschoss K, Heiskanen E (2017) Making it experimental in several ways: the work of intermediaries in raising the ambition level in local climate initiatives. J Clean Prod 169(15):85–93

Matschoss K, Repo P (2018) Governance experiments in climate action: empirical findings from the 28 European Union countries. Environ Polit 27(4):598–620

Mazzucato M (2018) Mission-oriented innovation policies: challenges and opportunities. Ind Corp Chang 27(5):803–815

Miedzinski M, Mazzucato M, Ekins P (2019) A framework for mission-oriented innovation policy roadmapping for the SDGs: the case of plastic-free oceans. UCL Institute for Innovation and Public Purpose, Working Paper Series (IIPP WP 2019-03)

Morisson A (2020) Framework for defining innovation districts: case study from 22@ Barcelona. In: Bougdah H, Versaci A, Sotoca A et al (eds) Urban and transit planning. Springer International, Cham

Morisson A, Bevilacqua C (2019) Balancing gentrification in the knowledge economy: the case of Chattanooga's innovation district. Urban Res Pract 12(4):472–492

Nielsen JS, Farrelly M (2019) Conceptualising the built environment to inform sustainable urban transitions. Environ Innov Soc Trans 33:231–248

OECD (2023) Science, technology and innovation policy for sustainability transitions. In: OECD Science, Technology and Innovation Outlook 2023—enabling transitions in times of disruption. OECD, Paris

Pfotenhauer S, Jasanoff S (2017) Panacea or diagnosis? Imaginaries of innovation and the "MIT model" in three political cultures. Soc Stud Sci 47:783

Rogge KS, Reichardt K (2016) Policy mixes for sustainability transitions: an extended concept and framework for analysis. Res Policy 45(8):1620–1635

Røpke I (2012) The unsustainable directionality of innovation—the example of the broadband transition. Res Policy 41:1631–1642

Schot J, Steinmueller WE (2018) Three frames for innovation policy: R&D, systems of innovation and transformative change. Res Policy 44:1554–1567

Sengers F, Berkhout F, Wieczorek AJ et al (2016) Experimenting in the city—unpacking notions of experimentation for sustainability. In: Evan J, Karvonen A, Raven R (eds) The experimental city. Routledge, New York, pp 15–31

Sovacool BK, Turnheim B, Martiskainen M et al (2020) Guides or gatekeepers? Incumbent-oriented transition intermediaries in a T low-carbon era. Energy Res Soc Sci 66:101490

Talmar M, Walrave B, Raven R et al (2022) Dynamism in policy-affiliated transition intermediaries. Renew Sustain Energy Rev 159:112210

United Nations (2015) Transforming our world: The 2030 Agenda for Sustainable Development. United Nations, New York

Uyarra E, Ribeiro B, Dale-Clough L (2019) Exploring the normative turn in regional innovation policy: responsibility and the quest for public value. Eur Plan Stud 27(12):2359–2375

van Veelen B (2019) Caught in the middle? Creating and contesting intermediary spaces in low-carbon transitions. Environ Plan C Polit Space 38(1):116–133

Victoria State Government (2021) Innovation Victoria

Wanzenbock I, Wesseling JH, Frenken K et al (2020) A framework for mission-oriented innovation policy: alternative pathways through the problem-solution space. Sci Public Policy 47(4):474–489

Weber KM, Rohracher H (2012) Legitimizing research, technology and innovation policies for transformative change: combining insights from innovation systems and multi-level perspective in a comprehensive "failures framework". Res Policy 41:1037–1047

Wolfram M (2016) Conceptualizing urban transformative capacity: a framework for research and policy. Cities 51:121–130

Zukin S (2020) New York tech dossier: "innovation districts" in New York: contentious geographies of growth. Metropolitics

Thi Minh Phuong Nguyen is a Laureate Postdoctoral Research Fellow at Fenner School of Environment and Society, Australian National University. Her research interests are at the intersections of sustainability experimentation, urban transitions, and climate governance. As an early-career researcher, she has written and co-authored a number of academic publications in flagship journals in the field of urban studies.

Kathryn Davidson is an Associate Professor at the University of Melbourne and works in the field of climate and resilience governance. Kathryn is an author of numerous international publications and has attracted funding from various bodies including the Australian Research Council and the Cooperative Research Centres.

Megan Farrelly is a Professor of Environmental Geography in the School of Social Sciences and the Associate Dean Graduate Research in the Faculty of Arts, Monash University. She has written and co-authored numerous academic publications and commissioned industry reports that focus on unpacking governance, policy, and experimentation dynamics associated with expediting urban sustainability transitions within the water and energy sectors in Australia and beyond. She currently leads a number of research projects supported by Australian and international funding agencies.

Part V
Healthy Cities

Chapter 12
Healthy Cities: Transitioning to Polycentric Cities Can Enhance Population Health

Manoj Chandrabose, Nyssa Hadgraft, Neville Owen, and Takemi Sugiyama

Abstract Cities around the world are recognising the need for innovative solutions to address the growing chronic disease burden in the context of urban population growth. In this chapter, we propose transitioning from monocentric to polycentric city models as a solution to this contemporary challenge, particularly in the context of Australia. Such transitions will bring fundamental changes to the way people move across the city: from car dependency to active mobility. Essential features of the polycentric city for enhancing population health include improving walkability and bikeability, reducing driveability, promoting public transport use, and integrating greenspaces. We discuss the importance of some key aspects of urban planning and public-health research for providing a strong evidence base to guide the implementation of the polycentric transition, including identifying specific thresholds for planning attributes that are optimal for population health and understanding community preferences for attributes of polycentric urban forms.

Keywords Urbanisation · City planning · Population health · Travel behaviors · Active living

12.1 Introduction

From the beginning of the industrial era to the present, city-planning strategies have been instrumental in shaping human health. For instance, during the industrial era, health threats such as the spread of waterborne diseases were addressed by

M. Chandrabose (✉) · N. Hadgraft · T. Sugiyama
Centre for Urban Transitions, Swinburne University of Technology, Hawthorn, VIC, Australia
e-mail: mchandrabose@swin.edu.au

N. Owen
Centre for Urban Transitions, Swinburne University of Technology, Hawthorn, VIC, Australia

Physical Activity Laboratory at the Baker Heart and Diabetes Institute,
Melbourne, VIC, Australia

© The Author(s) 2025   265
N. Frantzeskaki et al. (eds.), *Future Cities Making*, Theory and Practice of
Urban Sustainability Transitions, https://doi.org/10.1007/978-981-97-7671-9_12

constructing sewerage systems, and the implementation of zoning laws helped to reduce exposure to factory smoke. The growing burden of chronic diseases is another significant public health challenge that can be addressed by twenty-first-century city planners. Since the mid-twentieth century, Australia and many other countries have created sprawling cities to accommodate their growing populations. This has been accompanied by the expansion of road networks, including the construction of highways, that prioritize car-based transportation. This imposed reliance on cars has contributed to the development of sedentary and obesogenic lifestyles that increase the risk of chronic disease. To counteract this impact, it is essential that health is considered as an integral part of planning strategies, rather than an afterthought.

This chapter begins by describing the link between city planning and health, with a focus on city planning's capacity to influence people's travel behaviors, which contribute to the growing burden of chronic diseases. It considers the rapid population growth in Australian cities as a critical background issue with significant implications for public health and outlines the planning strategies to manage it. We then discuss some of the key evidence and public-health initiatives, which can provide a foundation for transitioning towards polycentric city models as a mission-oriented solution to enhance population health in the context of urban growth.

12.2   The Burden of Chronic Diseases in Australia

Australia is experiencing an increasing burden of chronic disease. In 2017–2018, nearly half (47%) of the Australian population had one or more chronic diseases, an increase from 42% in 2007–2008 (Australian Institute of Health and Welfare 2021). The most common chronic diseases in Australia are heart disease, stroke, type 2 diabetes, kidney diseases, and some cancers, which together are responsible for a large proportion of illnesses, poorer quality of life, disability, and premature death (Australian Institute of Health and Welfare 2021). People living with chronic diseases are also more vulnerable to communicable diseases, such as COVID-19, and at a higher risk of complications (Geng et al. 2021). The economic burden of chronic disease in Australia is also substantial, with an estimated cost of more than $100 billion during 2017–2018. This includes direct costs such as hospital expenditure, as well as indirect costs such as lost productivity (Australian Institute of Health and Welfare 2021).

To reduce the chronic disease burden, prevention strategies should address the contributing behavioral risk factors, including physical inactivity, poor diet, and tobacco smoking. Traditional approaches to modify these risk factors have often involved motivation-based interventions, such as providing incentives to encourage individuals to adopt healthier behaviors. However, achieving long-term behavior change remains a challenge (Marcus et al. 2006). This is especially relevant in the context of promoting physical activity, because, particularly in Australia, many people live in areas that do not support active living. In such contexts, motivation alone may not be sufficient to ensure sustainable healthy behaviors.

Given the challenge of promoting healthy, physically active lifestyles, there is increasing recognition of the need to target the social, physical, and economic conditions in which people live—known as the social determinants of health. Such "upstream", non-medical factors are considered to be the underlying causes of poor health (Australian Government Department of Health 2021). To reduce the burden of chronic diseases, addressing their causes by engaging non-medical sectors is of paramount importance.

12.3 City Planning as a Determinant of Health

One of the key non-medical sectors relevant to public health is city planning. It shapes the form and function of the built environment, which influences how people move (being either physically active or sedentary) for their daily routines such as commuting and shopping. It is well-known that physically active travel behaviors such as walking and cycling have numerous health benefits (Riiser et al. 2018), whereas sedentary travel (prolonged car use) is a risk factor for chronic diseases (Sugiyama et al. 2020). Since individuals' travel behaviors are regular and maintained over time, decisions made by city planners in designing neighbourhood-built environments can have important ramifications for their health. It is thus important to understand how planning-related attributes can affect residents' travel behaviors. Such knowledge can inform the development of upstream prevention strategies that can reduce chronic disease risk at a population level.

In the city-planning literature, the fundamental built-environmental characteristics of urban forms relevant to travel behaviors are defined as the 3Ds: "density", "diversity", and "design" (Cervero and Kockelman 1997). "Density" refers to the population and residential densities of an area; higher densities can facilitate more investment in creating infrastructures and local destinations (such as shops and services) that can be conducive to active travel. An area's diversity is determined by how different types of land use (i.e. residential, commercial, recreational, and industrial) are distributed within the area. The design features of an area typically refer to street design, including the connectivity of streets within an area for supporting direct (easier) trips to local destinations and availability of pedestrian and cyclist-supportive infrastructures (such as footpaths and bike lanes).

Figure 12.1 illustrates an archetypal comparison between two types of urban areas, featuring different levels of density, diversity, and design characteristics. Compact neighbourhoods (a) are characterised by higher densities, more diverse land uses, and better street connectivity. In contrast, dispersed neighbourhoods (b) have lower densities, predominantly single land use, and less well-connected streets. Easy access to various types of destinations, such as public transport stops, retail shops, services, and recreational facilities, makes compact neighbourhoods more walkable, that is, easier for residents to engage in active travel. In contrast, residents

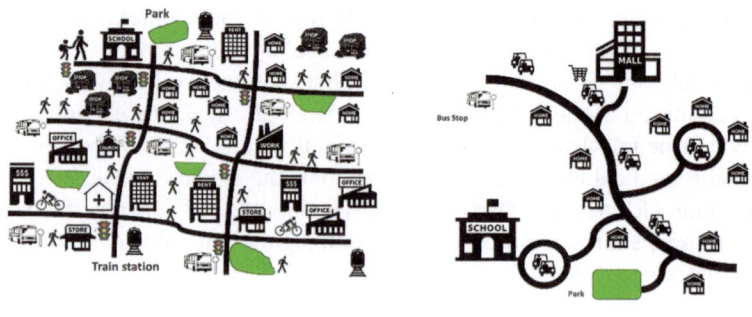

(a) Compact area

(Higher population density, mixed land use

and better street connectivity)

(b) Dispersed area

(Low population density, single land use and

limited street connectivity)

Fig. 12.1 Archetypal comparison of (**a**) compact and (**b**) dispersed urban areas. (Source: Authors)

of a dispersed neighbourhood would depend largely on cars for their day-to-day travel.

Many empirical studies over the past two decades have examined the relationships between aspects of urban forms and risk factors of chronic diseases, such as obesity and hypertension. A systematic review of 36 longitudinal studies published in 2019 found strong evidence that living in walkable neighbourhoods can be protective against the development of chronic diseases (Chandrabose et al. 2019b). These findings were further supported by a recent systematic review of Australian studies (Hadgraft et al. 2022).

12.4 Health Implications of Urban Population Growth: A Challenge for Australian Cities

Urbanisation is a global issue: by 2050, nearly 70% of the world's population will live in cities, compared to 56% in 2018 (United Nations 2019). The Australian population has been growing at a rapid rate since the beginning of the millennium, and nearly 90% of the population currently live in urban areas. Overseas migration has been the key driver of the rapid urban population growth in Australia, which had a slight pause during the COVID pandemic, but has already resumed and is expected to accelerate (Australian Bureau of Statistics 2022b). Governments, policymakers, and city planners now need to make strategic decisions in order to manage the needs of growing urban populations such as providing basic services, infrastructure, and housing.

The two fundamental and contrasting planning approaches to accommodate population growth are sprawling and densification. Sprawling is the process of expanding city boundaries into previously undeveloped greenfield lands, which creates low-density, single-use, residential developments on the outskirts. In contrast, densification involves accommodating the population in existing urban areas via medium- to high-density infill housing and encouraging mixed-use development. As shown earlier, sprawling would lead to the development of more dispersed neighbourhoods, whereas densification would add more compact neighbourhoods within a city.

It has been demonstrated that living in sprawling, car-dependent neighbourhoods can have a range of detrimental impacts on health. In such areas, access to essential services and job opportunities can be highly limited, and public transportation is infrequent or non-existent. In a recent study using Victorian Household Travel Survey data, those living 20 km away from major city centres reported, on average, spending very little time in active travel (less than 10 min/day) but longer durations (more than an hour per day) sitting in cars (Chandrabose et al. 2023). This is of concern because prolonged sedentary travel and low levels of active travel both contribute to the risk of developing chronic diseases. Moreover, a longitudinal study conducted in Adelaide found that those living in sprawling areas (20 km or more from the city centre) increased their waist circumference at a faster rate than those living closer to the city (Sugiyama et al. 2016).

Urban densification has the potential to enhance residents' health, provided it is implemented effectively. Recent longitudinal research has found that densifying existing neighbourhoods was associated with increased physical-activity levels and reduced obesity risk (Chandrabose et al. 2019a, 2021). Densifying urban areas, which can bring more local destinations to serve residents, is likely to facilitate residents' active travel and therefore reduce reliance on cars. Thus, it can be a health-enhancing planning strategy in response to urban population growth. However, research has also found that densification has some adverse health effects. For instance, an increased risk of hypertension was identified among those residing in densifying areas (Chandrabose et al. 2019a). Overcrowding and traffic congestion (if transport infrastructures are inefficient) in higher-density areas can generate noise and air pollution, which can elevate psychological and physiological stress, resulting in a greater risk of cardiovascular and respiratory diseases (Giles-Corti et al. 2012). Also, compact neighbourhoods typically have a higher concentration of retail establishments, including those with potential negative health impact, such as fast-food outlets, that make it easier to access unhealthy food. Moreover, some densification initiatives can reduce the amount of urban green spaces (Newton et al. 2022), to the detriment of local residents' health (Van den Bosch and Sang 2017).

This evidence highlights the need to develop city-planning strategies that prioritise human health as a key element. As Australia's major cities continue to grow steadily in terms of population size, it is more important than ever to identify optimal planning strategies that can support healthier living for urban residents.

12.5 Transition from Monocentric to Polycentric Cities to Enhance Population Health

The spatial structure of Australian cities provides some clues as to why population growth poses a major challenge in addressing health-related issues in an effective and sustainable manner. Figure 12.2 shows the population density of Statistical Area Level 2 units (SA2s; equivalent to suburbs) by their distance from the central business district (CBD) for the five largest cities: Sydney, Melbourne, Brisbane, Perth, and Adelaide. This clearly demonstrates that Australia's major cities are monocentric. There are some higher-density suburbs within 5–10 km of the CBDs in Sydney and Melbourne. However, it is noteworthy that these cities predominantly have lower-density suburbs about 20–30 km away from their respective CBDs. Employment opportunities, typically located in high-density areas, are far from where most people live. Such a monocentric urban structure without sufficient public transport reinforces car dependency; this is evident from recent census data, which reports that about 85% of the Australian working population who commuted on the census day used a car to travel to work (Australian Bureau of Statistics 2022a).

If Australian cities continue to remain monocentric and spread horizontally, the future could present an escalated burden of chronic diseases. Densifying a single CBD and its inner-ring suburbs is unlikely to accommodate the projected population growth and does not improve the situation for those who live in the middle and outer suburbs. A polycentric city—where multiple sub-centres of employment, economic, and social activity with surrounding residential areas exist throughout the metropolitan region—could provide a mission-oriented solution to this predicament. In this arrangement, there would be more people residing closer to workplaces and retail areas, making it convenient for them to use active travel modes and reduce their reliance on cars.

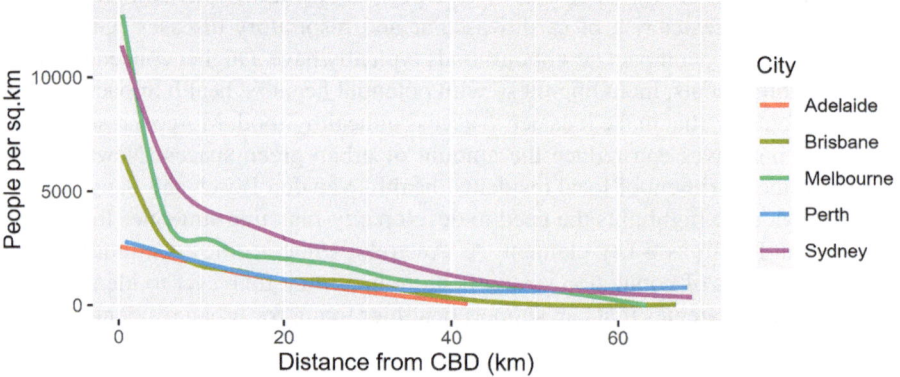

Fig. 12.2 Statistical Area 2 (SA2) population density by distance from CBD. (Source: Authors, using 2021 census data)

The aspiration of home ownership—specifically, owning a detached house on a sufficiently sized plot of land—is the "Great Australian Dream". This has contributed to the political and social acceptance of peri-urban, low-density residential development as the solution to Australia's growing urban population. It is also a model that has been strongly reinforced by the property-development industry, which has undertaken significant land banking of rural property, awaiting opportunities for rezoning to broadacre residential subdivisions. However, there is a growing recognition that this monocentric urban model is inadequate to address the interrelated challenges of chronic disease burden and environmental sustainability in the context of rapid growth in urban populations.

The polycentric city model is not new—such cities (e.g. London, Amsterdam, and Tokyo) already exist in many European and Asian countries. Australia's major cities have also started to understand the potential of such a model, with the Greater Sydney Commission aiming to create three "CBDs" in Sydney over the next 40 years (Greater Sydney Commission 2018). Similarly, the Victorian government has developed its own strategy, "Plan Melbourne 2017–2050", which seeks to promote polycentric areas in Melbourne (Victoria State Government 2017). However, the benefits of polycentric cities have been considered mainly from a perspective of environmental sustainability and productivity. Having strong evidence that polycentric urban forms can also help to address the growing chronic-disease burden can be a compelling element in supporting the transition towards polycentric cities in Australia. It is important to note that there is strong inertia to maintain the status quo in planning (Newton et al. 2022). To disrupt this planning inertia, it is essential to build a vision and roadmap that are grounded in public-health evidence and embraced by those who are involved in city planning. Creating such a vision and identifying key advocates who can promote it across all levels of government is likely to be crucial to mobilising stakeholders and driving positive change. The following section presents evidence that can support the realisation of this vision.

12.6 Policy-Relevant Health Evidence Supporting Transitioning to Polycentric Cities

Development of polycentric urban forms typically requires a top-down approach driven by a political climate for change. It requires concerted efforts from multiple sectors, including planning, transport, commerce, environment, education, sports/recreation, and healthcare, along with their sustained commitment. Urban planning and public health research can support this process by generating new and supportive evidence that adds momentum to the transition to polycentric city models and offers guidance for optimal planning and implementation.

For example, research can identify specific thresholds for urban-growth-related environmental attributes, such as local population density and residential proximity to the CBD, where health benefits are observed. Such thresholds could be used to inform guidelines regarding target density and size of sub-centres. Much of the existing research has oversimplified such relationships by investigating linear associations, which do not provide any information on thresholds of density and city sizes above (or below) which chronic disease risk increases. Further research that explores the non-linear associations of such environmental attributes with health-related behaviors and outcomes, as shown in Fig. 12.3 (hypothetical examples), could provide more helpful and policy-relevant evidence to assist planners in moving towards polycentric city models.

The process of developing polycentric cities will take time. Incremental steps need to be taken to support a gradual shift from monocentric to polycentric urban forms. A potential intermediate stage is the "20-minute neighbourhood", which is now widely recognised as a future direction in planning-policy documents (Victoria State Government 2017). This policy directive can be used as a springboard to make a transition to polycentric cities. However, there is also resistance to the idea of 20-minute neighbourhoods. Some people simply prefer to have suburban lifestyles with car-based mobility; these preferences support (and are supported by) the real-estate market, which favours peri-urban residential developments. Evaluating the health impact of implementing compact neighbourhoods would be an important step for further convincing stakeholders in city planning and other government sectors of the merits of this approach.

It is also important to understand the level of public support for different aspects of polycentric urban forms in Australia, as this will be an important factor underpinning successful implementation. Research can help by identifying community perceptions and preferences for different urban-design scenarios within polycentric models. For example, choice experiments can be carried out to assess community preferences for different urban design typologies that include different ways of arranging residential areas, commercial areas, and greenspaces. This can help to identify those aspects of polycentric cities that may be more or less likely to receive public support. In this context, providing relevant information and engaging in appropriate community consultations of different kinds is important. Legitimate concerns and potential unintended consequences can be identified, and as the benefits of innovations become apparent, people will be more willing to adapt their

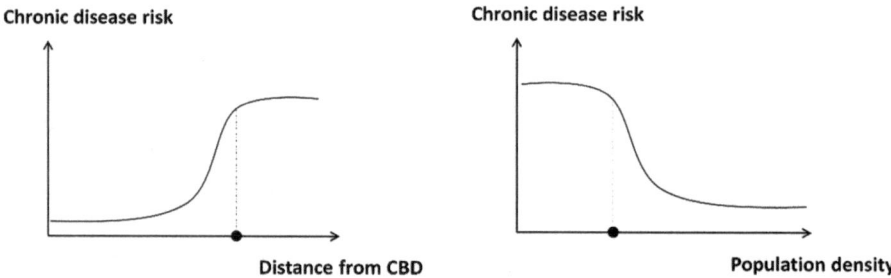

Fig. 12.3 Hypothetical non-linear associations of local population density and distance to the central business district (CBD) with chronic disease risk. (Source: Authors)

ways of living. Notably, a recent survey conducted in Sydney and Melbourne has revealed a significant shift in household preferences towards living in medium-density housing close to public transport, to the point where it now equates with living in a detached house in a car-dependent suburb (Newton et al. 2017). Evidence of public support for higher-density planning models may also provide decision-makers with the rationale and endorsement to restrict monocentric developments. Identifying features of polycentric urban strategies that both promote health and have strong public support can be informative for planners.

To empirically confirm the health benefits of polycentric cities (over monocentric cities), research needs to develop indicators of polycentricity. There appear to be diverse instruments for measuring the level of polycentricity (Natalia and Heinrichs 2020); for example, it can be assessed as the spatial distribution of sub-centres, which are often defined using employment density. A simple-to-use indicator of polycentricity could help public-health researchers to examine how the level of polycentricity can be related to health behaviors and outcomes of interest. This can provide compelling policy-relevant evidence and can be employed to underpin and guide initiatives towards polycentric city transitions.

International comparative studies in which monocentric and polycentric cities are compared in terms of residents' behaviors and health outcomes can be also informative for decision-makers. A strong contrast between Australian cities and Asian or European cities in their behavior patterns due to different urban structures may help city leaders in Australia to realise that their cities are far behind and need proactive and sustained efforts to catch up.

12.7   Enhancing Health Through Sub-Centre Design in Polycentric Cities

Polycentricity is a city-scale phenomenon, and the transition to polycentric urban forms can be a challenging task due to the necessary involvement of many stakeholders, who are likely to have competing interests. However, achieving polycentric urban forms may not always require a top-down approach at the city-wide macro scale. Urban design approaches at the meso (i.e. neighbourhoods and local suburbs) and micro (i.e. blocks and streets) scales can be employed to establish distinct sub-centres gradually, which can be considered as a complementary bottom-up approach. It is important to note that there is already a solid evidence base on the health impacts of neighbourhood-scale built-environment design that can be used to guide this approach. Based on the existing knowledge in this area, we recommend five health-related environmental features that should be considered in identifying and developing future sub-centres in polycentric cities: improving walkability, improving bikeability, reducing driveability, promoting public transport, and integrating greenspaces (Fig. 12.4).

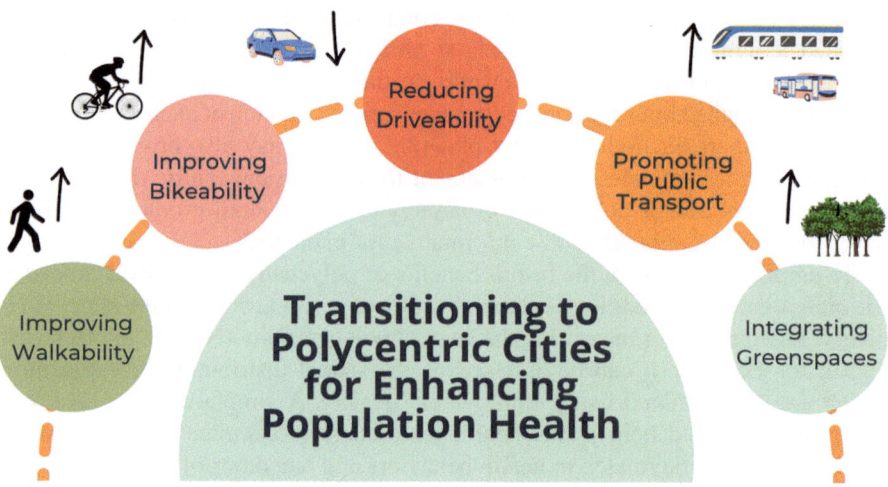

Fig. 12.4 Key health-related environmental features of future polycentric cities. (Source: Authors)

12.7.1 *Improving Walkability*

Walkability is the most fundamental concept linking urban forms and travel behaviors (Cervero and Kockelman 1997); it should be a key feature of sub-centres. The concept of walkability is essentially a combination of the 3Ds (density, diversity, and design). The synergy of these 3Ds would yield several health benefits. To improve walkability, it is recommended that the local areas have medium to higher residential densities (mostly with townhouses and mid-rise apartments), accompanied by diverse land uses and well-designed streets. A diverse mix of land uses in an area helps to ensure that residents have convenient access to essential services and amenities. Such services must include retail stores, health facilities, schools, childcare, and places for social interaction such as libraries. Having such destinations within short distances is crucial for promoting self-sufficient, walkable communities. This can positively influence residents' health not only by encouraging walking but also by fostering a sense of community and belonging, which can contribute to better mental health and well-being. Well-designed pedestrian infrastructure is another key aspect of walkability. Features such as wider footpaths, more frequent pedestrian crossings, and tree canopies along streets can provide a safer and more accessible and aesthetically pleasing experience for pedestrians. It is important to ensure that such pedestrian infrastructures are supportive for everyone, including children, older adults, and those with limited abilities to walk.

12.7.2   Improving Bikeability

Cycling is an underused mode of transportation in Australian cities, with less than 2% of all trips being undertaken by bicycle. Since it is possible to travel longer distances by cycling than by walking, improved bikeability between sub-centres and the CBD could be an important feature of a polycentric city. To improve bikeability, it is crucial to build bike paths that are separated from motor traffic, as safety concerns are a major deterrent for bicycle use (Castañon and Ribeiro 2021). Other supportive infrastructures, such as providing secured parking at workplaces and public transport hubs, and other end-of-trip facilities (e.g. showers and lockers), may also assist with greater uptake of cycling. Additionally, citywide schemes for bicycle sharing (including e-bikes) could also encourage more people to choose cycling as a means of transportation.

12.7.3   Reducing Driveability

Previous research has demonstrated that the socio-demographic and environmental determinants that underlie car use and active travel are distinct from one another (Chandrabose et al. 2023). This implies that, even if cities implement strategies to improve walkability and bikeability, car use will not necessarily be reduced without further actions to reduce the drivability of areas (den Braver et al. 2020). Examples of such measures include reducing the speed limit for cars, limiting car parking, and implementing car-free streets. Prioritising such initiatives could help Australian cities to reduce car use and promote active modes of transportation. It is worth noting that some of the current trends towards low-carbon mobility, such as electric cars and ride-sharing services, have the potential to reduce air pollution, thus contributing to improved health outcomes. However, they will not diminish the effects of sedentary travel, as sitting for prolonged periods can still be detrimental to health.

12.7.4   Promoting Public Transport

The monocentric city model has the disadvantage of requiring longer car commutes to access employment, particularly for those living in peri-urban areas that are located at a considerable distance from radial rail-transport networks. A key advantage of the polycentric city model is that it would bring people closer to employment opportunities. However, to maximise the benefit, it is important that sub-centres have easy access to public transportation. Using public transportation instead of cars can offer significant health benefits to individuals because using public transport typically requires individuals to walk before and after the ride, which can contribute to the recommended amount of daily physical activity. Decreasing the number of vehicles on the road, which reduces greenhouse-gas and air-pollutant emissions, can also help to lower the risk of respiratory and cardiovascular diseases.

12.7.5 Integrating Greenspaces

The health benefits of access to greenspaces are well established. They provide not only a setting for physical activity but also opportunities to connect with nature, facilitate social interactions, and offer cognitive restoration, thus contributing to improved physical and mental health (Sugiyama et al. 2018). Additionally, integrating access to greenspaces in neighbourhoods can assist in enhancing biodiversity, improving air quality, and mitigating the urban heat-island effect, all of which can have positive impacts on human health. However, the development of polycentric cities involves densification of sub-centres, using strategies such as infill development, which can lead to a decrease in the quantity of greenspaces (Haaland and van den Bosch 2015). For the polycentric city to be more effective in promoting population health, it is important to retain public greenspaces. "Nature-based solutions" have become an increasingly popular strategy to address the challenges that are accelerating in the context of growing urban populations (Frantzeskaki et al. 2020). This involves both traditional and innovative approaches, such as planting trees along streets, establishing community gardens, and integrating greenery into buildings' rooftops and walls. Additionally, greenbelts, which can be used to demarcate urban boundaries, are known to curb urban sprawl (Pourtaherian and Jaeger 2022). Such green infrastructure can be integrated into the design of new sub-centres.

12.8 Conclusion

This chapter has presented the transition from monocentric to polycentric urban forms as a vision and vehicle for enhancing population health in the context of urban population growth. The need for such transitions is being recognised in policy documents as a future direction for Australian cities. However, the current debate on polycentric cities lacks the perspective of population health. We have presented evidence derived from public health that can inform urban design and planning policies towards achieving healthy cities, which can also be used to propel the transition to polycentric urban forms. However, for such evidence to be applied to the city-planning decision-making process, robust coordination between the health sector and many other sectors is needed. Currently, Australia lacks governance mechanisms or formal structures that enable such coordination (Breadon et al. 2023). This is a structural barrier that is difficult to overcome by the good intentions of the health sector alone. Institutional reform, which creates a new way of developing planning policies, will be needed. Since it will take time to implement such reform, collaboration between researchers, practitioners, policymakers, and community groups is an important avenue to make progress. Identifying grassroots advocates and champions capable of influencing political decision-makers is also crucial to facilitate successful implementation of polycentric-city development initiatives. Sharing information, activities, capacities, and resources among these stakeholders can help them to make this vision more focused, to understand differences between them, and to build trust. Capacity-building and professional development in this interdisciplinary field are also needed for improved coordination between the sectors. We emphasise that the synergy between the health and planning sectors, at various levels, will play a pivotal role in driving the successful transition to more sustainable polycentric urban forms, which can be beneficial for enhancing population health.

References

Australian Bureau of Statistics (2022a) Australia's journey to work. Canberra

Australian Bureau of Statistics (2022b) National, state and territory population. Canberra

Australian Government Department of Health (2021) National preventive health strategy 2021–2030. Canberra

Australian Institute of Health and Welfare (2021) Chronic condition multimorbidity. Canberra

Breadon P, Fox L, Emslie O, Richardson L (2023) The Australian Centre for Disease Control (ACDC): highway to health. Grattan Institute, Melbourne

Castañon UN, Ribeiro PJG (2021) Bikeability and emerging phenomena in cycling: exploratory analysis and review. Sustainability 13(4):2394. https://doi.org/10.3390/su13042394

Cervero R, Kockelman K (1997) Travel demand and the 3Ds: density, diversity, and design. Transp Res Pt D Transp Environ 2(3):199–219. https://doi.org/10.1016/S1361-9209(97)00009-6

Chandrabose M, Owen N, Giles-Corti B, Turrell G, Carver A, Sugiyama T (2019a) Urban densification and 12-year changes in cardiovascular risk markers. J Am Heart Assoc 8(15):e013199. https://doi.org/10.1161/JAHA.119.013199

Chandrabose M, Rachele JN, Gunn L, Kavanagh A, Owen N, Turrell G, Giles-Corti B, Sugiyama T (2019b) Built environment and cardio-metabolic health: systematic review and meta-analysis of longitudinal studies. Obes Rev 20(1):41–54. https://doi.org/10.1111/obr.12759

Chandrabose M, Owen N, Hadgraft N, Giles-Corti B, Sugiyama T (2021) Urban densification and physical activity change: a 12-year longitudinal study of Australian adults. Am J Epidemiol 190(10):2116–2123. https://doi.org/10.1093/aje/kwab139

Chandrabose M, Forkan ARM, Abe T, Owen N, Sugiyama T (2023) Joint associations of environmental and sociodemographic attributes with active and sedentary travel. Transp Res Pt D Transp Environ 116:103643. https://doi.org/10.1016/j.trd.2023.103643

den Braver NR, Kok JG, Mackenbach JD, Rutter H, Oppert J-M, Compernolle S, Twisk JWR, Brug J, Beulens JWJ, Lakerveld J (2020) Neighbourhood drivability: environmental and individual characteristics associated with car use across Europe. Int J Behav Nutr Phys Act 17:8. https://doi.org/10.1186/s12966-019-0906-2

Frantzeskaki N, Vandergert P, Connop S, Schipper K, Zwierzchowska I, Collier M, Lodder M (2020) Examining the policy needs for implementing nature-based solutions in cities: findings from city-wide transdisciplinary experiences in Glasgow (UK), Genk (Belgium) and Poznań (Poland). Land Use Policy 96:104688. https://doi.org/10.1016/j.landusepol.2020.104688

Geng J, Yu X, Bao H, Feng Z, Yuan X, Zhang J, Chen X, Chen Y, Li C, Yu H (2021) Chronic diseases as a predictor for severity and mortality of covid-19: a systematic review with cumulative meta-analysis. Front Med 8:588013. https://doi.org/10.3389/fmed.2021.588013

Giles-Corti B, Ryan K, Foster S (2012) Increasing density in Australia: maximising the health benefits and minimising the harm. National Heart Foundation of Australia, Melbourne

Greater Sydney Commission (2018) A metropolis of three cities: the greater Sydney region plan. Sydney

Haaland C, van den Bosch CK (2015) Challenges and strategies for urban green-space planning in cities undergoing densification: a review. Urban For Urban Green 14(4):760–771. https://doi.org/10.1016/j.ufug.2015.07.009

Hadgraft N, Chandrabose M, Bok B, Owen N, Woodcock I, Newton P, Frantzeskaki N, Sugiyama T (2022) Low-carbon built environments and cardiometabolic health: a systematic review of Australian studies. Cities Health 6(2):418–431. https://doi.org/10.1080/23748834.2021.1903787

Marcus BH, Williams DM, Dubbert PM, Sallis JF, King AC, Yancey AK, Franklin BA, Buchner D, Daniels SR, Claytor RP (2006) Physical activity intervention studies. What we know and what we need to know: a scientific statement from the American Heart Association Council on nutrition, physical activity, and metabolism. Circulation 114(24):2739–2752. https://doi.org/10.1161/CIRCULATIONAHA.106.179683

Natalia VV, Heinrichs D (2020) Identifying polycentricism: a review of concepts and research challenges. Eur Plan Stud 28(4):713–731. https://doi.org/10.1080/09654313.2019.1662773

Newton P, Meyer D, Glackin S (2017) Becoming urban: exploring the transformative capacity for a suburban-to-urban transition in Australia's low-density cities. Sustainability 9(10):1718. https://doi.org/10.3390/su9101718

Newton PW, Newman PW, Glackin S, Thomson G (2022) Greening the greyfields: new models for regenerating the middle suburbs of low-density cities. Palgrave Macmillan, Singapore

Pourtaherian P, Jaeger JAG (2022) How effective are greenbelts at mitigating urban sprawl? A comparative study of 60 European cities. Landsc Urban Plan 227:104532. https://doi.org/10.1016/j.landurbplan.2022.104532

Riiser A, Solbraa A, Jenum AK, Birkeland KI, Andersen LB (2018) Cycling and walking for transport and their associations with diabetes and risk factors for cardiovascular disease. J Transp Health 11:193–201. https://doi.org/10.1016/j.jth.2018.09.002

Sugiyama T, Niyonsenga T, Howard NJ, Coffee NT, Paquet C, Taylor AW, Daniel M (2016) Residential proximity to urban centres, local-area walkability and change in waist circumference among Australian adults. Prev Med 93:39–45. https://doi.org/10.1016/j.ypmed.2016.09.028

Sugiyama T, Carver A, Koohsari MJ, Veitch J (2018) Advantages of public green spaces in enhancing population health. Landsc Urban Plan 178:12–17. https://doi.org/10.1016/j.landurbplan.2018.05.019

Sugiyama T, Chandrabose M, Homer AR, Sugiyama M, Dunstan DW, Owen N (2020) Car use and cardiovascular disease risk: systematic review and implications for transport research. J Transp Health 19:100930. https://doi.org/10.1016/j.jth.2020.100930

United Nations (2019) World urbanization prospects: the 2018 revision (ST/ESA/SER.A/420). United Nations, Department of Economic and Social Affairs, Population Division, New York

Van den Bosch M, Sang ÅO (2017) Urban natural environments as nature-based solutions for improved public health—a systematic review of reviews. Environ Res 158:373–384. https://doi.org/10.1016/j.envres.2017.05.040

Victoria State Government (2017) Plan Melbourne 2017–2050. Metropolitan Planning Strategy, Melbourne

Manoj Chandrabose is a Research Fellow at the Centre for Urban Transitions, Swinburne University of Technology. He is a spatial epidemiologist with expertise in investigating the health implications of urban environmental settings. With a PhD in public health and an MPhil in statistics, he uses cutting-edge data-science techniques to explore the link between urban planning and public health. His research focuses on both the built and natural environmental settings, and their influence on physical activity, sedentary behaviors, travel patterns, and cardiovascular disease risk. With over 20 peer-reviewed publications, his work has informed the development of public health policies.

Nyssa Hadgraft is a Research Fellow at the Centre for Urban Transitions, Swinburne University of Technology. She holds a Master of Public Health and PhD in Public Health and has published over 40 peer-reviewed journal articles. Her research focuses on understanding the correlates of the cardiometabolic health outcomes associated with active living behaviors, with the aim of informing public health and planning initiatives to enhance population health.

Neville Owen is a Distinguished Professor at the Centre for Urban Transitions, Swinburne University of Technology and a Senior Scientist in the Physical Activity Laboratory at the Baker Heart and Diabetes Institute. His studies include epidemiological and experimental examinations of the biological correlates and consequences of time spent sitting, workplace behavior-change trials, and linkages of large-scale health, transportation, and social and spatial data sets to identify environmental and policy implications. He received a 2021 American College of Sports Medicine Citation Award and is a Clarivate Highly Cited Researcher; his 620+ peer-reviewed publications have garnered over 117,000 citations.

Takemi Sugiyama is a Professor of Healthy Cities at the Centre for Urban Transitions, Swinburne University of Technology. With an extensive interdisciplinary research background in architecture, urban design, and epidemiology, his research explores the nexus between urban form and population health. His research currently focuses on urban built environments that support adults' active lifestyles and health, urban greenspace and health, the health impacts of active and sedentary transport, and environmental approaches to reduce health inequalities. He has published over 150 peer-reviewed journal articles and book chapters. He serves on the editorial boards of Landscape and Urban Planning, Environment and Behavior, the Journal of Transport & Health, and Wellbeing, Space and Society.